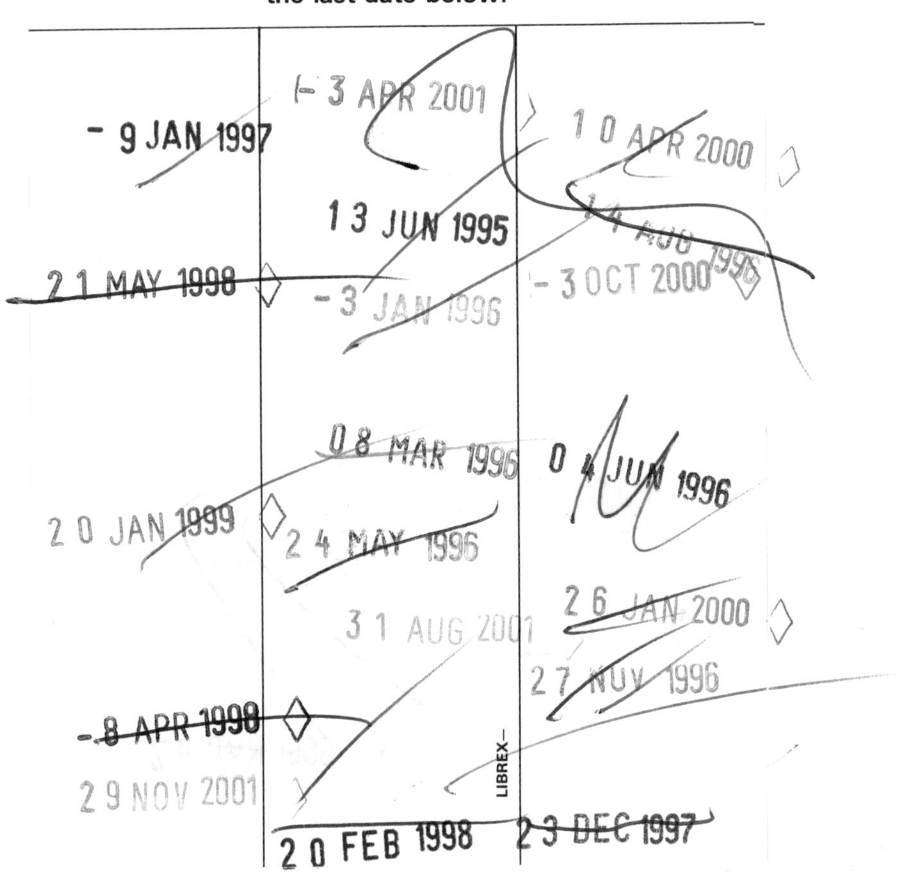
- 9 JAN 1997
- 3 APR 2001
1 0 APR 2000
13 JUN 1995
14 AUG 1996
2 1 MAY 1998
- 3 JAN 1996
- 3 OCT 2000
0 8 MAR 1996
0 4 JUN 1996
2 0 JAN 1999
2 4 MAY 1996
3 1 AUG 2001
2 6 JAN 2000
2 7 NOV 1996
- 8 APR 1998
LIBREX
2 9 NOV 2001
2 0 FEB 1998
2 3 DEC 1997

OPTICAL NETWORK TECHNOLOGY

BT Telecommunications Series

The BT Telecommunications Series covers the broad spectrum of telecommunications technology. Volumes are the result of research and development carried out, or funded by, BT, and represent the latest advances in the field.

The series includes volumes on underlying technologies as well as telecommunications. These books will be essential reading for those in research and development in telecommunications, in electronics and in computer science.

1. *Neural Networks for Vision, Speech and Natural Language*
   Edited by R Linggard, D J Myers and C Nightingale
2. *Audiovisual Telecommunications*
   Edited by N D Kenyon and C Nightingale
3. *Digital Signal Processing in Telecommunications*
   Edited by F A Westall and S F A Ip
4. *Telecommunications Local Networks*
   Edited by W K Ritchie and J R Stern
5. *Optical Network Technology*
   Edited by D W Smith

OPTICAL NETWORK TECHNOLOGY

Edited by

D.W. Smith

Manager

Networks Research Unit

BT Laboratories

Martlesham Heath, UK

CHAPMAN & HALL

London · Glasgow · Weinheim · New York · Tokyo · Melbourne · Madras

Published by Chapman & Hall, 2–6 Boundary Row, London SE1 8HN, UK

Chapman & Hall, 2–6 Boundary Row, London SE1 8HN, UK

Blackie Academic & Professional, Wester Cleddens Road, Bishopbriggs, Glasgow G64 2NZ, UK

Chapman & Hall GmbH, Pappelallee 3, 69469 Weinheim, Germany

Chapman & Hall USA, One Penn Plaza, 41st Floor, New York NY 10119, USA

Chapman & Hall, Japan, ITP-Japan, Kyowa Building, 3F, 2-2-1 Hirakawacho, Chiyoda-ku, Tokyo 102, Japan

Chapman & Hall Australia, Thomas Nelson Australia, 102 Dodds Street, South Melbourne, Victoria 3205, Australia

Chapman & Hall India, R. Seshadri, 32 Second Main Road, CIT East, Madras 600 035, India

---

First edition 1995

Printed in Great Britain at the University Press, Cambridge

ISBN 0 412 60120 6

A catalogue record for this book is available from the British Library

∞ Printed on acid-free text paper, manufactured in accordance with ANSI/NISO Z39.48-1992 (Permanence of Paper)

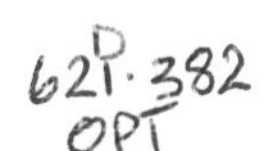

Contents

Contributors

M J Adams	Optoelectronic Devices, BT Laboratories
B J Ainslie	Advanced Technologies, BT Laboratories
J R Armitage	Network Concepts and Applications, BT Laboratories
P E Barnsley	Network Concepts and Applications, BT Laboratories
C J Beaumont	Network Technology, BT Laboratories
L T Blair	Network Concepts and Applications, BT Laboratories
L C Blank	Photonics and Quantum Processing, BT Laboratories
K J Blow	Photonics and Quantum Processing, BT Laboratories
M C Brierley	Linear Waveguides, BT Laboratories
J D Burton	Network Transport, BT Laboratories
R J Campbell	Linear Waveguides, BT Laboratories
S A Cassidy	Service Systems, BT Laboratories
S T Davey	Materials & Technology Application, BT Laboratories
D A O Davies	Optoelectronic Devices, BT Laboratories
P Eardley	Wireless Systems, BT Laboratories
A D Ellis	Photonics and Quantum Processing, BT Laboratories
P J Fiddyment	Laser Amplifiers, BT Laboratories
M A Fisher	Optoelectronic Devices, BT Laboratories
T G Hodgkinson	Network Evolution, BT Laboratories
C A Jones	Advanced Technologies, BT Laboratories
R Kashyap	Advanced Technologies, BT Laboratories
D A Mace	Service Creation Engineering, BT Laboratories
F MacKenzie	Network Concepts and Applications, BT Laboratories
J F Massicott	Linear Waveguides, BT Laboratories
G D Maxwell	Advanced Technologies, BT Laboratories
M McCullagh	Wireless Systems, BT Laboratories
P F McKee	Network Mobility and Intelligence, BT Laboratories
C A Millar	Network Concepts and Applications, BT Laboratories
P S Mudhar	Information Network Applications, BT Laboratories

M Nield Advanced Technologies, BT Laboratories
A W O'Neill Network Evolution, BT Laboratories
D B Payne Fibre Systems, BT Laboratories
S J D Phoenix Photonics and Quantum Processing, BT Laboratories
S Ritchie University Research Programme, BT Laboratories
M J Robertson Semiconductor Optoelectronics, BT Laboratories
J D Rush Optical Processing, BT Laboratories
G Sherlock Semiconductor Optoelectronics, BT Laboratories
J Singh Formerly Laser Amplifiers, BT Laboratories
D W Smith Networks, BT Laboratories
I C Smith Network Transport, BT Laboratories
K Smith Photonics and Quantum Processing, BT Laboratories
P P Smyth Network Mobility and Intelligence, BT Laboratories
D M Spirit Future Undersea Networks, BT Laboratories
D Szebesta Network Transport, BT Laboratories
A Thurlow Advanced Technologies, BT Laboratories
J R Towers Optical Processing, BT Laboratories
P D Townsend Photonics and Quantum Processing, BT Laboratories
D Wake Advanced Lasers, BT Laboratories
N G Walker Future Network Infrastructure, BT Laboratories
L D Westbrook Advanced Lasers, BT Laboratories
T J Whitley Network Concepts and Applications, BT Laboratories
H J Wickes Laser Amplifiers, BT Laboratories
M R Wilkinson Network Concepts and Applications, BT Laboratories
D L Williams Advanced Technologies, BT Laboratories
D Wisely Wireless Systems, BT Laboratories
D Wood Network Mobility and Intelligence, BT Laboratories
R Wyatt Advanced Technologies, BT Laboratories

Foreword

In these heady days of competition, deregulation, market forces and all-consuming focus on the customer, it is easy to overlook that the world is driven by raw technology. One of the most fundamental, and remarkable, of these drivers is optical fibre. Without it the current telecommunications revolution would have been a minor event, with the prospect for the information society greatly diminished. Trying to support the exponentially growing global demand for telecommunications with satellite, terrestrial radio and copper cable alone would have stalled the emerging information revolution. The limited transmission capacities of these elemental bearers would have precluded the widespread availability of bandwidth and connectivity for all at a price we can all afford.

Realizing the telecommunications infrastructure we enjoy today would have been prohibitively expensive with copper and radio technologies. In addition, the concatenation of the huge amounts of necessary electronics would have seen lamentably poor levels of reliability. In less than 20 years optical fibre emerged as the saviour — a technology that looked like the transmission engineer's dream:

- effectively unlimited bandwidth;
- near-zero attenuation;
- near-zero signal distortion;
- near-zero power requirement;
- near-zero material usage;
- near-zero space requirement;
- near-zero cost.

Today, the BT network employs three million kilometres of single-mode optical fibre to transport over 85% of its traffic. Internationally, it supports over 60% of the global traffic. This is staggering stuff! Without it the human race would not have been able to satisfy its insatiable desire to communicate, and would not have sustained the continuing exponential growth.

I have personally been fortunate to have enjoyed an intimate involvement in the development of optical transmission systems and networks since 1979. I saw the deployment of TAT-8, PTAT-1 and other undersea cable systems, long- and short-haul terrestrial and, more recently, access networks. In contrast to my earlier experiences with copper cable and radio, where developments were constrained by all-too-evident practically limiting bounds to capacity and distance, fibre never offered such a restriction. As we approached the fundamental limits predicated by electronic repeaters in the fibre transmission path, optical amplifiers emerged to remove the bottleneck. They were first realized in a discrete chip form, shortly after as short lengths of doped fibre, and ultimately as an entire cable span in the form of a distributed amplifier. Similarly, the present limitations posed by the electronics required for time division multiplexing can now be swept aside by wavelength division multiplexing. More recently, the emergence of solitons (perfect waves) has sidelined issues of pulse distortion build-up due to nonlinear effects accrued on future international routes using optical amplifiers.

If the physicists, materials scientists, chemists, mathematicians, engineers, systems and network designers involved in the development of fibre technology have committed any crime, it is that of giving the human race technological indigestion — too much, too fast, and at an accelerating rate. For we are now, apparently, well ahead of any user demand in our ability to transport and process information. From any commercial standpoint, the optical community perhaps seems out of step, and well ahead of the game. But, as we look to computing, education, entertainment, physical travel and many other human activities there is a rapidly growing, and pent up, demand for more transmission and network capacity. Any explosion in demand for information and communication can only be satisfied by fibre technology. We need, and fortunately have, a storehouse of technology ready to go. What is required is the courage to leapfrog lesser technologies and implement radically new solutions. The decision to do this will require the same commitment, and management bravery, that was in evidence in the early 1980s when it was decided to 'go with single-mode fibre' despite all the risks, difficulties and unknowns.

Optical technology has already revolutionized telecommunications — and is now set to do it again. Just as the world has gone digital and electronic, fibre technology looks set to realize transparency. Networks without electronics in the transmission and switching path are now entirely feasible — as is photonic signal processing. In this book you will catch a fascinating glimpse of the new and developing technologies being researched at BT Laboratories (BTL): fibres that amplify and filter, networks without electronic switches, quantum cryptography, radio over fibre, optical wireless, picosecond logic, invisible-to-visible light conversion, new forms of signal routeing and processing, and much more. The many contributing authors to this volume go well beyond the bounds of standard optical fibre texts dealing with established system, network and application thinking and explore a cornucopia of technology that is still neither complete nor remotely near exhaustion. Of necessity this book had to be limited in its scope and treatment as the field of options is vast. What have purposely been selected are those developments that are currently working in the laboratory and show a clear possibility for application within the next decade. On this basis I have no hesitation in recommending the text to students, educators, workers and researchers in the field.

A tantalizing excerise for you, the reader, is to contemplate, as the teams at BTL do, the possibilities offered by the interconnection and interaction of these technologies, plus the cumulative impact on telecommunications, services and markets. Further, you may wish to contemplate the changes that will be engendered in network signalling, control and management, software, billing systems and operations. The technologies discussed will invoke radical change and some interesting and demanding challenges to the *status quo*.

It is really quite stunning to reflect that the situation today looks as it did 20 years ago — we can see no slow down, nor limitation to the continuing development of optical fibre technology. We can see no end to this story — the dream continues.

Peter Cochrane
Advanced Applications and Technologies
BT Laboratories

1

NERVES OF GLASS

D W Smith

1.1 A BRIEF HISTORY OF FIBRE

In a period of just 20 years, optical fibre communication has progressed from primitive laboratory bench-top experiments to become a multibillion dollar international business. It is estimated that there are already more than 40 million kilometres of single-mode fibre installed around the planet; indeed, silica glass has now become the fabric of the global nervous system. The huge information-carrying capability of this technology will fundamentally change the way we all conduct our business and take our leisure.

Before outlining the new concepts and opportunities that are discussed in the chapters that follow, it is worth reflecting on the factors that spurred on the fibre pioneers of the early 1970s. At that time there was a significant pressure of demand to increase the transmission capacity of the coaxial cable systems that formed the major information highways. However, the repeater spacing of a coaxial line system has an inverse power law relationship with system capacity. Repeater spacings for 140 Mbit/s systems were already down to about 2 km and rising costs and maintenance liabilities could not easily be contained. The leading contender for a replacement technology was considered to be circular millimetric waveguide — a rather elaborate solution that had been made to work by some courageous and innovative engineering. Some researchers had also previously considered free-space optical propagation in hollow pipes but the practical difficulty of periodically refocusing and steering the light round corners was immense. The application of glass optical waveguides, as first proposed by Kao and Hockham [1] in 1966, was seen to be a more elegant but rather unlikely long shot. However, some strong-willed visionaries, in particular the late F F Roberts of the British Post Office (the forerunner to BT), initiated speculative but well-focused research programmes. These eventually led to experimental field trials of fibre systems [2]. It was only after those and other trials during the late 1970s that fibre was taken seriously by the wider communications community.

The first operational systems were based on multimode fibre, which, although offering the possibility of practical cost advantages over coaxial cable, was rather limited in its ultimate transmission capacity. The next major step was to develop and then extensively install single-mode fibre. The early adoption of single-mode fibre was a bold decision because it placed increased engineering demands on laser coupling, connectors and fibre splices since it required micron-level tolerances to be maintained within a field environment. With the benefit of hindsight, no one today would question the rightness of the single-mode decision. For the first time we have a cable infrastructure that is proof against almost any foreseeable capacity demand. The advanced optical technology discussed in the chapters that follow illustrates many of the ways that the capacity could be harnessed to meet the needs of future bandwidth-hungry services, such as videoconferencing, video-on-demand, interactive networked multimedia, telepresence and virtual reality.

1.2 HIGH-CAPACITY TRANSMISSION

Single-mode fibre technology has led to such a rapid reduction in the intrinsic cost of bit transport that we could expect a future where the cost of providing basic telephony would be distance independent. Experience has shown that the cost of optical transmission equipment approximately doubles for each fourfold upgrade in capacity. Today, several equipment suppliers are already developing transmission equipment for 10 Gbit/s operation and, within research laboratories, experiments at up to 100 Gbit/s have recently been reported [3]. Figure 1.1 shows the rapid progress in increasing the transmission capability of fibre systems; this is given in terms of the product of bit rate × distance and has increased by nine orders of magnitude since the first field trials. With optical amplification it could now be possible to send the same signals, which in 1974 required electronic regenerators every 2 km along coaxial cable, about 1000 times around the planet without any electronic intervention!

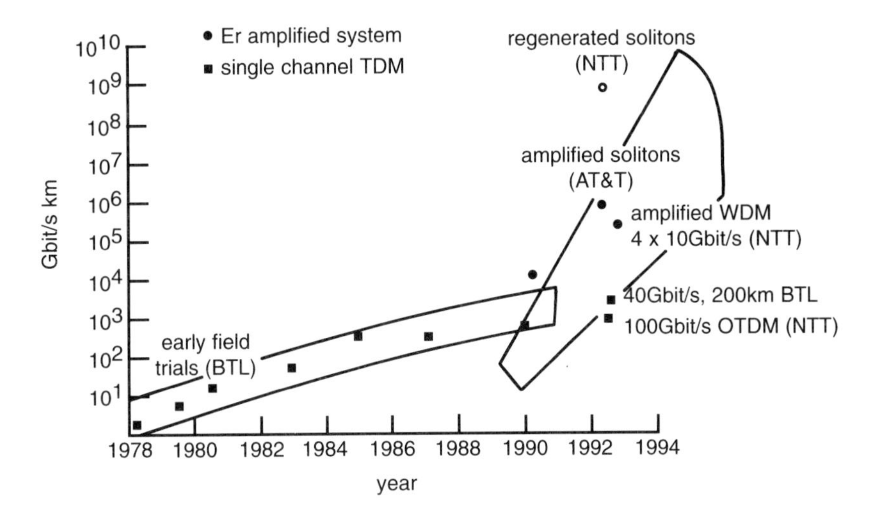

Fig. 1.1 Progress towards ultra-high capacity transmission.

But the future is video and multimedia, not just plain telephony. Even after taking account of the recent advances in video bandwidth compression, there is every expectation that the capacity required of the core network by these new services will significantly increase. However, the full range of these new broadband services is still largely undetermined and the potential growth rate is less than certain. What is sure is that in today's increasingly competitive environment the network operator must be able to respond rapidly and economically to the addition of any new service. The use of advanced optical technology, such as fibre amplifiers (Chapters 10 and 11) and advanced optical multiplexing techniques (Chapters 3 and 4), will enable a much leaner physical transmission network to provide the huge capacity necessary to meet the more futuristic demands. By stripping out excess complexity, such a network would not only be more flexible but could also be easier to manage and maintain. Of course, the established network operator does not start out with the 'ideal', but must now evolve towards it gracefully, and to this end there are now an increasing variety of upgrade strategies being debated at international conferences. Figure 1.2 outlines the wide range of choices available and includes the options of either increasing data rates of single-wavelength channel systems or the use of wavelength-division multiplexing to achieve a high aggregate data rate. One important issue is 1.3 μm wavelength working versus 1.5 μm operation. Whilst the latter can give the best performance in terms of fibre loss and ready availability of fibre amplifiers, most installed links are designed to operate at 1.3 μm. Therefore the recent demonstration of practical fibre amplifiers operating at 1.3 μm (Chapter 11) is an important milestone achieved.

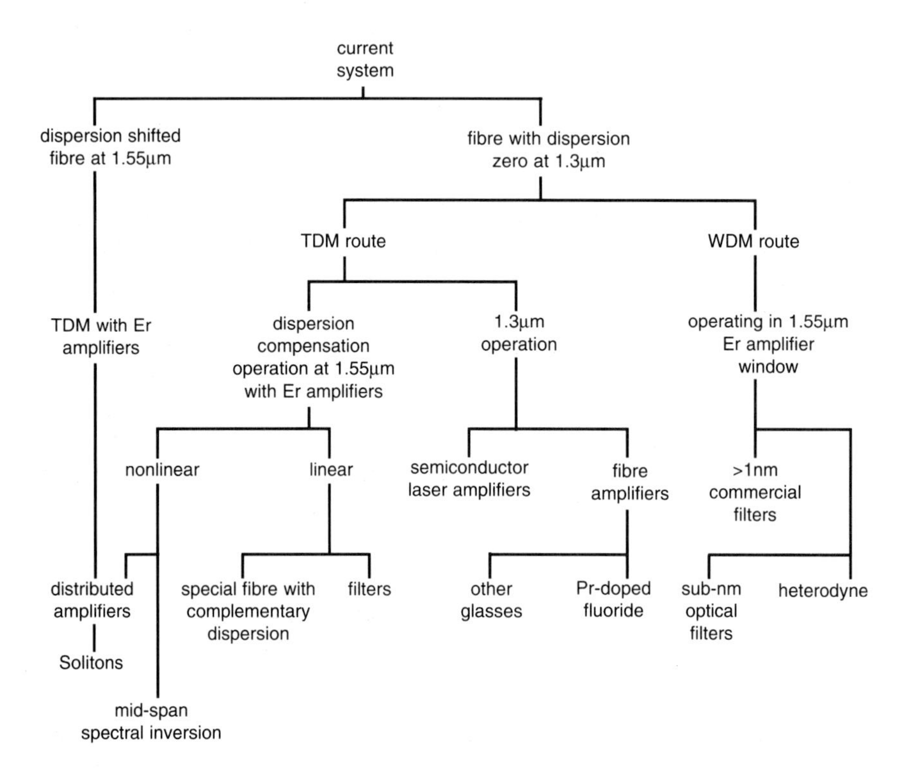

Fig. 1.2 Options for ultra-high capacity transmission.

1.3 OPPORTUNITIES FOR NEW OPTICAL TECHNOLOGY IN THE ACCESS NETWORK

For some time it has been widely agreed that the next significant breakthrough for optical communications is within the access network or local loop. Although point-to-point direct fibre feeders are already finding their way into the larger urban business customer it is unlikely that this approach will scale economically to meet the needs of the smaller business user or residential customer. A key step towards breaking out of the cost/applications vicious circle is optical networks, which share fibre plant between several end users via passive optical components [4].

This theme of cost reduction by resource sharing is taken a stage further (Chapters 2 and 5) through some novel approaches for delivering broadband services over fibre direct to the customer. These new architectures can use wavelength-division multiplexing (WDM) to increase flexibility by providing a useful degree of bit rate and service independence. It is also possible to use WDM techniques to provide a means of passive routeing to increase network resilience.

Clearly, for local loop applications the cost of optoelectronic components is always likely to be of paramount concern. Chapter 15 describes a new, potentially low-cost, method to generate optical signal-processing functions, such as wavelength selection and single-frequency lasers. The technique involves the direct writing of periodic gratings within the fibre core by u.v. light. This has recently been one of the fastest-developing new areas of optical technology research around the world. More complex functions can be created when this writing method is combined with new advanced computer-based optimized optical design techniques based on simulated annealing (Chapter 16). For some applications it will be necessary to integrate several signal-processing functions on to one easily manufacturable optical circuit board, or 'motherboard', to obtain a cost advantage — currently silica planar integrated optics looks to be a very attractive option for this (Chapter 14).

1.4 OPTICS IN FUTURE SWITCHING AND INFORMATION-PROCESSING MACHINES

To access the new services and manage the available network capacity, more advanced switching nodes will be required. It is felt that many of the emerging broadband services could be handled by cell-based switching technologies, such as asynchronous transfer mode (ATM). Whilst it is anticipated that any initial introduction of ATM will rely mainly on powerful electronics, at some future time new hardware solutions will be required. The precise node capacity where a new technology is required cannot be exactly predicted nor can the time when that capacity will be required (see Fig. 1.3). However, it is generally agreed that, when aggregated node capacities go much beyond 100 Gbit/s, optics will increasingly be required within the switch fabric, firstly for interconnect and eventually for signal routeing. Optical switching can also play other useful roles, e.g. network reconfiguration to improve resilience to faults. A range of optical switching techniques are considered in Chapters 9 and 10. In addition to optical space switching it is possible to consider wavelength switching (Chapter 2) and time switching. The true power of optical switching probably only comes in when the three dimensions of switching are combined to make machines of huge capacity that can permute signals in space, wavelength and time [5, 6].

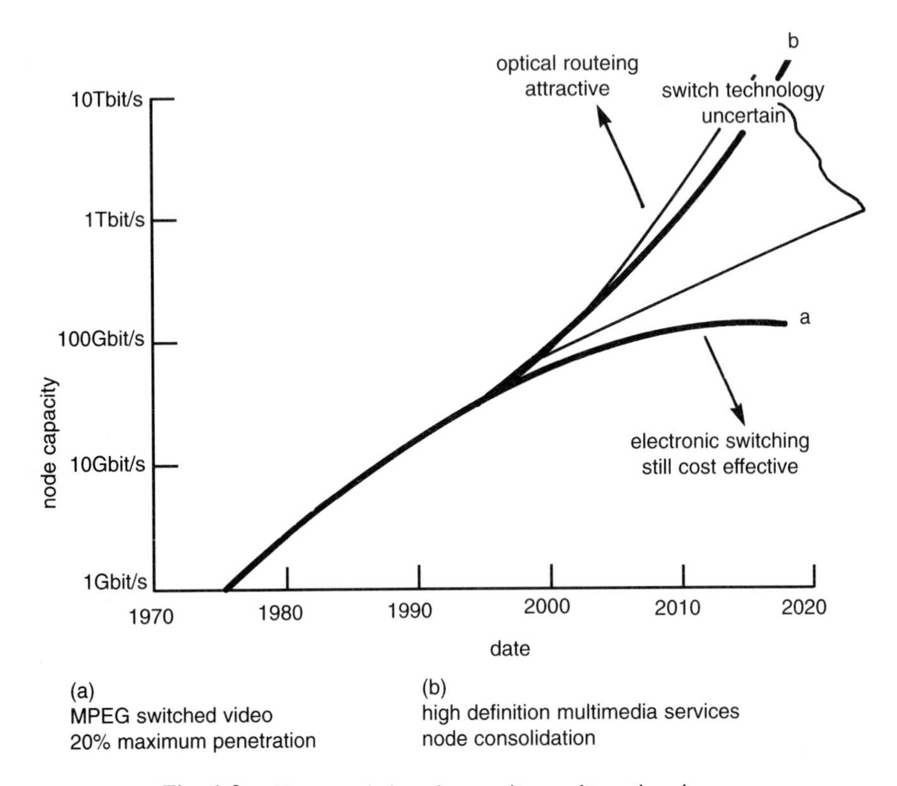

Fig. 1.3 Future switch node capacity — alternative views.

Ultimately, some of the fastest switching and logic operations could be performed directly by nonlinear optical elements, i.e. light controlling light. Chapter 10 describes a promising ultra-fast switch based on a nonlinear loop mirror constructed out of silica single-mode fibre. These and similar devices could form the basis of optical regeneration circuits, clock recovery and multiplexer circuits operating at speeds that cannot easily be accessed by electronics; sub-picosecond recovery time indicates that operation beyond 100 Gbit/s is a practical possibility.

1.5 END-TO-END PHOTONIC NETWORKS — THE VISION

The combination of fibre amplifiers, wavelength and time division multiplexing, optical switching and signal processing will lead to a wide range of new network concepts with the long-term possibility of end-to-end optical transparency. Some of the expected desirable characteristics from these networks include:

- transparency to service type and protocols;
- huge capacity for future needs and with fast response to difficult-to-predict broadband service growth;
- lean — efficient resource sharing and lower capital cost with less to manage and maintain;
- better reliability from fewer elements and able to use surplus bandwidth to build in resilience;
- graceful growth with low upfront costs as it is possible to add network equipment and terminals only when needed;
- dumb, with intelligence located around the periphery to exploit trend towards low-cost and high-functionality silicon in customer premises equipment;
- removal of interfaces, gateways and optical energy conversions since a significant part of network costs is from interfaces.

1.6 ADVANCED OPTICAL TECHNOLOGY ENABLING NEW SERVICES

Whilst much of the rationale for the further development of optical communications has been directed at the capacity likely to be generated by new services, there are other important physical attributes that remain to be developed further.

Information security is likely to be a vital issue to both individuals and business as they become increasingly dependent on the network for transactions that involve both financial and personal data. Chapter 7 describes a technique to improve data security on a fibre link by applying the laws of quantum physics towards achieving a fundamentally secure cryptographic key distribution system.

The wireless local area network (LAN) is a growing area for the future office where people will demand tetherless communications from their personal digital assistants — phones and file servers. Chapter 6 outlines the possibility of a local area network based on free-space optical links to provide a wideband interconnect to office equipment. Chapter 8 discusses the role of optoelectronic technology at the interface between fibre and radio networks. The new services which will be carried by the future network will be dependent on advances in terminal equipment, particularly at the man/machine interface. The creation, processing and display of visual images will be of increasing importance — large and bright displays are likely to be a prerequisite of the advanced terminals of the next decade. Three-dimensional displays, based on solid state lasers covering the full visible spectrum, are considered by some to offer the ultimate solution. Until now the existence of efficient, room-temperature operation, solid-state blue lasers has proved to be elusive. The recent achievement of light emission over a range of visible wavelengths in lasers made from doped fluoride fibre (Chapter 13) brings us one step closer to that goal.

1.7 SUMMARY

There continue to be advances in techniques to increase the transmission capacity of the embedded base of single-mode fibre well beyond any currently projected service demand. What is of greater significance is the new possibility to trade the reduced network complexity from having fewer network elements for improved reliability and lower maintenance costs. Advanced low-cost optical technology is still a crucial issue for the access network and further developments in component integration and packaging are still required.

Further into the future the end-to-end all-photonic network is clearly becoming a technical possibility. However, the need for such end-to-end transparency is only a virtue if it provides good value to the user and greater flexibility for the operator. Electronic switching still has significant potential from which to draw, and it is considered highly unlikely that optical switching will ever compete at the granularity, for example, of a compressed MPEG (Moving Picture Experts Group) video service. What is more likely is a network where photonics is used more and more extensively but not exclusively and the optimized solution uses the best features of both.

The photonic networks discussed in this chapter, although lean in complexity, would have huge capacity for bandwidth-hungry services. They would also be fairly dumb with most of the intelligence residing in electronics and software around the periphery, which is somewhat counter to the current evolution path of telecommunications networks. It is by no means certain which direction, if either, is likely to be the natural order of things and which, in the longer term, may eventually turn out to be a blind alley.

In conclusion, a very exciting future is seen for advanced optical technology not only in its crucial role in the main transmission pipes but also for the feeders, local area networks and within the powerful node processors necessary to deliver the new services.

REFERENCES

1. Kao C and Hockham G: 'Dielectric fibre surface waveguides at optical frequencies', Proc IEE, 113, p 1158 (1966).

2. Midwinter J E: 'Optical-fibre transmission systems: overview of present work', Post Office Electr Eng J, 70, No 3 (1977).

3. Kawanishi S et al: '100 Gbit/s 50 km optical transmission employing all-optical multi/demultiplexing and PLL timing extraction', OFC/IOOC 93 Post deadline paper, San José (February 1993).

4. Payne D B and Stern J R: 'Transparent single-mode fibre optical networks', IEEE, J Lightwave Technol, LT-4, No 7, pp 864-869 (1986).

5. Healey P, Smith D W and Cassidy S A: 'Photonic switching and interconnect for future network nodes', BT Technol J, 9, No 4, pp 19-29 (1991).

6. Hill G R: 'A wavelength routeing approach to optical communications networks', BT Technol J, 6, No 3, pp 24-31 (1988).

2

OPPORTUNITIES FOR ADVANCED OPTICAL TECHNOLOGY IN ACCESS NETWORKS

D B Payne

2.1 INTRODUCTION

Over the next ten years there is likely to be a significant penetration of fibre technology into the access networks of most major Telcos throughout the world. The initial penetration will be against plain old telephony services (POTS) and cable television (CATV) (not necessarily together) but with a growing demand for two-way multimegabit broadband services in the latter part of the decade.

Future networks will need to exist in an increasingly competitive environment. The networks that survive will be those that have the greatest flexibility and can respond to change and new service demands quickly and efficiently. The major potential advantage of optical networks over metallic systems is the inherent flexibility to carry any conceivable service package to any mix of customers over a very large geographical range. The use of optical networks as the bearer infrastructure will enable one network platform to carry all service types with common management and operations and maintenance (O&M) systems. Very rapid service provisioning could be possible with many services being 'dial up — on demand'. The huge geographical reach of optical networks can also aid node consolidation and at the same time benefit the implementation of low-penetration services by concentrating traffic from sparce and widely distributed customers back to highly centralized servicing nodes.

There is now a wide consensus that shared access passive optical networks (PON) will provide the basis for cost-effective penetration of fibre systems into the local network. The PON architecture is an immensely flexible network platform capable of delivering narrowband and broadband interactive services. It also has an inherent capability for distributive service provision.

2.2 EXPLOITING FIBRE CAPACITY

An advantage of a fibre transport layer in the access network is the potential for further exploitation of the fibre capacity for new and enhanced services. One attractive method of accessing the fibre capacity is to use wavelength multiplexing techniques.

Optical wavelength multiplexing offers several unique features that can be exploited to advantage in passive optical networks:

- the ability to add services/capacity without disturbance to existing customers;
- each added wavelength is a new transparent optical channel and can be independent of any other wavelength in terms of bit rate, modulation format and services carried;
- wavelength multiplexing allows access to the enormous capacity of the fibre without requiring new advanced electronic technologies — 'existing' electronics and optoelectronics can be used;
- if the correct initial PON architecture is employed only terminal equipment needs to be added to utilize additional wavelengths — if wavelength-tuneable devices are used in terminal equipment, access to new services, carried on additional wavelengths, can be achieved by software control without additional terminal hardware;
- the optical multiplexing and demultiplexing components can be passive and therefore placed at any convenient point in the network, allowing greater flexibility in network layout and implementation.

All early PON systems will carry at least POTS and narrowband digital services and it is generally agreed that these services will be carried in the 1300 nm optical window. To minimize the cost of the optical network termination unit (ONU) these early systems will have all of the usable spectrum of the 1300 nm window allocated for use by the ONU laser. This allows a simple uncooled F-P laser to be utilized over a wide temperature range without compromising yields with tight operating wavelength specifications. Unfortunately it also means that no service enhancement, requiring extra capacity, is possible for dialogue services within the 1300 nm window. In the exchange to ONU transmission direction there is more scope for use of a laser with better wavelength stabilization. This could be a temperature-controlled F-P device combined with a tighter operating wavelength specification or even a low-cost DFB laser. The use of such lasers could enable a few additional wavelengths within the 1300 nm window to be added to the system (assuming of course that at the outset a suitable wavelength-blocking filter is fitted to the ONU receiver). These additional wavelengths can then be used to carry distributed services from the exchange to the ONUs.

2.3 UPGRADING IN THE 1300 nm WINDOW

An experimental system, which demonstrated few-wavelength exploitation of the 1300 nm window, was constructed and reported by BT Laboratories in December 1990 [1]. A simplified schematic of the experimental system is shown in Fig. 2.1. A range of services were supported including telephony at 1300 nm and two distributive video services at 1320 and 1340 nm. A

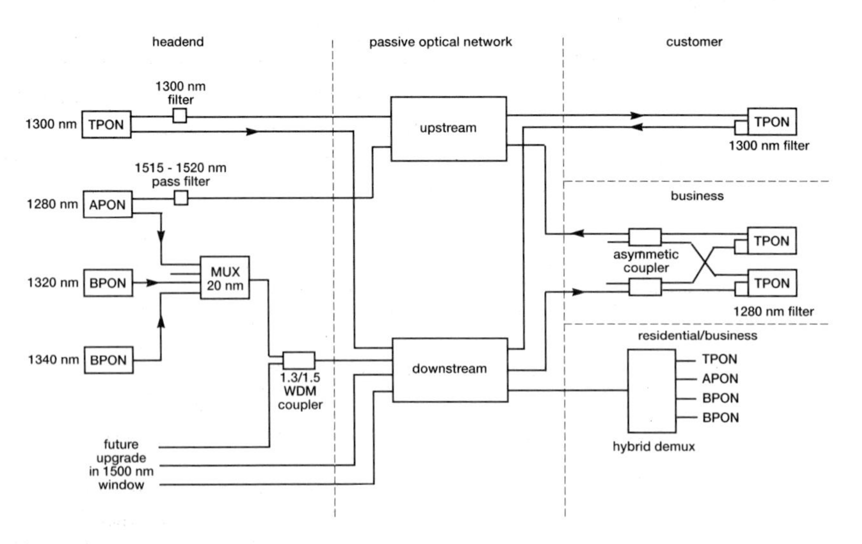

Fig. 2.1 Experimental passive optical network.

second two-way or dialogue service was also added at 1280 nm. This was an asychronous transfer mode (ATM) based system (APON) [2] operating at 155 Mbit/s with a pay load of 140 Mbit/s. Because of the use of uncooled, simple F-P lasers for the telephony service the return channel for the APON system was placed in the lower part of the 1500 nm window at 1515 nm. For exchange to ONU transmission distributed feedback (DFB) sources were used at the exchange and a range of filter and demultiplexer technologies were used for channel selection at the ONUs.

To enable fuller exploitation of the 1300 nm window a laser source with much greater wavelength stability and accuracy will be required for the telephony service ONU. The increased stability and accuracy needs to be achieved without the need for thermo-electric cooling, as this will significantly increase power consumption and increase packaging complexity and hence cost. Several options are possible that could achieve increased wavelength stability over a wide temperature range.

One interesting and potentially low-cost approach is the use of a UV written grating within the pigtail fibre [3]. A simple schematic of a possible device is shown in Fig. 2.2. An anti-reflection coating is placed on the output facet of a standard F-P chip and this is mounted in a conventional low-cost package. The back facet could have a high-reflectivity coating, if this did not increase costs prohibitively, as this would raise output power and lower threshold currents. The second facet is now formed by the narrowband reflection filter produced by the UV written grating in the core of the fibre pigtail. The lasing wavelength is defined by the spectral shape of this filter. For moderate bit rates (< 100 Mbit/s) the grating position is not critical and can be written several centimetres from the chip facet. Because this filter is written in silica the temperature stability is excellent with a temperature coefficient of 0.016 nm/°C and the absolute wavelength can be defined very precisely. This type of solution could therefore offer very precise and stable wavelength sources with the potential of only a marginal increase in cost compared with low-cost, uncooled F-P laser modules.

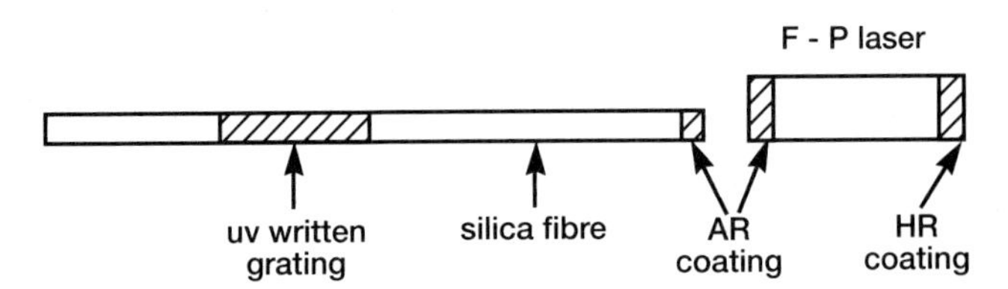

Fig. 2.2 Wavelength stable laser utilizing UV written fibre grating.

It is important that a low-cost wavelength-stable source is available for operational use as early as possible before a large number of 'basic' ONUs have been installed. The cost of retro-fitting a large installed base of simple

F-P laser ONUs could become prohibitive, increasing the pressure to use early but very wasteful upgrade options in the 1500 nm window.

2.4 EXPLOITATION OF THE 1500 nm WINDOW

The ability to provide early upgrade in the 1300 nm window will allow more strategic applications for the 1500 nm window. This would include exploiting high-density wavelength-division multiplexing (HDWDM ~1 nm) or even ultra-high density WDM (UHDWDM ~0.1 nm) and high-power erbium-doped fibre amplifiers to realize new and significantly more powerful network solutions.

To exploit fully the available spectrum of the 1500 nm window and also to increase the networking opportunities it will be necessary to move to higher-density multiplexing techniques. The use of HDWDM raises some important questions concerning wavelength accuracy and stability and the operational problems of managing networks with significant numbers of wavelengths. An immediate implication is the necessity of different wavelength sources being required in ONU equipment. If these are nominally fixed wavelength sources (e.g. DFB lasers) then problems associated with manufacturing and operational tolerances and the maintenance and installation issue of placing the correct source in a specified ONU begin to look prohibitive. A solution to this problem is to use tuneable components for both sources and receivers such that the operating wavelength of any ONU is set up and controlled by the network management system. The opportunity for universal components then arises, resulting in cost advantages from volume production of a common type of device for source or receiver.

Once tuneable components are installed in ONUs the increased functionality allows new networking opportunities to be considered. The wavelength agility provided at the network periphery can be utilized as a distributed switch with point-to-point and broadcasting capability. If this is combined with electrical multiplexing and channel selection at the ONU (as in current PON systems) then a very powerful distributed switching stage could be realized. A simplified schematic of a distributed network is shown in Fig. 2.3. ONUs connected to PONs are brought together via an amplifier and splitter chain. In this system all traffic is collected together, amplified and broadcast to all terminals where wavelength and time slot assignment protocols enable channel selection, authentication, etc. A ranging and synchronization protocol is required, as in conventional PON systems, to avoid data collision between the multiple users sharing a wavelength at any one time. One possible approach, using a TDMA protocol with timeslots scrambled with unique identification sequences, is described in Smith [4].

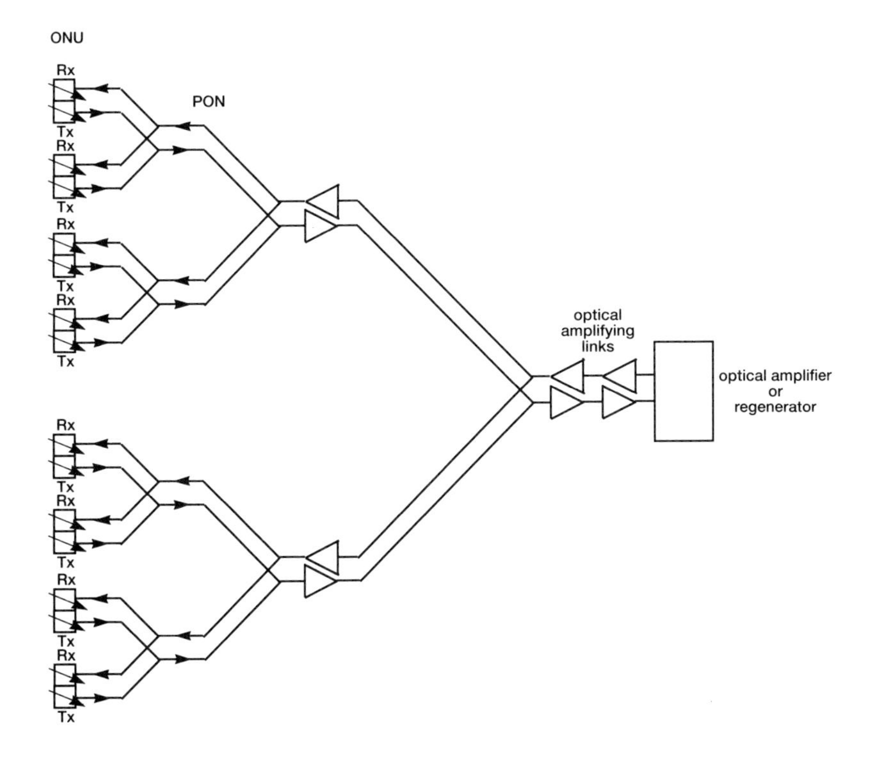

Fig. 2.3 Distributed switching network.

Although this network structure is very simple with no switching hardware outside the ONU, modest bit rates and numbers of wavelengths can realize a large broadband switch structure. Table 2.1 shows the approximate number of ONUs supported as a function of the bit rate/ONU and the bit rate/wavelength for a 40-wavelength system. Even at a modest 150 Mbit/s on each wavelength with a protocol efficiency of only 50% and assuming 10:1 concentration (comparable with concentration in the digital public switched telephone network (PSTN)), ~12 000 ONUs could be serviced. Of course to access such a large number of terminals in practice will require significant use of optical amplifiers but such numbers, and indeed significantly greater, have already been demonstrated in experimental systems with geographical distances of 500 km [6]. The figures in Table 2.1 are not meant to be definitive but serve to illustrate the potential of upgraded PON systems when wavelength agile techniques are combined with electrical multiplexing to provide switched broadband services.

Table 2.1 Number of ONUs supported on distributed switch network.

Mbit/s wavelength	*Example service bit rate*			
	ISDN 30 2 Mbit/s	CATV 8 Mbit/s	High quality interactive video 30 Mbit/s	HTDV 140 Mbit/s
34	3000	750	—	—
155	12 000	3000	750	—
622	48 000	12 000	3000	750
2400	192 000	48 000	12 000	3000

Note: 40 wavelengths, 10:1 concentration, 50% protocol efficiency.

The simple network shown in Fig. 2.3 may cause concern from a network reliability and disaster recovery standpoint because all the network traffic is ultimately routed through a single path. However, the network can be readily provided with alternative routes by using passive WDM components to partition the wavelength spectrum. Figure 2.4 shows a network with passive

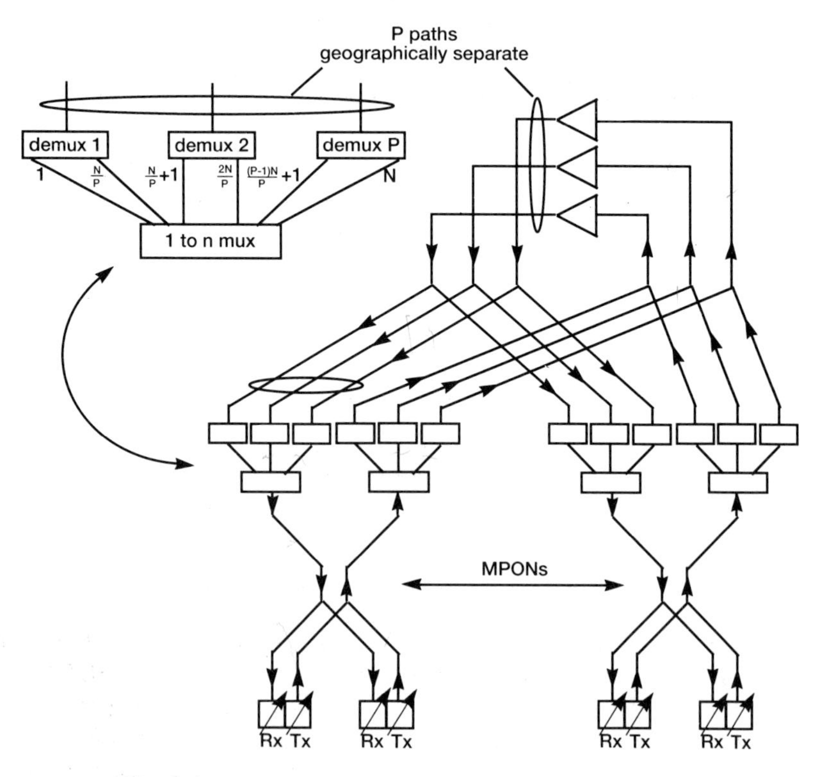

Fig. 2.4 Distributed switch network with multiple routes.

WDM components providing multiple transmission paths for traffic routeing. All routes are in use at all times carrying different subsets of wavelengths. The effect of a route failure is to reduce the total number of wavelengths available and therefore to increase the blocking probability within the network. It does not stop customers receiving service, although calls in progress over the failed route would be lost in this simple system until new connections were re-established. An alternative approach would be to provide a switch within the passive multiplexer-demultiplexer partitioning units as shown in Fig. 2.5. The wavelengths are distributed across the alternative routes as before but in the event of a route failure the switch enables the wavelengths carried on the failed routes to be added to the wavelengths on the surviving routes. This approach fully restores network capacity at the expense of non-passive nodes and additional management overhead.

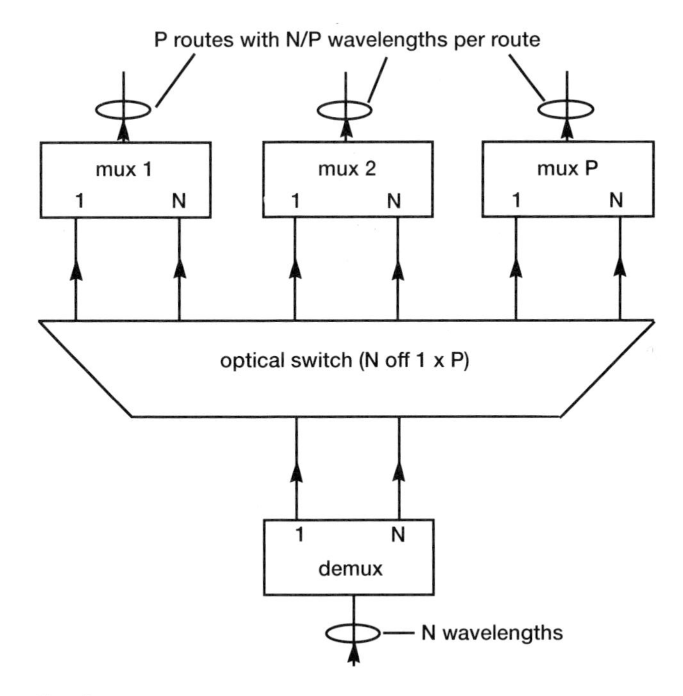

Fig. 2.5 WDM wavelength partitioning unit with optical switch.

An additional network management issue associated with this type of network arises from the use of erbium-doped fibre amplifiers (EDFA). In a simple EDFA the spectral gain profile is not flat and is also dependent on the input power level. To make the amplifiers suitable for use in HDWDM systems some technique to flatten the gain profile will need to be employed.

Although techniques for wavelength-flattening amplifiers have progressed well, the spectral control of amplifiers in dynamic systems is still a research topic requiring further work.

A further feature of wavelength-partitioning is that it can enable wavelength reuse via a hierarchical wavelength structure. If there is a sufficient community of interest between the customers on PONs connected to a wavelength-partitioning unit, a subset of wavelengths can be utilized for 'local' connections. The remaining wavelengths can be further partitioned for onward-transmission over geographically separate routes. The 'local' wavelengths can be used by all the PON groups connected to wavelength-partitioning nodes. The onward-transmitted wavelengths are common to, and therefore shared between, all PONs connected to the network. A simplified scheme for wavelength reuse is shown in Fig. 2.6.

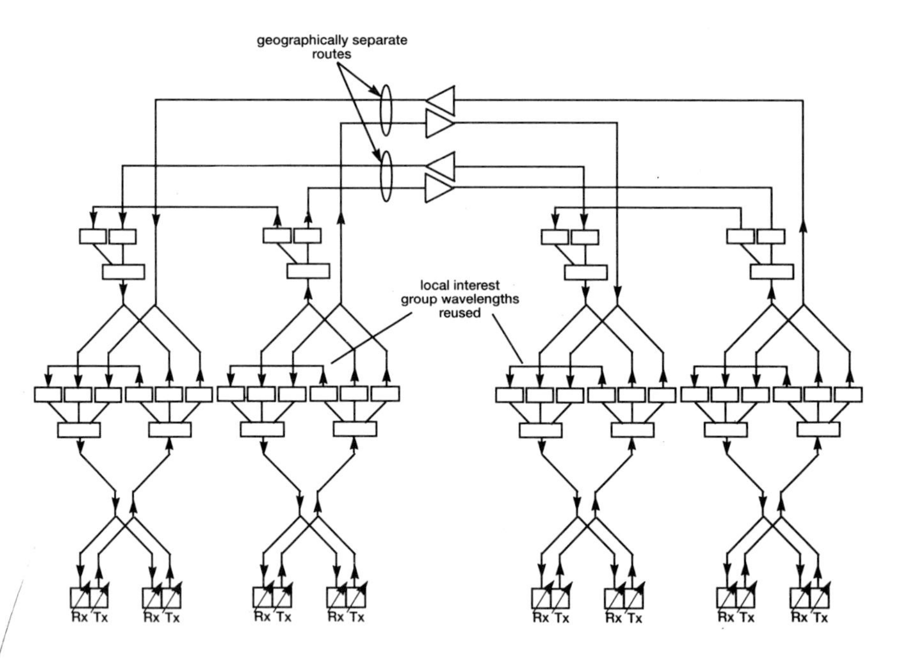

Fig. 2.6 Distributed switch network with wavelength reuse and multiple routes.

This simple distributed network still suffers from loss of service for customers connected to a wavelength-partitioning node if the node fails. Node bypass protection can be arranged by including additional input mux/demux units in the wavelength-partitioning unit as shown in Fig. 2.7. PONs could then be provided with two or more routes to additional geographically separate wavelength-partitioning nodes. Traffic would be switched to the alternative nodes in the event of first-choice node failure. It is important to note that the node protection and route protection is provided with only a very small percentage overhead in total installed equipment in these types of network. Most of the capital investment resides in the periphery of the network. A few additional routes in the centre of the network and additional inputs to the wavelength-partitioning nodes do not have a significant impact on the total network cost but enormously enhance the capability of the network.

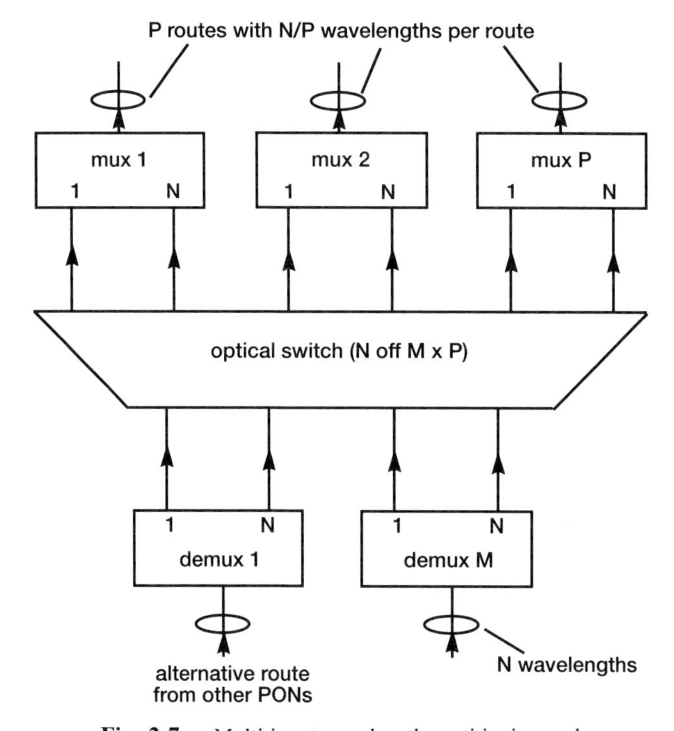

Fig. 2.7 Multi-input wavelength-partitioning node.

2.5 BROADBAND OVERLAY UTILIZING DISTRIBUTIVE AND NODE SWITCHING

As the demand for broadband capacity increases, the distributive networks described above will ultimately become capacity-limited even if UHDWDM with several hundred wavelengths is exploited. The solution is to enable complete wavelength reuse within different geographical regions. This can be achieved by combining distributive switching with central node switching.

At this node, space switching, and possibly wavelengths translation, will be performed to enable interconnection to further switching nodes and distributive switch stages. An outline of such a network is shown in Fig. 2.8. PONs connected to a switch node (which can be expanded versions of the switched wavelength-partitioning node shown in Fig. 2.7) form the input and output stages of a three-stage switching network [5]. Internode trunks are connected via a wavelength translation unit. This could consist of a back to back tuneable receiver and tuneable transmitter and can be considered as PON ports connected back to back. Because the PONs have inherent broadcast capability, the network readily lends itself to distributive services such as CATV. This is shown in Fig. 2.8 by the remote head end feeding the downstream PONs. Such a head end could be very remote; indeed one CATV head end could serve a country the size of the UK with thousands of video channels [6]. The whole network would be locked to a centralized reference comb of wavelengths derived from absolute standards and broadcast to all switch nodes [7].

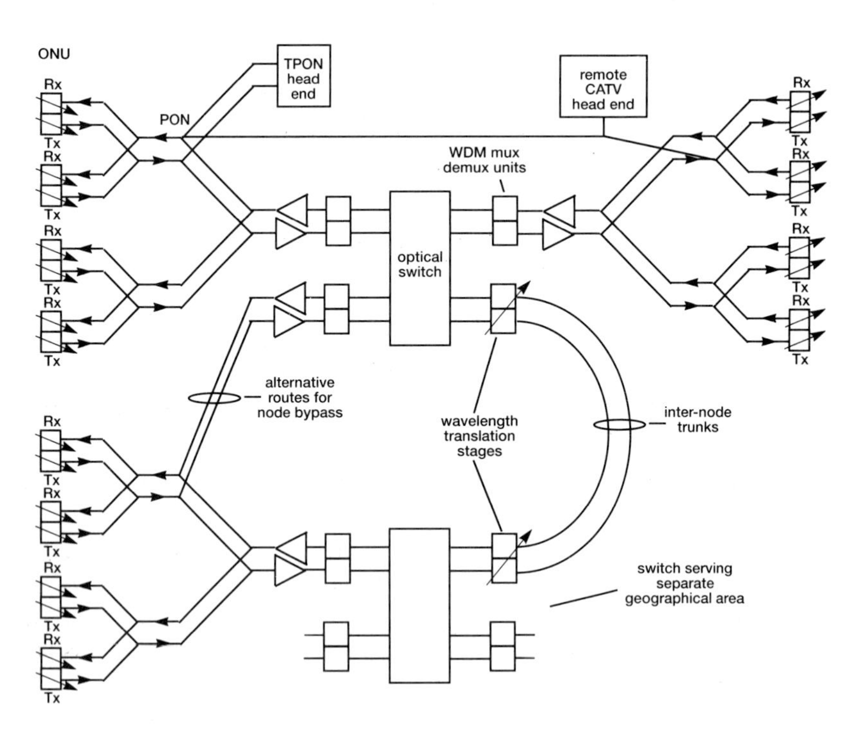

Fig. 2.8 Distributed and centralized switching network.

2.6 CONCLUSIONS

The fuller exploitation of the optical wavelength domain within a PON architecture will inevitably require the use of wavelength-tuneable components. These components will be an integral part of the optical network termination units at the periphery of the network. Such functionality, initially provided for channel or service selection, can form the basis of a distributed switching structure and can have significant impact on the design of the overall network-switching architecture. When combined with optical amplifier technology radically new approaches to whole network design can be considered.

Such networks can be fully compatible with an evolving PON infrastructure and can co-exist side by side with the existing PSTN which for many years to come will carry the majority of narrowband traffic. The new network structure will, however, be able to provide switched broadband connections for all types of service, with capacity available on demand. The broadband switching capability will also enable movement of large amounts of traffic about the network to meet unpredictable demand with minimal, or no necessity for, additional network infrastructure. The networks will also have the potential to offer a high degree of fault tolerance and significant immunity to disaster situations providing major gains in network and service reliability.

REFERENCES

1. Hill A M et al: 'An experimental broadband and telephony passive optical network', Proc Glob Comm '90, San Diego, USA (1990).
2. Ballance J W et al: 'ATM access through a passive optical network', Electron Lett, 26, No 9 (April 1990).
3. Bird D M et al: 'Narrow line semiconductor laser using fibre grating', Electron Lett, 27, No 13 (June 1991).
4. P J Smith et al: 'A high speed optically amplified TDMA distributive switch network', Proc IOOC-ECOC '91, Paris (1991).
5. Hill A M and Payne D B: 'A distributed wavelength switching architecture for the TPON local network', Proc 13th International Switching Symp, ISS 90, Stockholm, 3, Paper P4 (June 1990).
6. Forrester D S et al: '39.81 Gbit/s, 43.8 million-way WDM broadcast network with 527 km range', Electron Lett (1991).

7. Blyth K J et al: 'Wavelength applications in the local access network', IEE Colloquium on Wavelength Standards for Fibre Optic Systems, Savoy Place, London (February 1992).

3

FUTURE CORE NETWORKS USING NOVEL BUT SIMPLE OPTICAL TECHNOLOGY

P E Barnsley

3.1 INTRODUCTION

Any surplus capacity present in today's networks may be quickly removed with the expected customer demand for future feature-rich broadband services. This growth in demand has required in the past, and is likely to require in the future, exponential growths in line and switch capacities. The optical fibre currently installed in telecommunications networks has the capability of providing almost unlimited bit-rate transmission, but existing electronic switch and control hardware solutions prevent full exploitation of this potential. The present strategy of introducing standardized synchronous digital hierarchy (SDH) transmission equipment will have two major beneficial effects. Firstly, it will offer network operations the management capability needed to provide the feature-rich services and, secondly, it offers network operators a procurement advantage by standardizing the network interfaces.

But what of the future? Any network strategy for coping with future demands must be evolutionary and all technology developments must be mutually compatible — an integrated approach. The present strategy of upgrading the existing node technology to higher and higher bit rates may not offer the most cost-effective long-term solution and this approach could cause problems if broadband service demand grows exponentially, as the technology becomes cheaper. But the bottleneck, not necessarily in the speed of discrete electronics but in the ability to provide the management interfaces as electronic speeds and complexity increase, still remains, and what must be avoided is the possibility of having to confront a technology problem similar to that which prompted the introduction of SDH. Future-proofing the network is extremely important. Transparent optical networks are one possible strategic approach, which can offer a large degree of future-proofing to the network upgradability. The selective use of optical rather than electrical components to provide routeing within the communications networks of the future may result in significant operational improvements. The use of this advanced optical technology will be dependent on its cost relative to existing electronic plant available from many suppliers world-wide; however, the added value associated with the service, format and capacity-insensitivity demands strong consideration.

This chapter will discuss why such optical technology may be of such importance to the future networks in terms of its ability to assist improving network flexibility, management and maintenance. A possible upgrade strategy will be outlined showing how both wavelength and time multiplexing may be used to increase link capacities and reduce switching plant requirements at the network nodes. Experimental results will be given, summarizing the activity at BT Laboratories in the area of optical switching and routeing. These results will show that optical switch technology can prove a major benefit to implementing networks that the future may require.

3.2 THE BENEFITS OF OPTICAL TECHNOLOGY

As the penetration of fibre into the access network increases to satisfy the capacity requirements of present business customers and to allow future residential users access to advanced services, the capacity demands placed on the 'metropolitan' and 'trunk' networks will increase significantly. Field studies, such as the BT Bishop's Stortford TPON trials, have shown that optical hardware can be used to link customer premises equipment [1]. By providing this broadband pipe, future customers may gain access to an array of services provided by network and database operators, such as home-working, shopping and entertainment, multimedia and database interrog-ation. In this scenario any spare capacity present within today's networks will be very quickly depleted.

Accompanying the increased access capacity will be a drastic increase in the 'metropolitan' and 'trunk' network capacities. The following situation is just such an example of what could be facing network planners in the future.

By the year 2010 it is predicted that the total traffic in the core transmission network could be 16 times that at present [2] and this figure does not include the introduction of broadband services. The growth in demand is not easy to predict. In times of economic growth the demand may be higher than today (perhaps ~3% per annum) as demonstrated by the mobile communications market in the 1980s. If the mean access bandwidth used by network customers were to increase from 64 kbit/s to 2 Mbit/s in the same period, then a 500-fold increase in network capacity would be expected. Node consolidation, being a key cost-reduction driver, will mean that very large node capacities will be required to cater for this future demand. Some estimates suggest that link capacities in excess of 10 Gbit/s and switch capacities of approximately Tbit/s may be required in the trunk network by around the year 2010 [3, 4]. As depicted in Fig. 3.1, optical switch technology offers a potentially integrated solution to this problem. Optical switches within a transparent optical network structure can potentially offer the improvements in network operations and maintenance management while providing the platform to satisfy the customer capacity demand.

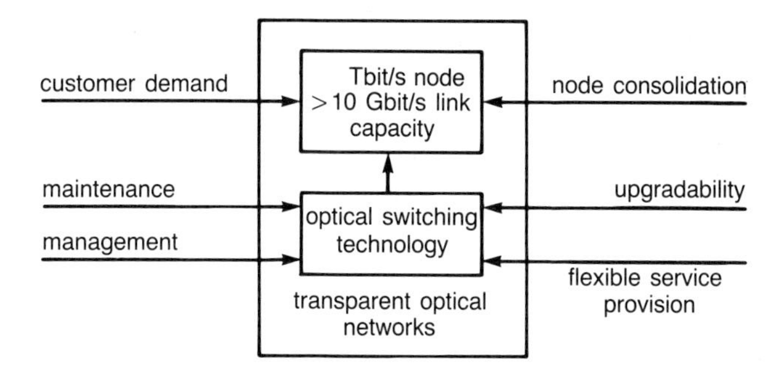

Fig. 3.1 A schematic representation of how transparent optical networks can provide the network platform on which any future customer demand for feature-rich services could be accommodated.

As the number of subscriber services increases, the networks providing the transport and switching for these services will need careful design. Already the move away from PDH towards SDH/Sonet networks is prevalent worldwide, with BT and other network operators installing 'switching' equipment capable of synchronous operation. In addition to the move towards synchronously switched networks, line capacities are increasing towards 2.5 Gbit/s and beyond, with commercial systems at these line rates already being procured in Canada, France, Spain and the USA. Furthermore, 10 Gbit/s systems have already seen field trial in Japan [5] and are currently under development by system suppliers.

To be efficient, network operators need to maximize the utilization of duct and terminal space. Better utilization of the inherent bandwidth potential of the existing transmission fibre is an obvious way of gaining further capacity without the need to pull in more fibre. In metropolitan and trunk networks the termination of multifibre cables with switching equipment implies large network installations with large electronic cross-connect switches. Since much of the trunk traffic reaching the network nodes is for onward transmission, a significant network advantage can be gained from developing optical systems capable of spans much larger than the majority of existing links (≤ 100 km in the BT inland network), and using coarse optical routeing to reduce the switch requirements at the nodes. This is illustrated schematically in Fig. 3.2. A transparent optical network carrying multi-gigabits per second data, multiplexed in either the wavelength or time domains, passes a network node. Optical switching technology (circuit or packet) centred at this node routes channels ascribed to that node down from the all-optical layer into the electronically switched SDH layer. At the same time, trunk traffic, from the node's vicinity, would be multiplexed on to the optical network. The electronic layer would provide the traffic grooming, management and end-to-end monitoring [2], with the optical switch acting as a semi-dumb bandwidth management device.

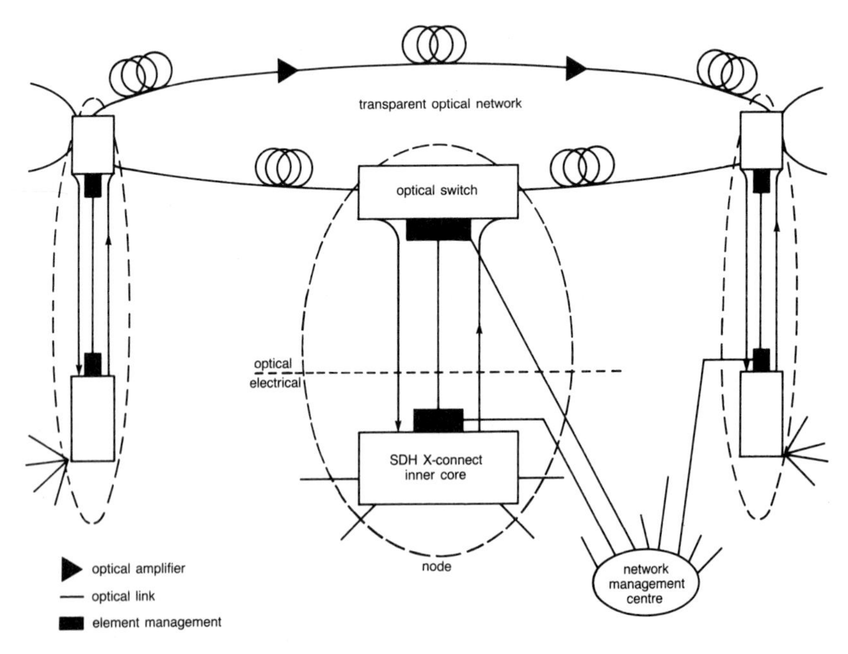

Fig. 3.2 A schematic diagram showing the integration of a transparent optical network with an SDH electrically switched network and the links to the management centre.

In order to achieve the desired line capacities alternative optical multiplexing techniques are being investigated, such as wavelength-division multiplexing (WDM) [6, 7] and optical time-division multiplexing (OTDM) (see Chapter 4), with systems already being demonstrated in the laboratory with capacities in excess of 20 Gbit/s, which is also discussed in more detail in Chapter 4. As the capacity of a network increases it is vitally important to maintain flexibility and resilience. Optical switching within the time, space and wavelength domains within the high-capacity network can provide improved protection strategies, reducing the need for $1+1$ protection and thereby providing further increased network utilization. Such strategies are potentially simple, offer a fast response time and the equipment does not need upgrading but is driven from the SDH management platform.

The optical switch depicted in Figs. 3.1 and 3.2 can offer four further areas where improved network performance can be expected.

- Flexible service provision — the introduction of transparent optical hardware, such as optical amplifiers, optical switches and optical clock recovery devices, can provide a physical platform that can accommodate virtually any type of service that customers could want. With correct choice of transport format (packet, circuit) the network operator could charge the customer for the services used. The use of fast optical switches can also allow fast (~ns) network reconfiguration with gigabits per second channel granularity.
- Network upgradability — since the hardware platform is broadband and virtually transparent to bit rate, further capacity demand can be quickly accommodated without the need for the procurement of whole systems, for example, by the addition of a further wavelength channel or by change of the transmission equipment at the main exchange nodes.
- Network maintenance — error activity should be reduced as high-speed optical components are inherently less susceptible to factors such as static damage, crosstalk and lightning strikes. This reduction of error activity in transparent routes could significantly reduce maintenance costs. In addition, the lifetime of optical may also be significantly longer than electronic hardware. Thus the mean time between failure for all-optical systems is higher, reducing operating costs further, and at the same time improving quality of service for the customers.
- Network management — as transmission capacity increases, the complexity of the software managing the switch may increase significantly. The risks of software failure, due either to a single event or a combination of events, and the prospect of the propagation of failures throughout the network, make software reliability a critical issue.

The introduction of optical components could reduce the amount of electronic equipment, and thereby reduce the required management overhead needed, so easing the demands placed on the software tools. However, the optical switching devices would also need to offer 'hooks' on to which the network management systems can be attached. These hooks are discussed in more detail in section 3.4.6 and in Hill [8] and Hawker [2].

3.3 A POSSIBLE NETWORK UPGRADE STRATEGY USING OPTICAL TECHNOLOGY

The evolution of optical networks is governed by many factors including the service demand, regulatory restrictions and the available technology. One possible proposal for network upgrading, based on the technology and operating drivers described in the previous section, is shown in Fig. 3.3, which schematically illustrates some of the potential node upgrades using optical technology that could result in the desired future networks. The first level of the core network (level 1 in Fig. 3.3) will be the synchronous (SDH) network, with optical transport and electronic switching at VC1 (2 Mbit/s) and VC4 (155 Mbit/s) with a central clock synchronizing the switch operations [9]. Each multiplex level can be mapped into the others, avoiding the need for

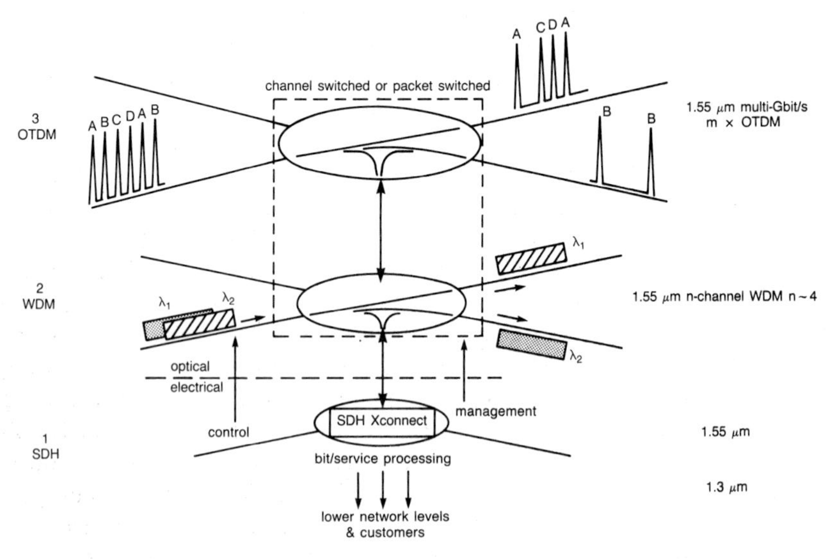

Fig. 3.3 A schematic representation of a possible evolution strategy for networks using optical technology to advantage.

bit-stuffing and multiplexer mountains inherent in PDH networks, and can provide increased network flexibility due to the add/drop capability. Networks of this type are expected to be operational within the BT inland network by the latter part of this decade.

Initially, further capacity upgrades are expected to be obtained by increasing the line rate towards 2.5 Gbit/s, at both the 1.3 μm and 1.55 μm transmission wavelengths, and upgrading the terminal switching equipment. The replacement of regenerator technology with broadband erbium-doped fibre amplifiers (EDFA) [10] for 1.5 μm systems, and perhaps praseodymium amplifiers currently under development (see Chapter 12) for 1.3 μm systems, will reduce the amount of electronic plant within the network, and will have a major impact on the timescales for further upgrade strategies. Further capacity may then be gained by implementing WDM on the transmission fibre and upgrading terminal equipment. However, this will result in a large increase in the electronic switching plant. Since much of the traffic entering the node is for retransmission, wavelength routeing [8] at the network node could then be used to reduce the switching plant required and increase the network flexibility (see layer 2 in Fig. 3.3). More details of this approach to network design are given in Hill [11], which also describes field trials of this technology within the London network [12].

In brief, the use of selective filters to route specific wavelength data signals across the network allows direct transfer of information between two points without the need for intermediate optical-to-electrical conversion and processing. Optical cross-connection fabrics allow the connectivity and wavelength converters to reduce blocking probability [13, 14]. The concept of self-healing SDH rings can be extended into the optical WDM ring architecture [15] to provide reliable, flexible network structures.

However, by about the year 2010, if present growth continues as the result of widespread take-up of broadband services, the trunk network may be operating with line rates greater than 20 Gbit/s (see layer 3 in Fig. 3.3). The distances involved, the number of wavelengths needed to fully interconnect the network nodes, and the nonlinear characteristics of both the optical amplifiers and fibre, mean that a multiwavelength transmission solution may be difficult to implement. OTDM systems are potentially ideal for long-span, high bit-rate transmission, with recent laboratory demonstrations including a 20 Gbit/s 205 km four-channel system (see Chapter 4) and a 20 Gbit/s 1020 km system [16]. Such an OTDM network may only link a few nodes, located at the largest demand centres, which for the BT network may number around twenty. In order to gain maximum advantage of such a network, dispersion-shifted fibre (DSF) could be used since the lower fibre loss at 1.5 μm, compared with 1.3 μm, allows longer transmission spans and the low dispersion allows for a wide wavelength range over which transmission is possible (see Chapter 4). The topology of this OTDM network may therefore be very different from that currently employed within existing networks. Ring and star topologies can be meshed to provide the necessary connectivity between the nodes while maintaining flexibility and simplicity.

Optical technology can therefore successfully exploit the bandwidth of fibre and allow network capacity to grow with demand. As access demand increases, WDM wavelength-routeing technology can be installed in the metropolitan area. OTDM technology can then be used to interlink the high-capacity metropolitan networks at the highest demand centres (see Fig. 3.4).

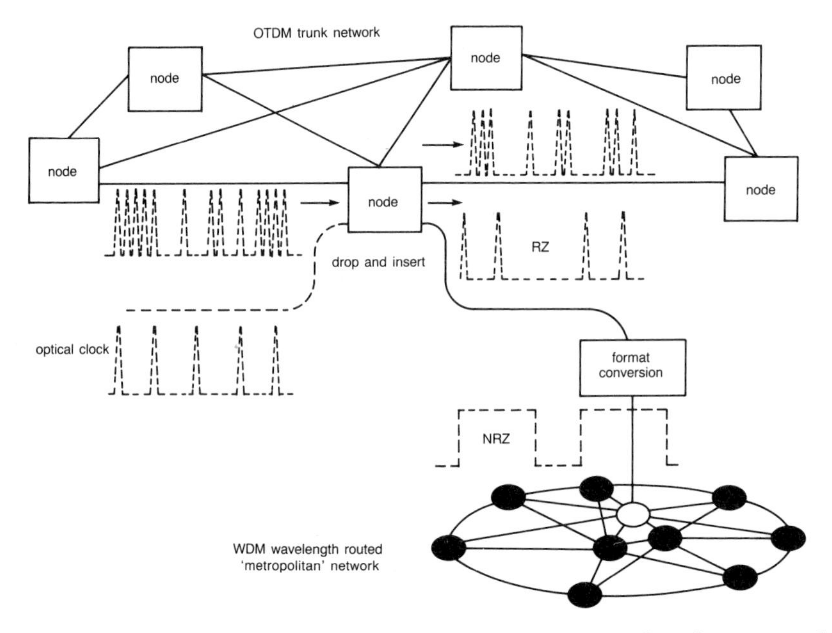

Fig. 3.4 A schematic illustration showing how an OTDM trunk network can be transparently linked to a WDM metropolitan network using an optical drop and insert switch coupled with a format converter.

3.4 OPTIONS FOR OPTICAL SWITCHING IN ALL-OPTICAL NETWORKS

3.4.1 Wavelength networks

As discussed earlier, wavelength routeing will probably be the first all-optical network upgrade solution. However, if the interconnections between inputs and outputs at the nodes within the wavelength-routed network are fixed, the network remains inflexible due to the manual reconfiguration time for changing the interconnection pattern (see Fig. 3.5(a)). Space switching, under management control, at each system wavelength within the interconnection field in the wavelength-routed network nodes (Fig. 3.5(b)) can regain the added network flexibility, improving the utilization factor by reducing the need for one-to-one protection strategies. If the input signals at each wavelength are time-multiplexed either optically or electrically, time switching (described in more detail in section 3.4.2) can also be used, resulting in further improvement in network flexibility.

The limitation on the optical power within the fibre due to the effects of fibre and amplifier nonlinearities (e.g. gain saturation) will probably limit the number of wavelengths able to be employed in such wavelength-routed networks to ≤ 10. Even with node consolidation the inner core of the BT network will probably require more than 20 nodes. For multinode connection, reuse of wavelengths between different regions of the network [2] can reduce the total number of wavelengths used in any link. However, wavelength reuse reduces the network flexibility by isolating network regions by specific wavelength sets. This flexibility can be recovered by using wavelength switching at the gateways between network regions. In addition, wavelength switching between the WDM channels within a wavelength-routed network would result in improved flexibility due to the increased node interconnectivity [14] (see Fig. 3.6), as wavelength routeing effectively produces many independent parallel networks. Wavelength switching can also be used to integrate networks operating at different wavelengths, e.g. the present 1.3 μm network to future 1.55 μm WDM wavelength-routed networks (Fig. 3.5). It has been shown [13, 14] that wavelength switches are imperative if high availability of network connections is required. This is particularly important in networks covering a large number of nodes with a few wavelengths [13].

Wavelength switching has been demonstrated in fibre and semiconductor devices. In general, the fibre devices (see Chapter 10) can operate at high bit rates, but require higher optical powers and are physically larger, requiring many metres (kilometres) of fibre, and stabilization circuits may be required to reduce thermal and vibrational effects from the exchange environment. Semiconductor devices (see Chapter 9) are not as fast but are small, require less optical power and have the potential of being integrated with control and drive electronics. Recent reports of 20 Gbit/s wavelength conversion suggest that semiconductor devices, integrated with planar waveguide structures, also offer such high speeds [17].

The linking of networks operating at 1.3 μm with wavelength-routed networks at 1.55 μm, as depicted in Fig. 3.5(b), has been demonstrated using two-contact semiconductor optical amplifiers (see Chapter 9). Here,

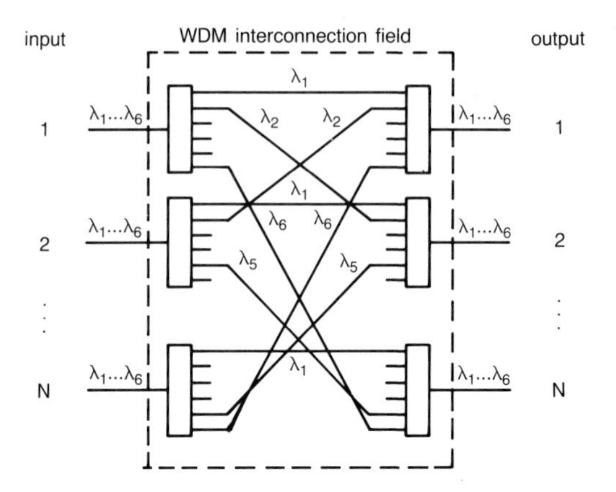

(a) Inflexible without reconfiguration switching.

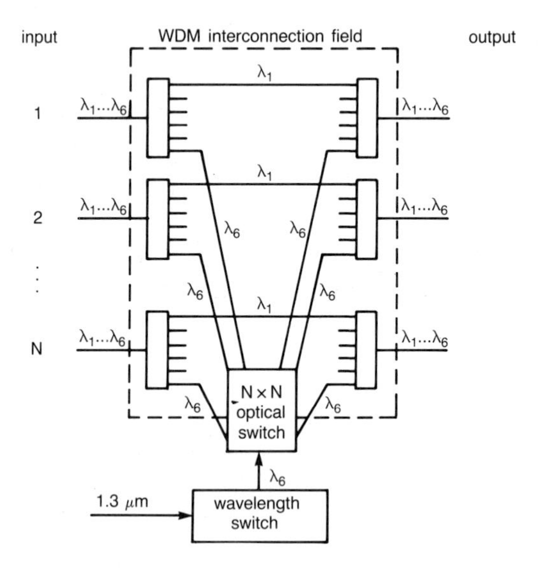

(b) With flexibility achieved by optical switching technology in time, space and wavelength.

Fig. 3.5 A WDM interconnection field with a wavelength-routed network node.

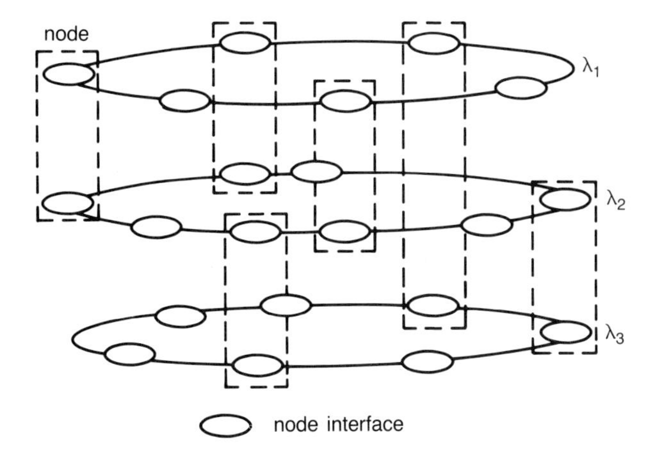

(a) Unconnected networks at different wavelengths.

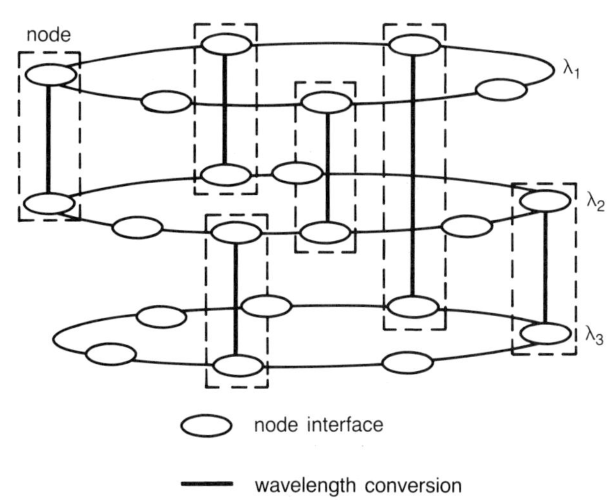

(b) An integrated network using wavelength conversion to provide linkage between the different wavelengths.

Fig. 3.6 The effects of wavelength routeing without wavelength conversion.

saturation of an absorbing region within the semiconductor device results in increased signal amplification at 1.55 μm [18]. Data at 1.3 μm (or 0.85 μm) incident on the saturable absorber modulates the gain observed by a second, local, 1.55 μm signal. Wavelength conversion at speeds up to 1 Gbit/s has been demonstrated [19]. Extinction ratios of >10 dB were obtainable using optical filtering. The local 1.55 μm source can be tuned across the gain bandwidth schematic of the device offering conversion to wavelengths between 1.53 μm and >1.56 μm [20]. Figure 3.7 shows a schematic of the experiment where the output from the wavelength converter was first amplified before being routed via an optical routeing switch and a WDM multiplexer on to 63 km of standard fibre, simulating a metropolitan scenario [12]. BER measurements, shown in Fig. 3.8, showed only a small, ~2 dB, system penalty across the whole network [21]. With further device improvements, operation at 2.5 Gbit/s is expected (see Chapter 9) which will allow upgrades to the present 1.3 μm networks to be integrated to 1.5 μm networks at a later date, thereby future-proofing today's investments in 1.3 μm switching plant. Other researchers have utilized this type of technique, with a single contact device, to demonstrate 2.5 Gbit/s [22] and 20 Gbit/s [17] wavelength conversion between wavelengths within the 1.5 μm wavelength window, proving the high value of semiconductor technology. Since tuneable semiconductor lasers have been developed that can be reconfigured within nanoseconds [23] this type of switching can offer fast network configuration at the granularity of the maximum SDH transport rate.

3.4.2 OTDM networks

Within the OTDM layer there is need for both space and time switching. If WDM and wavelength routeing are also utilized [24] (see also Chapter 4) then wavelength switching may also be required to provide added flexibility and routeing options as with wavelength networks. The different time channels on a particular input fibre should be capable of being linked to different output fibres (or space channels) and routeing at the node obtained by both optical cross-connect switches, based on multiplex and demultiplex operations, and timeslot interchangers will be required. The routeing of the different timeslots within the input frame depends on the configuration of the mux/demux switches for that timeslot with the cycle of configurations repeated for each successive frame. Different timeslots have different routes through the switch, either being remultiplexed with other channels for onward transmission (cross-connect) or dropped out of that particular layer (demultiplexed) and another data channel inserted (multiplexed). A timeslot interchanger switch, which reconfigures the order of the channels within the frame at the input to the switch, can be used to alter the routeing without having to reset the cycle of configurations at the mux/demux switches. The use of the timeslot interchanger is dictated by the size of the switch. For large high-speed switch fabrics, the increase in control complexity needed to cater for any reconfiguration of the switch elements may mean that a simple timeslot interchanger at each input is more cost-efficient.

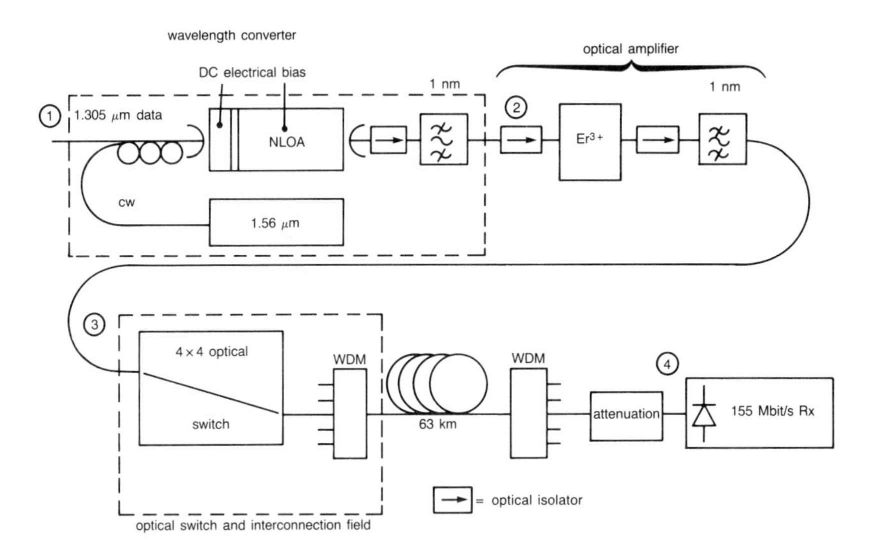

Fig. 3.7 A schematic diagram of the experimental set-up to investigate the proposal described in Fig. 3.5(b).

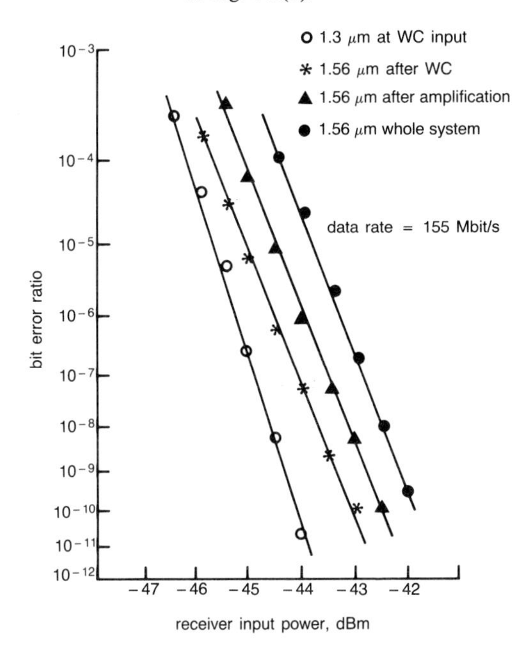

Fig. 3.8 The variation of BER with receiver input power at various points within the experiment shown in Fig. 3.7.

For both drop-and-insert and timeslot interchange the switch elements must have a fast rise-and-fall time and have a high repetition rate capability. If the channels are bit interleaved (OTDM) then the repetition rate is equivalent to the line-rate and all-optical techniques preferable, but for byte-interleaved systems (SDH/WDM), the repetition can be significantly lower than the data rate (typically ⅛) and electrically driven switches can be used. Other factors that affect the choice of switch are cascadability (which for active devices is linked to the noise figure and for passive devices is linked to insertion loss and crosstalk extraction [25]), stability and size. Furthermore, the architecture must offer non-blocking characteristics to ensure efficient network flexibility. Although classical space switch architectures (Benes [26], Clos [27], etc) require established control algorithms, they use a large number of switch elements compared with time-domain multiplexed (TDM) switch architectures. TDM architectures that use optical storage in fibre delay lines offer a reduction in switch complexity and use the same basic control algorithm. They offer timeslot interchange, optical cross-connection and drop-and-insert functions and are modular and scalable, allowing for continuous upgradability [28, 29].

Many potential switching technologies are now emerging, including lithium niobate ($LiNbO_3$) [30], semiconductor InP amplifier and electro-absorption devices [31] (see also Chapter 9), all-optical fibre devices (see Chapter 10), and organic polymers [32]. Both the polymer and fibre devices are generally polarization-sensitive and require optical clocking powers currently obtainable from only the best laboratory erbium-fibre amplifiers. They are, however, potentially very fast and ideally suited to the fibre environment. The further development of the fibre devices (see Chapter 10) is expected to result in greatly improved performance. The lithium niobate devices are presently commercially available, are quite fast (20 GHz [30]), but suffer from high insertion loss and drive voltages, and 10 Gbit/s devices have been successfully used in 40 Gbit/s OTDM systems [33] showing that the switch speed can be significantly lower than the optical line-rate and the drive electronics compatible with the highest speed SDH layer. The semiconductor devices are potentially as fast as the lithium niobate devices, potentially offer lossless operation, and are presently being investigated by commercial switch suppliers [34]. Recent results with travelling wave semiconductor amplifiers suggests that demultiplexing 10 Gbit/s channels from 40 Gbit/s data streams is possible when the device is used in a loop-mirror configuration [35].

Two types of technology have been successfully employed to demonstrate the feasibility of using optically transparent components to implement optical time-slot-interchange and drop-and-insert functions. One class of architectures, which offers the desired characteristic, have been theoretically investigated by Strathclyde University and recently demonstrated experimentally at BT Laboratories [29]. The switch fabric is built up of many stages of crosspoint switches where one output from each crosspoint is linked by optical delay lines (lengths of fibre) to an input of the next stage. The delays feed forward to produce even attenuation and good crosstalk performance. More detail of the switch fabric performance can be found in Hunter and Smith [28]. The length of the delay between stages may vary depending on the size of the overall switch fabric. The size of the delay is matched to the switch rate, i.e. block/packet/byte or bit. Delays of ~1 ns (1 Gbit/s) upwards are possible using fibre, but using other waveguide techniques such as silica-on-silicon (see Chapter 14), delays suitable for >10 Gbit/s operation are possible and, in addition, are able to be integrated and should produce more compact and stable switch fabric modules. Crosstalk effects must be carefully considered in designing this type of switch fabric [25].

These non-blocking fabrics are functionally identical to Benes networks and can therefore use standard control algorithms. Fabrics with and without frame integrity (all timeslots entering on one frame leave the fabric on the same frame) can be configured and dilation can be used to reduce the crosstalk coefficient of the fabric. Reduction in control complexity can be achieved by increasing the number of crosspoints. This trade-off between control complexity and switch count is currently under more detailed investigation. The number of switches required for an $N \times N$ switch fabric can in some cases equal the theoretical minimum, demonstrating the potential power of this type of switch fabric architecture. With some fabrics, the number of switches is less than the number of channels being switched.

A sample schematic illustration of this type of switch fabric is illustrated in Fig. 3.9, with the principle being extendable to much larger switch fabric designs using substrates containing many 2×2 crosspoints to reduce

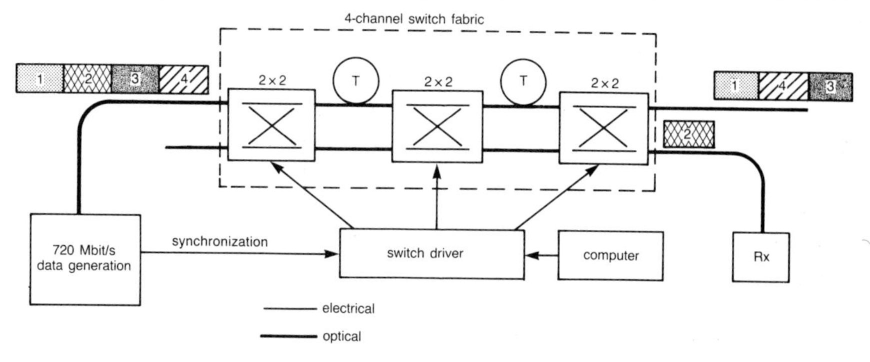

Fig. 3.9 A schematic diagram showing the type of switch fabric architecture described in Hunter and Smith [28] along with the experimental configuration used to evaluate the approach.

interconnection losses. This fabric consists of three stages and can accommodate four channels, whereas a switch consisting of 11 stages could accommodate 64 channels. The experimental configuration shown in Fig. 3.9 was built using lithium niobate 2×2 optically transparent switches (broadband electrical drive bandwidth ~4 GHz), driven using a computer-controlled programmable pattern generator. The limited rise and fall time of this generator meant that the switch configuration time was ~2-2.5 ns. The computer automatically derived the control signals required to obtain any desired mapping of input to output channels (assignments). In this demonstration [29] the optical delays were chosen for 45 Mbit/s switching. Four different optical blocks/packets, each 16 bits long and at a data rate of 720 Mbit/s, were injected into one input of the first switch. All possible configurations of the four timeslots were attainable at the output, an example of which is shown in Fig. 3.10. In addition a drop function could be implemented with one block switched to the second output from the fabric, as shown in Fig. 3.10. The variation in bit error ratio (BER) with mean input power to an optical receiver for the dropped channel is shown in Fig. 3.11. These results indicate that this fabric architecture is indeed suitable for implementing optical time-slot-interchange functions. A slight difference in BER performance is observed between different switch fabric assignments (lines 1 and 3) due to the differing switch configurations possible to implement a given assignment. In fact slight improvements can be obtained, since the switches can be configured so as to reduce unwanted noise on the input line — see line 1 in Fig. 3.11 [29].

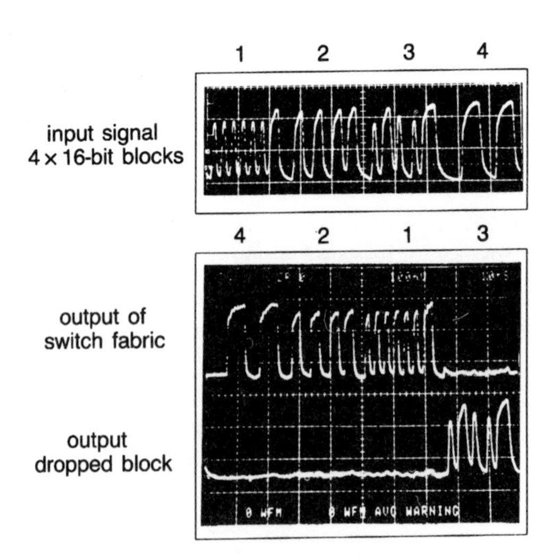

Fig. 3.10 Typical experimental results of the switch fabric shown in Fig. 3.9 demonstrating the time slot interchange and drop-and-insert capability of the switch fabric.

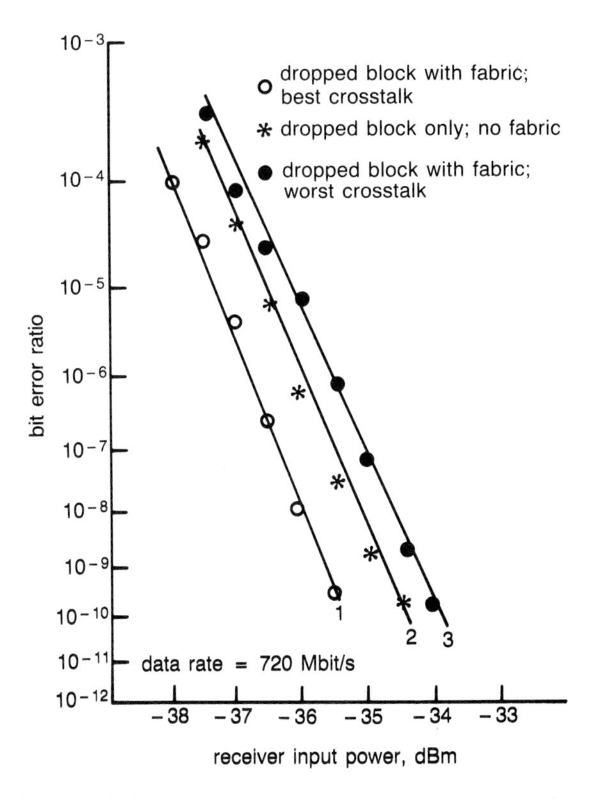

Fig. 3.11 Typical BER measurements for the case shown in Fig. 3.10.

Although lithium niobate devices are commercially available they tend to have a high insertion loss. The devices used in the above experiments had an average loss of ~6.5 dB. This high loss, even with the use of optical amplifiers, limits the number of switch stages that can be concatenated, due to reductions in signal-to-noise ratio. Another alternative switch technology is indium phosphide, with devices based on semiconductor optical amplifiers whose loss is dependent on the electrical bias applied. When unbiased, the device is blocking (loss typically 30 dB) but when biased with ~100 mA, a fibre-to-fibre gain of ~10 dB can be obtained. These devices can be combined with fibre couplers to produce 1×2 or 2×2 switch units, which can then be used to implement the same type of function as described for the lithium niobate devices (see Fig. 3.12). The coupler losses are offset by the inherent gain of the amplifier switch. Experiments have shown that these gated amplifier devices can operate at speeds up to 5 Gbit/s [36]. Other experiments have demonstrated optical drop-and-insert and optical

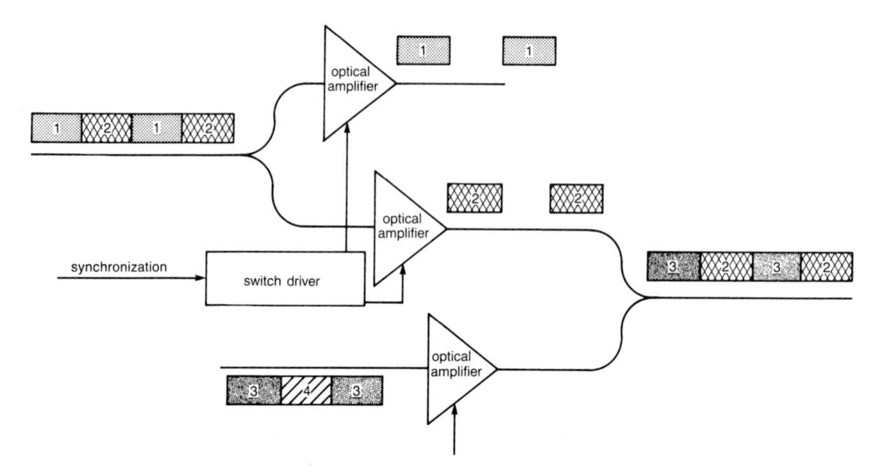

Fig. 3.12 Schematic representation of drop-and-insert switching using gated semiconductor optical amplifiers.

multiplexer and demultiplexer of 2.5 Gbit/s data packets [37]. Some typical results of these experiments are shown in Fig. 3.13, where the multiplexer and demultiplexer functionality for two 2.5 Gbit/s data streams is shown, and Fig. 3.14 shows the BER characteristics of a demultiplexed 8-bit packet/byte, revealing that slightly improved system performance (line 1) can be achieved due to the high extinction ratio of this technique, reducing unwanted noise [30].

The advantages of the semiconductor approach are the potential of lossless operation and the lower drive voltages required. The disadvantages of the semiconductor approach are that each discrete amplifier gate needs to be linked using fibre. A more compact and stable solution is to use the inherent

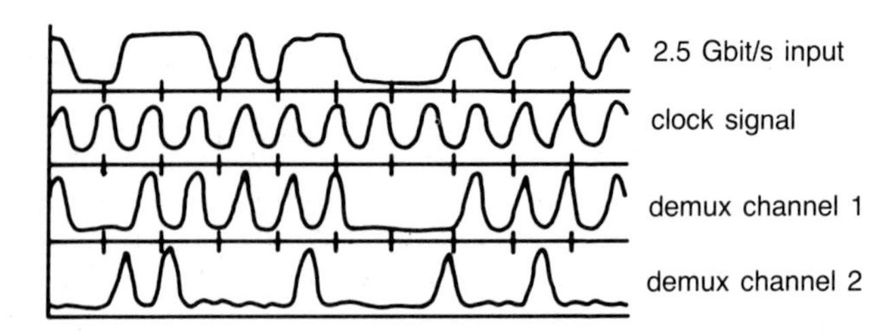

Fig. 3.13 Typical results showing the demultiplexing of two TDM channels from a 2.5 Gbit/s optical time-multiplexed data signal.

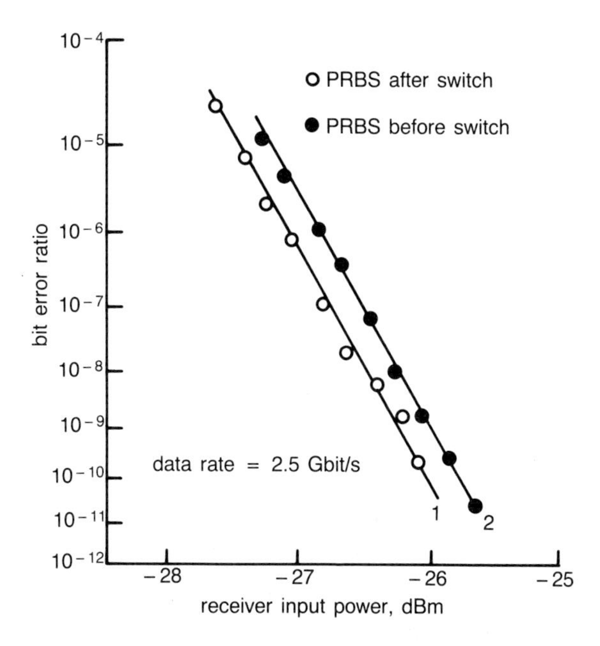

Fig. 3.14 Typical BER results when demultiplexing 8-bit long data packets from a 2.5 Gbit/s data stream.

integration potential of InP technology to produce 2×2 switch arrays on a single substrate (see Chapter 9). Rapid progress has been achieved world-wide in this field with 8×8 devices already in the research prototype phase. More details of this technology are given in Chapter 9. Eventually the drive and control electronics should be able to be integrated on the same substrate [38] thereby producing a highly functional, compact component with on-board element manager interfaces.

3.4.3 Linking between wavelength and time networks

The linking between the optical OTDM and WDM layers will probably be very important, especially if the WDM options are used primarily in future metropolitan network (MN) configurations and OTDM options are used in the trunk network to link many MNs as described in Fig. 3.4. The drop-and-insert function therefore presents a problem since in the WDM and SDH layers the data format is typically non-return-to-zero (NRZ), which is incompatible with the OTDM return-to-zero (RZ) format, and transfer between these formats is therefore required. In both cases there may also be a requirement (benefit) arising from a change in data wavelength.

Using the nonlinear saturation characteristics in semiconductor devices (see Chapter 9), this conversion process has been demonstrated experimentally at 5 Gbit/s. RZ data from the OTDM system described earlier was injected at ~ 1.547 μm into a two-contact semiconductor device. The optically filtered output from the device showed that the data had been transferred to the 1.547 μm signal. From Fig. 3.15 it is clear that along with the wavelength conversion the data format had been changed to NRZ. This is due to the limiting response time of the semiconductor material. Similar operation was observed when the CW input wavelength was tuned to 1.53 μm.

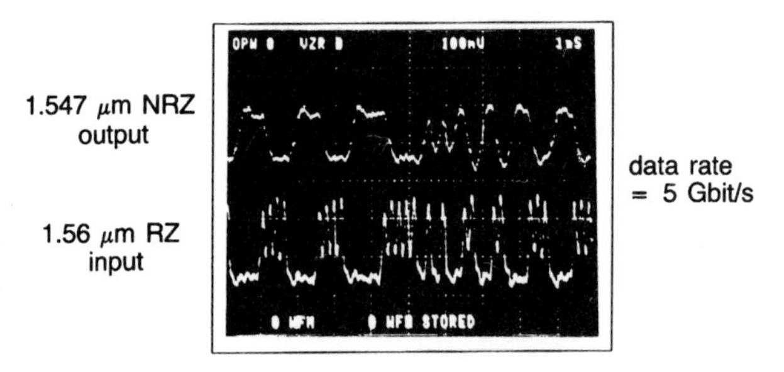

Fig. 3.15 Demonstration of both RZ to NRZ format conversion and wavelength switching of 5 Gbit/s data using a semiconductor device.

3.4.4 Packet switching

Packet-based switching fabrics have attracted much interest and form the subject of many research projects within RACE and other joint ventures. If optical packet switching is chosen, then the switch fabric architecture will be slightly different, as each data packet has an address, or header, associated to it. Buffering of arriving packets is also required to allow for contention resolution between packets destined for the same output. The optical switch depicted in Fig. 3.1 needs to be able to recognize this specific header code associated with that node and then switch the packet out of the transport layer into the lower bit-processing switch layers. A schematic diagram describing this implementation is shown in Fig. 3.16. The header can be separated from the data, and a header decoder determines whether the packet is destined for that node by comparing (correlating) address codes. The switch is then configured to route the packet either down to the electronic processing layer, or on to the next node after reattaching the header address. The header address can be coded in time or in wavelength in a variety of ways [39]. Time decoding using correlation techniques (Fig. 3.17) can be greatly helped by using integrated waveguide delays (see Chapter 14). If the address is coded using wavelength then this controlling signal can overlap the whole data

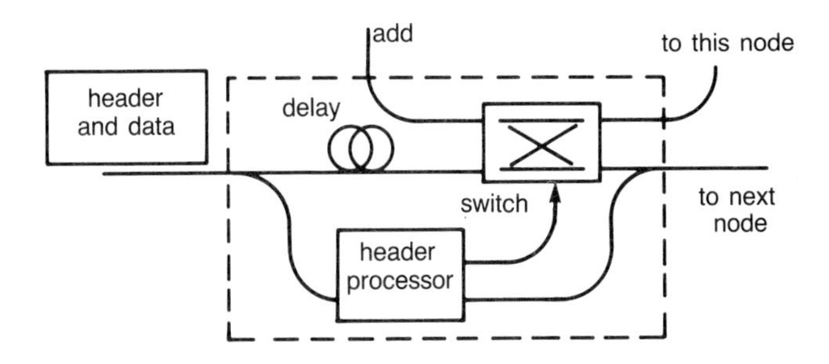

Fig. 3.16 A schematic diagram of a possible packet switch.

packet. The multiplexing of header and data information into the same time window allows better usage of the fibre capacity by allowing more packets per time slot (Fig. 3.17), but problems associated with group delay dispersion need careful consideration. Using wavelength instead of time as the address code makes decoding 'simple', using optical filtering technology (see Chapter 15), and also increases the compatibility with wavelength-routed networks. Figure 3.18(a) and (b) shows how both star and ring topologies can be used with combined header and data wavelength WDM interconnection fields (Fig. 3.5) to interlink a large number of network nodes. Group delay dispersion can be accommodated by introducing a transmission delay between header and data depending on the route length held in a look-up table. This technique could also be extended to offer a fast circuit-switching approach

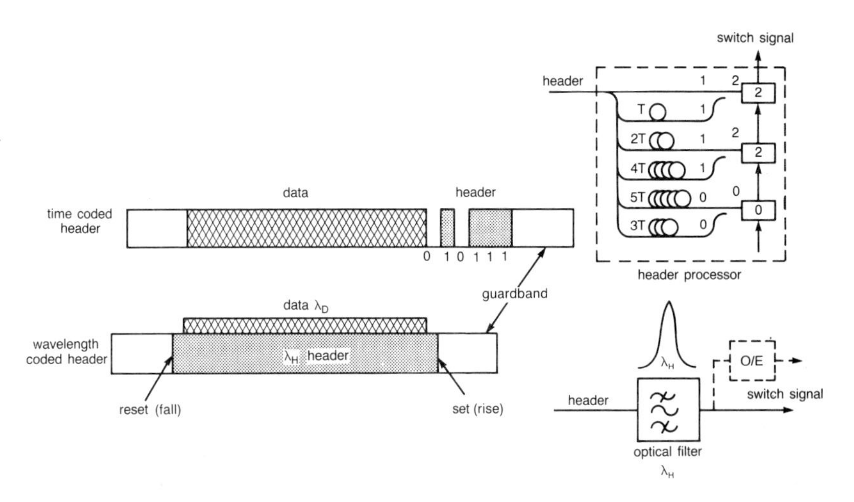

Fig. 3.17 A schematic of how the packet address code could be implemented using time or wavelength and a possible all-optical decode circuit for each approach.

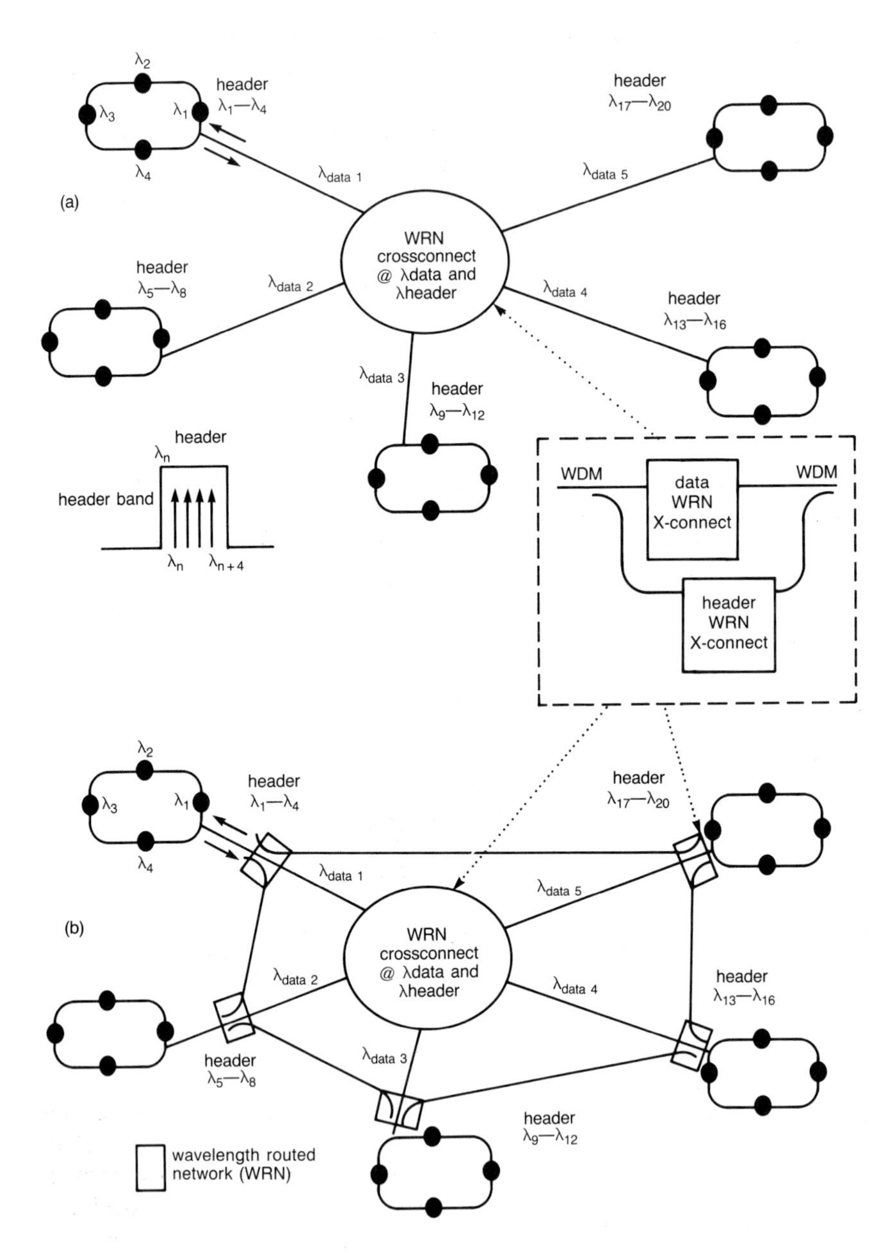

Fig. 3.18 Possible network architectures that could be used to link a number of nodes using wavelength routeing of both address and data with decoding and encoding at the network nodes — (a) star topology connecting many local rings, and (b) a star/ring interconnection topology for added resilience and flexibility.

using the multiplexed control signal to both set (rising edge) and reset (falling edge) the switching nodes along the route (see Fig. 3.17). Further work to investigate this technological approach is planned.

In an optical packet network, header recognition can be achieved electronically or optically using either parallel or serial processing. All-optical header recognition (Figs. 3.16 and 3.17) for time-coded header addresses has attracted some attention but the optical logic gates based on fibre (see Chapter 10) [40] and semiconductor devices (see Chapter 9) [41] are in an early stage of development. Wavelength-addressed header decoding is potentially simpler since it only requires narrowband filters (Fig. 3.17), which can be fabricated in a number of ways (see Chapters 15 and 16) [41].

Such an all-optical gating function has been demonstrated using the same two-contact device technology used for the wavelength-switching results described in section 3.4.1. In this configuration, however, the data was at the 1.5 μm wavelength and the 1.3 μm signal acted as a gating, or control signal, which selected packets of 16 bits (see Fig. 3.19). Extinction and contrast ratios, for the switched data, of >13 dB and ~10 dB respectively were obtained with open eye diagrams at data rates of 2.5 Gbit/s. These results show the potential of this type of gating technology for packet- and fast circuit-switched applications.

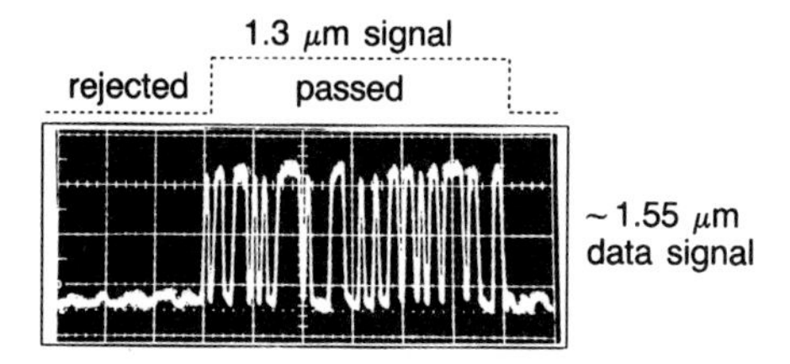

Fig. 3.19 A typical experimental result showing the principle of all-optical decoding of a wavelength-addressed packet where a block of 16 bits of 1 Gbit/s data was gated using a superimposed 1.3 μm control signal and a two-contact semiconductor device.

3.4.5 Synchronization

Any optical switch will require synchronization to both the information on the line and the start of a frame in channel-switched networks. The problem of clock extraction is therefore a very important one. Since the network must integrate with the electronically switched SDH network the all-optical network will have to operate synchronously. Clock jitter is therefore a great cause for concern and the standards associated with the present SDH systems operating at 155 Mbit/s are likely to be inappropriate for operation with 20 Gbit/s OTDM networks. It may become necessary to synchronize the network from the all-optical layer, utilizing the greater clock purity at the higher speeds to improve the operation of the whole integrated network. Full bandwidth flexibility at the switching node is only achieved if the data clock frequency has a wide operating range. While advances in YIG oscillators have shown electrical clock recovery operation >10 GHz [42] and electronic phase-locked loop circuits can be operated at 40 Gbit/s [33], for all-optical demultiplexing an optical clock extraction technique is potentially more desirable. Many methods of optical clock recovery in fibre (see Chapter 10) and semiconductor-based devices (see Chapter 9) [43, 44] have been demonstrated. Again, the fibre devices require kilometres of fibre at present and this results in a large latency. Improvements in silica waveguide technology (see Chapter 14) may reduce this size limitation in the future. Semiconductor devices are small, compact and have the potential of integration with electronic components.

The same two-contact device as described earlier for wavelength conversion can be used to produce all-optical clock recovery circuits. When operated above threshold, the devices can be made to self-pulsate at a specific frequency (from ~ 100 MHz to >5 GHz depending on the device [43, 44]). It has been shown [45] that injection of an optical data signal can lock the pulsations to the clock frequency and operation at data rates from 1 Gbit/s to 5 Gbit/s has been successfully demonstrated [43, 44]. When locked, the resulting optical clock pulse train has similar purity to the clock signal used to generate the optical data (<10 Hz for 5 GHz clock). Only 10 μW of optical power was needed to lock the optical clock circuit, resulting in clock output powers of milliwatts [43]. The extraction technique was independent of pattern, with operation for a $2^{31}-1$ PRBS being similar to that for a 2^7-1 sequence [46], implying that the technique should be suitable for real data where long strings of '0' bits can be accommodated [47]. Operation with NRZ data has also been demonstrated [48]. The missing optical clock component was first generated using a two-contact device (operated below threshold as a nonlinear optical amplifier) which then synchronized the pulsations of a second two-contact device above threshold.

A clock recovery circuit based on a semiconductor pulsating diode was used in a 20 Gbit/s OTDM system [43]. Figure 3.20 is a schematic illustration of the experimental configuration. RZ optical data at 20 Gbit/s was demultiplexed (see Chapter 4) to provide a 5 Gbit/s 2^7-1 optical data signal, which was then split. A small fraction of the signal was used to synchronize the self-pulsating laser clock recovery circuit while the main signal was incident on a 5 Gbit/s optical receiver. The BER measurements (Fig. 3.21) on the 5 Gbit/s channel with synchronization using both the transmitter clock and the extracted clock showed no appreciable difference, thereby confirming the system functionality of this type of all-optical clock

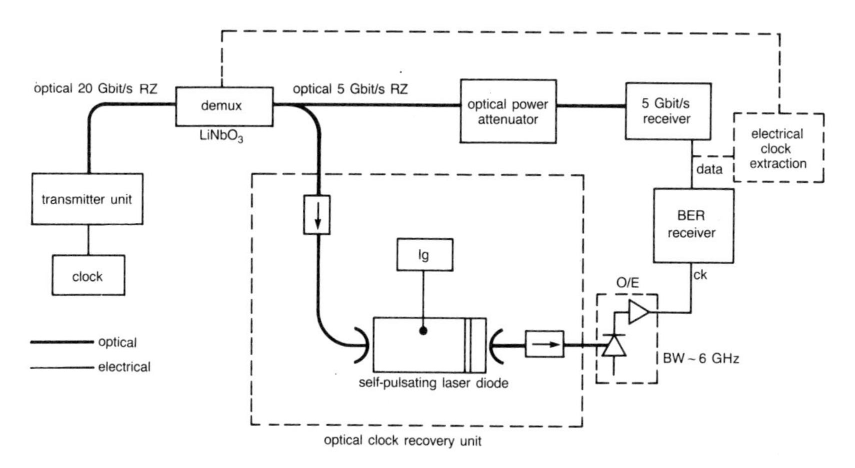

Fig. 3.20 Schematic diagram of the experiment to investigate the performance of a 5 Gbit/s all-optical clock recovery circuit using a 20 Gbit/s OTDM system.

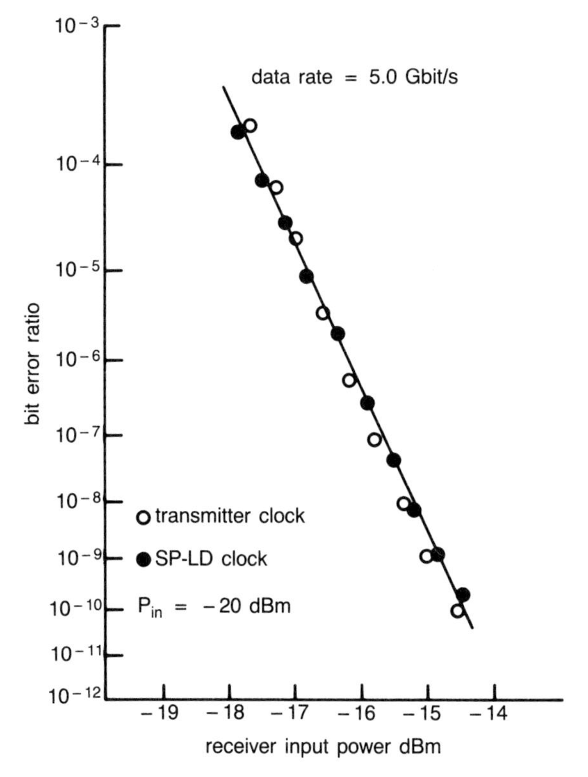

Fig. 3.21 BER results showing that the optical clock of Fig. 3.20 does not introduce any penalty.

technology [46]. Similar clock recovery operation has recently been demonstrated in a fibre device, where the data modelocks a fibre ring laser (see Chapter 10). This type of device can be operated at speeds up to 40 GHz [49]. Recent results have also shown that semiconductor SP-LD type devices can operate at >10 Gbit/s [50]. In all these devices only the principle has been established and consideration of the jitter transfer and tolerance need now to be studied.

All-optical clock recovery has a number of potential systems applications. Combined with a method of all-optical demultiplexing, such as the fibre loop mirror (see Chapter 10), it would allow all-optical demultiplexing of OTDM channels; more detailed network ideas are presented in Chapter 4. Another application area is in clock distribution in SDH systems. The 20 dB power gain coupled with the purity of the clock signal would allow a central clock signal to be distributed around a multi-Gbit/s SDH network, something other techniques may have problems achieving. If the clock recovery circuit was combined with an optical 'AND' gate, all-optical regeneration (see Chapter 10) may be possible, which could have a major impact on long-line submarine systems, especially those proposing to use nonlinear 'soliton' transmission.

3.4.6 Control and management

Any future all-optical network will have to integrate with the existing networks. The SDH managed network will be the base-line network and any overlay networks integrated to this must offer control points such that network management can be facilitated. This requires that the elements within the all-optical network, the optical amplifiers, switches and synchronization circuits, will need to have management interfaces able to communicate with the central management centres. This interface must be able to flag error events but need not necessarily add any extra control bytes to the optical overhead section of the transmitted STM-n frame, which could be handled by the element manager, perhaps at the optical line termination points. One of the key error events at the time-domain optical switch nodes is frame and channel misalignment. Any slow synchronization drift, such as those caused by thermal effects due to seasonal changes, could be accommodated by control signals from the management centre. The element manager need therefore only interrogate the element performance and not the data highway, and must be able to respond on a 'slow' timescale.

All-optical networks will require standard reference to synchronize factors such as wavelength and time across the different network regions. Clock distribution can provide a temporal synchronization reference for all the network elements as in SDH. Wavelength synchronization is also an important requirement for both multi-wavelength networks [11] and some OTDM applications, where Gordon-Haus jitter can be reduced by strict system wavelength control (see Chapter 4). A simple way of implementing wavelength standards throughout a network is to replicate a central wavelength source at each node. This standard could be a filter or laser whose wavelength is fixed. Atomic absorption or emission lines have been used to stabilize lasers but perhaps a cheaper and very effective approach is to use fibre-grating filters (see Chapter 15), which are stable, compact and highly reproducible using computer-aided fabrication techniques (see Chapter 16). These standards issues are discussed in more detail in Hill et al [11].

3.5 CONCLUSIONS

The introduction of SDH technology will yield a network providing greater flexibility, increased automation and improved management and monitoring compared with existing PDH telephony-based networks. However, the potential take-up of the new broadband services that can be offered using this technology and the growing penetration of broadband access poses problems for the future. The multi-gigabits per second link capacities and the terabits per second switch capacities required by a network supporting such a demand could pose major problems in the areas of control, planning, maintenance and operational costs. It has been shown how optical switching and routeing technology can be integrated into this network framework to provide a degree of future-proofing unavailable from other approaches. Transparent optical networks linked using this optical hardware to the inner-core electronically switched SDH layer offer upgradable, virtually unlimited, bandwidth. The network retains the monitoring capabilities and gains the enhanced flexibility provided by fast, simple optical technology.

Device technology research is now at a stage when these types of networks can really be tested. The recent rapid progress, a result of the fruition of fundamental research and development, has provided truly applicable optical solutions. Already, both fibre and semiconductor technology solutions are at hand, and attention is now turning towards the network implications. The potential rewards from pursuing these technology avenues are vast, with potential savings available at all stages. Although much is still to be done, there is now the opportunity of building an integrated network capable of encompassing virtually anything that the future may have in store. However, optical networks will only become strong contenders for deployment when they have been shown to work and when they show advantages over existing solutions. Engineering and systems trials of these ideas are necessary to support these conceptual ideas.

REFERENCES

1. Newman D and Stern J: 'Last link in the loop', Physics World, pp 45-48 (September 1991).

2. Hawker I: 'Evolution of digital optical transmission networks', BT Technol J, 9, No 4, pp 43-56 (1991).

3. Lalk G R et al: 'Potential roles of optical interconnections within broadband switching modules', SPIE Symposium on Advances in Interconnection and Packaging, paper 1389-31, Boston (November 1990).

4. Aoki I: 'Trends in electronic communications switching technologies', International Topical Meeting on Photonic Switching, Paper 12C3, Kobe (April 1990).

5. Nakagawa K et al: 'A bit rate flexible transmission field trial over 300 km installed cables employing optical fibre amplifiers', OSA Topical Meeting on Optical Amplifiers and their Applications, Post deadline Paper PD11, Snowmass (July 1991).

6. Toba H et al: '100-channel optical FDM transmission/distribution at 622 Mbit/s over 50 km utilising a waveguide frequency selection switch', Electron Lett, 26, No 6, pp 375-376 (1990).

7. Forrester D S et al: '39.81 Gbit/s, 43.8 million-way WDM broadcast network with 527 km range', Electron Lett, 27, No 22, pp 2051-2052 (1991).

8. Hill G R: 'A wavelength routeing approach to optical communication networks', BT Technol J, 6, No 3, pp 24-31 (July 1988).

9. Newall C (Ed): 'Synchronous transmission systems', Northern Telecom (1991).

10. Digonnet M F J (Ed): 'Selected papers on rare-earth-doped fibre laser sources and amplifiers', SPIE, MS37 (1991).

11. Hill G R et al: 'A transport network layer based on optical network elements', J Lightwave Technol, Sp Issue on 'Broadband Optical Networks', 11, No 5-6, pp 667-679 (May/June 1993).

12. Lynch T G et al: 'Experimental field trial demonstration of a managed multinode reconfigurable wavelength routed optical network', Proceedings 18th European Conference on optical communication ECOC, paper ThA 12.4, Berlin (1992).

13. Lowe E D and O'Mahony M J: 'Dimensioning and performance analysis of multi-wavelength optical transport networks', submitted to J Lightwave Technol (IEEE).

14. Lee K-C and Li V O K: 'Routeing and switching in a wavelength convertible optical network', Proceedings IEEE Infocom '93, paper 5b.4 (1993).

15. Elrefaie A F: 'Multiwavelength survivable ring network architectures', Proceedings of IEEE International Conference on Communications, Geneva, pp 1245-1251 (May 1993).

16. Nakazawa M et al: '20 Gbit/s, 1020 km penalty-free soliton data transmission using erbium doped fibre amplifiers', Electron Lett, 28, No 11, pp 1046-1047 (1992).

17. Mikkelsen B et al: '20 Gbit/s polarization insensitive wavelength conversion in semiconductor optical amplifiers', Proc 19th European Conference on Optical Communications, ECOC '93, Montreux, paper ThP 12.6 (September 1993).

18. Barnsley P E et al: 'Absorptive nonlinear semiconductor amplifiers for fast optical switching', SPIE Symposium on Optically Activated Switching, paper 1378-11, Boston (November 1990).

19. Chidgey P J, Barnsley P E and Westlake H J: 'Optical switching in wavelength division multiplexed networks', Proc EFOC/LAN, paper 3.8.4, London (1991).

20. Barnsley P E and Fiddyment P J: 'Wavelength conversion from 1.3-1.55 μm using split contact optical amplifiers', Photon Technol Lett, 3, No 3, pp 256-259 (1991).

21. Barnsley P E and Chidgey P J: 'All-optical wavelength switching from 1.3 μm to a 1.55 μm WDM wavelength for network system results', Photon Technol Lett, 4, No 1, pp 91-94 (1992).

22. Glance B et al: 'High performance optical wavelength shifter', Electron Lett, 28, No 18, pp 1714-1715 (1992).

23. Oberg M et al: 'A three electrode distributed Bragg reflector laser with 22 nm wavelength tuning range', Photon Technol Lett, 3, No 4, pp 299-301 (1991).

24. Mollenauer L F et al: 'Demonstration of error free soliton transmission over more than 15000 km at 5 Gbit/s single channel, and over more than 11000 km at 10 Gbit/s in two-channel WDM', Electron Lett, 28, No 8, pp 792-794 (1992).

25. Legg P J, Hunter D K, Barnesley P E and Andonovic I: 'Crosstalk effects in optical time division multiplexed switching networks', IEE Colloquium on Optical Switching, Digest 1993/137 (June 1993).

26. Benes V E: 'On rearrangeable three-stage connecting networks', Bell Syst Tech J, XLI, 5, pp 117-125 (1962).

27. Clos C: 'A study of non-blocking switching networks', Bell Syst Tech J, XXXII, pp 126-144 (March 1953).

28. Hunter D K and Smith D G: 'New architectures for optical TDM switching', J Lightwave Technol, 11, No 3, pp 495-511 (March 1993).

29. Hunter D K et al: 'Architectures for optical TDM switching', Proceedings of SPIE OE/Fibres, Paper 1787-18, Boston (September 1992).

30. Dolfi D W and Ranganath T R: '50 GHz velocity matched broad wavelength $LiNbO_3$ modulator with multimode active section', Electron Lett, 28, No 13, pp 1197-1198 (1992).

31. Wakita K et al: 'High speed InGaAs/InAlAs multiquantum well optical modulators with bandwidths in excess of 40 GHz at 1.55 μm', Proceedings Conference on Lasers and Electro-optics, Paper CTuC6, Anahiem (1990).

32. Westland D J et al: 'Degenerative four-wave mixing in polydiacetylene waveguides', Electron Lett, 27 , No 15, pp 1327-1328 (1992).

33. Ellis A D et al: 'Transmission of a true single polarization 40 Gbit/s soliton data signal over 205 km using a stabilized erbium ring laser and 40 GHz electronic timing recovery', Electron Lett, 29 , No 11, pp 990-991 (1993).

34. Gustavsson M et al: 'Monolithically integrated 4×4 InGaAsP/InP laser amplifier gate switch arrays', OSA Topical Meeting on Optical Amplifiers and their Applications, Paper PD9, Santa Fe (June 1992).

35. Ellis A D and Spirit D M: 'The use of GalnAsP amplifiers for 40 Gbit/s signal processing', Conference on Nonlinear Guided Wave Phenomena, Cambridge, Paper PD2 (September 1993).

36. Yao J et al: '5 Gbit/s transmission system using a MQW semiconductor optical amplifier as an amplitude modulator', IEE Colloquium on Sources for Very High Bit Rate Optical Communication Systems, Paper 6 (April 1992).

37. Yao J, Walker N and Sherlock G: 'High speed optical switching using MQW semiconductor laser amplifiers', IEE Colloquium on Optical Amplifiers for Communications, Paper 5 (May 1992).

38. Zylbersztejn A: 'Optoelectronic integrated circuits for fibre optic telecommunications', Critical Reviews, CR45, pp 320-340 (1992).

39. Boettle D et al: 'System and technology aspects for optical switching in broadband systems', IEE International Conference on Integrated Broadband Services and Networks, pp 270-275, London (October 1990).

40. Islam M: 'Ultrafast fibre switching devices and systems', Studies in Modern Optics, No 12, Cambridge University Press (1992).

41. Kawaguchi H: 'Progress in optical functional devices using two-section laser diode/amplifiers: review', IEE Proceedings Pt J, Optoelectronics, to be published.

42. O'Shea C D: 'A novel wideband 2-18 GHz clock extraction circuit for optical transmission systems', Electron Lett, 27 , No 25, pp 2324-2326 (1991).

43. Barnsley P E et al: 'All-optical clock recovery from 5 Gbit/s RZ data using a self-pulsating 1.56 μm laser diode', Photon Technol Lett, 3 , No 10, pp 942-945 (1991).

44. Barnsley P E and Wickes H J: 'All-optical clock recovery from 2.5 Gbit/s NRZ data using self-pulsating 1.58 μm laser diode', Electron Lett, 28 , No 1, pp 4-5 (1992).

45. Jinno M and Matsumoto T: 'All-optical timing extraction using a 1.5 μm self-pulsating multielectrode DFB LD', Electron Lett, 24 , No 23, pp 1426-1427 (1988).

46. Barnsley P E et al: 'A 4×5 Gbit/s transmission system with all-optical clock recovery', Photon Technol Lett, 4, No 1, pp 83-86 (1992).

47. Barnsley P E: 'All-optical clock extraction using two-contact devices', IEE Proc, Pt J, 140, No 5, pp 325-336 (1993).

48. Barnsley P E: 'NRZ format all-optical clock extraction at 3.2 Gbit/s using two-contact semiconductor devices', Electron Lett, 28, No 13, pp 1253-1254 (1992).

49. Ellis A D, Smith Y and Patrick D M: 'All-optical clock recovery at bit rates up to 40 Gbit/s', Electron Lett, 29, No 15, pp 1323-1324 (1993).

50. Erhardt A, As D J and Feiste U: 'All-optical clock extraction at 18 Gbit/s by a self-pulsating two-section DFB laser', Proc European Conf on Optical Communications, paper ThP 12.9, pp 85-88 (1993).

4

OPTICAL TIME DIVISION MULTIPLEXING FOR FUTURE HIGH-CAPACITY NETWORK APPLICATIONS

D M Spirit, L C Blank and A D Ellis

4.1 INTRODUCTION

To fulfil the demand for increasingly sophisticated and bandwidth-intensive services, telecommunications operators will require reconfigurable optical transmission networks which operate at higher data rates than those installed to date. Network flexibility will be significantly enhanced by the progressive roll-out of synchronous digital hierarchy (SDH) terminal equipment to replace existing plesiochronous digital hierarchy plant. Nonetheless, the maximum data capacity of optical systems will still be limited by the bandwidth of the electronics in the terminal and repeater equipment (currently around 2.5 Gbit/s). It is likely that, in the near future, advances in component design will allow electronic processing at data rates up to 10 Gbit/s in commercial terminal equipment. However, this may still be insufficient for services such as unencoded multichannel HDTV transport, which require bandwidths of around 1 Gbit/s per channel. Since all the capacity on any one set of terminal equipment in a link could be used completely to transport only a few channels, provision of sufficient transmission capacity would imply a fibre-rich trunk network. An alternative to this extension of 'traditional' electronic multiplexing, optical wavelength division (WDM) or time division (OTDM) multiplexing techniques may be used to access data rates substantially higher than 10 Gbit/s, offering a substantial increase in capacity on a single fibre. Both multiplexing techniques could be used in conjunction with switching in either the wavelength or time domains to allow increased network flexibility through, for example, the drop-and-insert function. Furthermore, optical fibre amplification is an excellent method of compensating for the loss of the transmission link, providing bit-rate-independent gain. This implies that increases in the capacity of individual sections of the network should be possible simply by upgrading the terminal equipment, provided this is allowed for during the planning and installation phases. However, the regenerative function of conventional optoelectronic repeaters is no longer provided in fibre amplifiers — linear dispersion of the optical pulses now becomes a major system constraint. The current availability of commercial fibre amplifiers restricts the signal wavelength of optically multiplexed systems to the range 1530-1570 nm, implying the use of dispersion-shifted transmission fibre to minimize dispersion penalties in large-scale networks. In principle, doping of optical fibres with alternative rare earth elements should enable similar systems in the 1300 nm region with standard step-index fibre (see Chapter 12). For moderate line rates requiring only a small number of wavelengths, WDM provides the simplest implementation of optical multiplexing, as shown in WDM and wavelength-routeing demonstrations in the installed London fibre network [1]. However, the provision of very high-capacity links using WDM may require a large number of closely spaced channels.

The operation of WDM networks with more than a few channels over fibre with low dispersion can lead to significant system degradation due to optical nonlinearities (four-wave mixing), even over fibre spans of less than 50 km [2]. An alternative, longer-term approach to ultra-high-speed transmission is to use OTDM, requiring a transmitter configuration based on short (picosecond) optical pulses, implying the use of return-to-zero (RZ) format optical data. Again, it is necessary to use dispersion-shifted fibre as the transmission medium to minimize linear dispersion penalties. However, these penalties may be significantly reduced by using optical nonlinearity to advantage, permitting transmission well beyond the linear dispersion limit. As an example, it has recently been demonstrated that 5 Gbit/s and 10 Gbit/s data can be transmitted over transoceanic (>10 000 km) distances when nonlinear pulse compression is used to balance the chromatic dispersion experienced by picosecond duration optical pulses [3]. This is known as soliton transmission and has formed the basis of many studies into the future prospects for ultra-long-span submarine systems. Optical amplifiers are a necessary part of such systems not only to compensate for fibre loss and to keep the signal amplitude above the system noise floor, but also to maintain the signal at a high enough level to sustain the nonlinear interaction between the pulses and the fibre. This chapter presents an overview of the current state of the art in circuit-switched optical time division multiplexing (illustrated with experience of 4×5 and 4×10 Gbit/s OTDM testbeds) and a survey of future opportunities for all-optical networking. The 4×10 Gbit/s testbed was funded by RACE project R2012, HIPOS.

It should be added that system capacities and network flexibility may be enhanced further by the use of optical multiplexing of OTDM bit streams (one or many channels), for example, in the wavelength or polarization domains.

4.2 OTDM CONCEPTS

Considering first a point-to-point OTDM link, Fig. 4.1 shows how n individual x Gbit/s optical data streams may be interleaved to $n \times x$ Gbit/s and subsequently demultiplexed back to the original channel. The multiplexing in circuit-switched OTDM systems is sequential in that single bits from each channel are interleaved in the same order, as opposed to packetized data, where blocks of bits are interleaved sequentially. The cornerstone underlying the implementation of OTDM networks is that the optical pulses produced by any transmitter are significantly shorter than the bit period at the highest multiplexed line rate in the network. The implication of this is that modulated data streams may be interleaved and demultiplexed with negligible crosstalk between channels. For the configuration shown in Fig. 4.1, the individual channels may be supplied from local terminal equipment or from remote transmitters and receivers. This issue is discussed further in section 4.5.

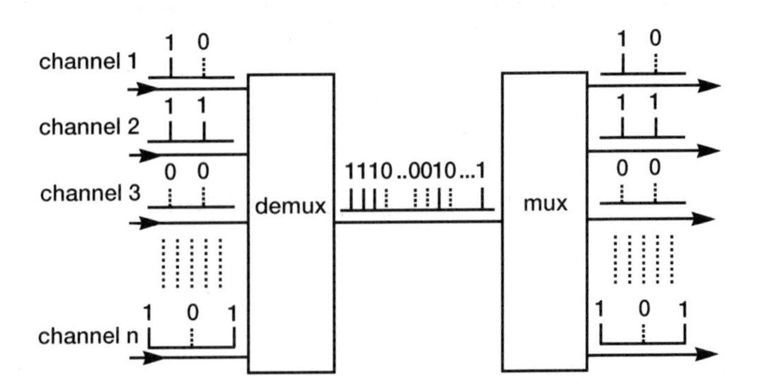

Fig. 4.1 Point-to-point OTDM trunk link.

Developing these ideas further into implementations of OTDM networks, a generalized OTDM network node (see Fig. 4.2) can be broken down into a number of functional parts. Traffic enters the node through a demultiplexing unit, which drops out (in this example) every fourth bit. The dropped-out channel may be detected at the network node or may be transported further to a remote location. The remaining channels of data are forwarded to the multiplexing unit, where one or more channels may be inserted to replace the dropped optical data streams. The inserted channel may originate from a transmitter local to the node or from a remote location. Clock recovery is necessary to provide a signal to drive both the demultiplexer and synchronization unit to ensure that inserted channels go into a vacant time slot in the multiplexed data stream.

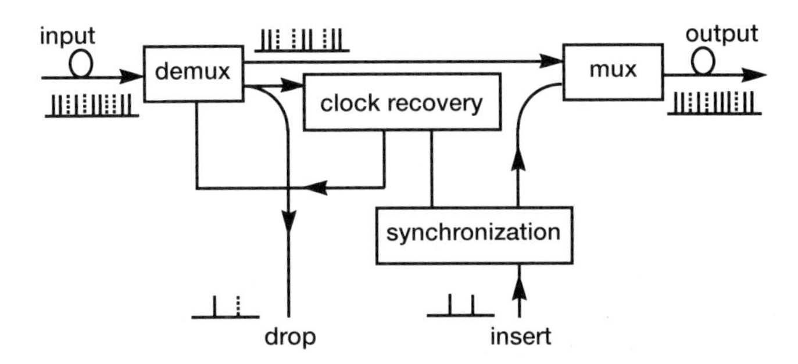

Fig. 4.2 Generalized OTDM network node based on four interleaved channels.

The functional units from which an OTDM network could be constructed are the same as for any optical system — transmitter, multiplexer and demultiplexer (optical in this instance), receiver and clock recovery. However, implementation of these functions occurs in a different manner from that for conventional optical systems, exceptions being the receiver and the clock recovery circuit, which may be common to both conventional NRZ (non-return to zero) format and OTDM RZ format. In the next section, the requirements of OTDM subsystems will be discussed and various technological solutions will be presented.

4.3 OTDM SUBSYSTEMS

4.3.1 Optical sources

In addition to the normal stability requirements of optical sources for transmission systems, there are additional requirements for an optical source for use in OTDM systems. Issues which require particular attention are:

- optical output consisting of a train of short pulses (data is RZ format);
- low cycle (typically 5:1) at highest multiplexed line rate in system;
- optical pulse characteristics close to transform limit for optimal nonlinear transmission;
- synchronization to line rate of terminal equipment.

In general, the requirements become more stringent as the size of a nonlinear transmission network is increased, in common with other solutions. At 20 Gbit/s, for example, although a duty cycle of 3:1 may be sufficient to transport data over a few hundred kilometres, it may require better than 5:1 for transmission over 1000 km. The requirements are such that it is currently not practicable to use directly modulated lasers — in all cases, external modulation of the optical source must be used.

Since OTDM is being considered for network applications, where the highest multiplexed line rate is in excess of 10 Gbit/s, say at least 20 Gbit/s, the bit period of the multiplexed data stream will be 50 ps at most. Efficient crosstalk-free multiplexing and demultiplexing will therefore be obtained for optical pulses of <25 ps duration (duty cycle requirement of only 2:1). In general, for a system with highest line rate of x Gbit/s, the maximum permissible optical pulsewidth will be $1000/2x$ ps. The requirement is for optical sources which generate trains of close to transform-limited pulses of much less than 25 ps width at repetition rates of at least a few gigahertz. Since synchronization with existing standard data rates is also required, active stabilization to a clock signal must be maintained. In all cases below, this is achieved in part by the supply of a high-quality electrical sinewave drive to an active electrooptic device, which forms a part of the pulse source. The remainder of this section will be taken up with a survey of existing options for meeting these requirements.

4.3.1.1 Gain-switched distributed feedback (DFB) laser

Gain switching is a versatile technique which requires the driving of a conventional DFB laser with a high-quality electrical sinewave, producing short pulses at gigahertz repetition rates. It is therefore a simple matter to synchronize this pulse source with a specific electrical data rate. However, the optical pulse trains produced by gain switching are highly chirped and, for simple devices, may possess unacceptable levels of jitter. Appropriate spectral filtering and pulse compression of the output from the laser can be used to produce pulses which are close enough to the transform limit to reduce linear dispersion effects and to permit soliton formation in system fibre if necessary. Sources of this type have been used for much of the Japanese work on soliton transmission [4].

4.3.1.2 Semiconductor mode-locked laser (MLL)

The external cavity semiconductor MLL is a development of the commercially available grating-tuned external cavity semiconductor laser, which provides a narrow linewidth source. The major new requirement is for the provision of a high quality RF bias connection to the laser chip which allows the laser to be driven with a single-frequency RF signal. The length of the laser cavity is adjusted so that the optical round trip time is equal to the period (or harmonics thereof) of the RF drive signal. The net laser output consists of a train of short optical pulses at a well-defined repetition frequency. Synchronization to a particular data rate requires an adjustment to match the cavity length to the RF frequency. Packaged MLLs have been in use in several groups at BT Laboratories for OTDM and nonlinear systems investigations over several years. For a 5 GHz sine wave drive, the output is typically a train of 15-30 ps full width half maximum (FWHM) pulses with a time × bandwidth product in the region of 0.4.

The integration of optoelectronic devices which otherwise consist of bulk optic components is clearly desirable to improve the long-term reliability and stability. Semiconductor MLLs can be integrated, consisting of a short active section in tandem with a long passive waveguide. For a 5 GHz pulse repetition frequency, the device length would be of the order of 8 mm. Initial integrated MLLs made by several laboratories around the world have produced encouraging results [5].

4.3.1.3 Mode-locked fibre laser

As the mode-locked fibre laser is discussed in detail in Chapter 10, only those issues of direct relevance to network applications will be presented here. The basic structure of a doped-fibre MLL consists of the same functional parts as a semiconductor MLL. The laser cavity consists of a closed loop of fibre, part of which is erbium-doped and optically pumped to provide optical gain. Mode-locking is performed in this case by driving an intra-cavity electrooptic modulator at a harmonic of the cavity round-trip time. Synchronization of a high harmonic of the laser cavity round-trip frequency to the gigahertz single-frequency RF drive to the intra-cavity modulator requires active control of the overall cavity length. This can be achieved by monitoring a portion of the laser output, which can be used to provide active control of the fibre length [6]. Very recently, active stabilization of a mode-locked fibre laser has enabled transmission at 2.5 Gbit/s over 12 000 km to be demonstrated in a recirculating fibre loop [7]. Alternative schemes based on intra-cavity interferometric elements to eliminate unwanted modes and improve amplitude stability have also been successfully demonstrated [8]. This has proved that the stabilization schemes which have been developed are sufficiently robust to permit use of these fibre lasers as transmitters for high-speed OTDM direct-detection systems.

4.3.1.4 External modulation

While mode-locked fibre lasers produce pulses which are transform-limited and so are ideal sources for OTDM systems, active synchronization of the fibre length will be required to ensure an acceptable degree of long-term stability. These issues may be avoided by the production of short pulses by the external modulation of a CW laser source. In principle, any external modulation technique with sufficient bandwidth may be used — electrooptic, electroabsorption or all-optical. In some cases, production of pulses of sufficiently short duration at the required data rate may require the use of more than one modulator in series. Perhaps the most promising option is the electroabsorption modulator, as it offers advantages over other techniques. The most significant of these is the potential for integration with semiconductor lasers to produce a monolithic picosecond pulse source, without the constraints imposed by mode-locking. Of secondary benefit is the moderately low drive voltage when compared with electrooptic devices. Such an integrated device has recently been used as a pulse source to enable 10 Gbit/s soliton transmission over 6700 km in a recirculating loop [9].

4.3.2 Multiplexers and demultiplexers

The main difference between OTDM and conventional optical systems is in the use of multiplexing and demultiplexing in the optical domain. The multiplexing may be active or passive — demultiplexing has to be performed as an active function. In this section, current and future alternatives for the mux/demux functions will be presented, with particular reference to an existing 4×10 Gbit/s demonstrator.

4.3.2.1 Multiplexing

Passive

A schematic of an OTDM transmitter is shown in Fig. 4.3 [10]. The optical source is a stabilized, mode-locked erbium-doped fibre ring laser, producing a continuous pulse train at a repetition rate of 10 GHz. The pulses have a full width half maximum (FWHM) of 6 ps, and are transform limited (time-bandwidth product ≈ 0.44). The signal level is boosted in an erbium-doped fibre amplifier before being split into four channels by passive fused-fibre devices. Each of the 4×10 GHz repetition rate optical pulse trains is modulated with 10 Gbit/s pseudorandom data using 4 GHz bandwidth lithium niobate Mach-Zehnder modulators. Unlike external modulation of a CW light source, the modulators are not being used to shape the pulses — they are being used as on/off switches. Before recombining the data streams using fused fibre couplers, each channel passes through a variable delay line (fibre stretcher) to allow interleaving with the correct delay to produce a single 40 Gbit/s RZ data stream. Tolerance to timing errors is maximized by active control of the fibre stretchers, which are aligned to minimize subharmonics of the drive frequency. This particular implementation is based around single-mode fibre components — a future semi-integrated version could be based on in-diffused waveguides on silicon substrates (see Chapter 14), offering more stable operation. Extensions to higher bit rates should, in principle, be relatively straightforward, requiring either electrooptic modulators of higher bandwidth and a higher repetition-rate optical source producing shorter pulses, or more modulators in parallel

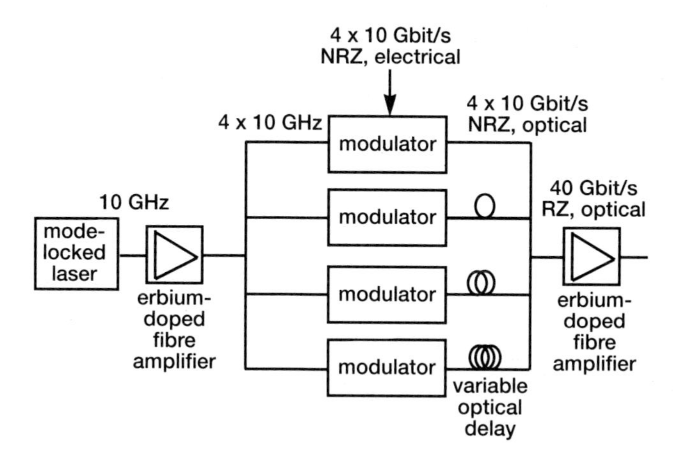

Fig. 4.3 4×10 Gbit/s OTDM transmitter using passive multiplexing.

in combination with a source producing shorter pulses at the same repetition rate. Recent results have demonstrated transmission at 80 Gbit/s over 50 km using similar methods in conjunction with polarization division multiplexing [11].

Active

While passive multiplexing is clearly simple to implement, there are circumstances in which active multiplexing may well provide additional functionality. A particularly important example of this is to provide the simultaneous functions of multiplexing and NRZ to RZ optical format conversion — i.e. the interface between 'conventional' NRZ systems and an ultra-high speed RZ OTDM network. An example of these concepts is summarized in Fig. 4.4, which shows how four optical NRZ data streams can be combined using electrooptic 2×1 switches [12]. The switches are driven with single-frequency RF signals, and act effectively as sampling switches. In this way, the 'top hat' NRZ data is converted into pulses more appropriate for RZ format transmission.

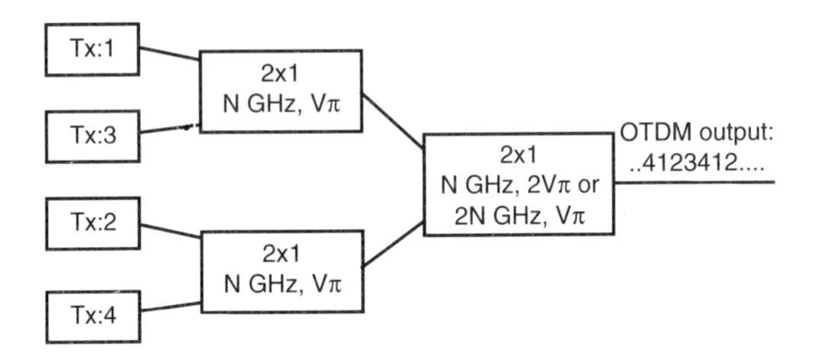

Fig. 4.4 Four-channel OTDM transmitter using active multiplexing.

4.3.2.2 Demultiplexing

While the multiplexing function may be implemented either actively or passively, the demultiplexing must be an active function. The functionality required from the demultiplexer will influence the design — e.g. will operation with circuit- and packet-switched data be supported, if required? In the first case, the demultiplexer should be gated with a period corresponding to that of the dropped-out channel. The gate duration should be approximately equal to the length of the individual data pulse. For packet-switched systems, the gate should be open for long enough to demultiplex a complete packet over several data periods. This may be achieved with a single, long gate pulse, or a sequence of short pulses (as in the circuit-switched case), configured to imitate a long gate pulse. A second, important issue is whether it will be necessary to separate all the channels individually or simply remove one channel from N. There is an analogy that can be made here with the multiplex hierarchies of plesiochronous and synchronous digital hierarchies (PDH and SDH). In electrooptic implementations, complete demultiplexing of all channels must normally take place to extract a single-channel ('multiplex mountains') if the advantages of a simple sinusoidal modulator drive are to be retained. By contrast, the all-optical solutions permit the drop of a single-channel (or more than one) and would permit network configurations similar to those permitted by the SDH add-drop multiplex. Demultiplexing of a single-channel using electrooptic methods would be possible in a device which supported sufficiently high frequencies, although this would, of course, negate many of the advantages of OTDM, where only relatively low electrical bandwidths are required.

Electrooptic

The design of an electrooptic demultiplexer will generally lead to a separation of all the channels in an OTDM system. The simplest conceptual form for an N-channel system would be a $1 \times N$ active switch array (see Fig. 4.5), or alternatively, a passive split followed by N copies of a single-channel demultiplexer in parallel. As an example of a single-channel demultiplexer, consider the receiver for the 4×10 Gbit/s testbed shown in Fig. 4.6 [10]. This consists of an erbium-doped fibre preamplifier, an optical demultiplexer to retrieve one 10 Gbit/s data stream, and a wideband PIN receiver. The finite bandwidth of the receiver and following electronic amplifiers converts the single 10 Gbit/s RZ optical data stream ($<$ 10 ps FWHM) into 10 Gbit/s NRZ electrical data. The data output from the receiver is used as the input to a 10 Gbit/s error detector. The output from the clock recovery circuit drives both the optical demultiplexer and the error detector. The demultiplexer consists of two lithium niobate Mach-Zehnder modulators in tandem. The first modulator is driven with a 10 GHz sinewave of 2 V_π amplitude, to demultiplex from 40 Gbit/s to 20 Gbit/s [12]. The second modulator is also driven with a 10 GHz sinewave, but of V_π amplitude to demultiplex from 20 Gbit/s to 10 Gbit/s. Different channels may be selected by changing either the electrical phase (delay) or d.c. bias of the electrical drive to the demultiplexers. Just as at the transmitter, the optical demultiplexer is not being used to shape the output and is acting only as an on-off switch. Hence the entire 40 Gbit/s system operates using components usually appropriate for system operation at 10 Gbit/s, with the addition of a small number of narrowband components at higher frequencies in the clock recovery circuit. In the current laboratory system, only a single 10 Gbit/s data stream

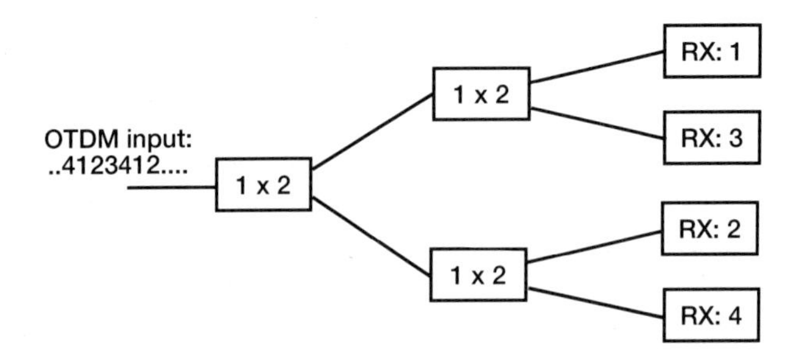

Fig. 4.5 4-channel OTDM receiver using active 1×2 electrooptic switches. Switches may be integrated into a 1×4 array on a single substrate.

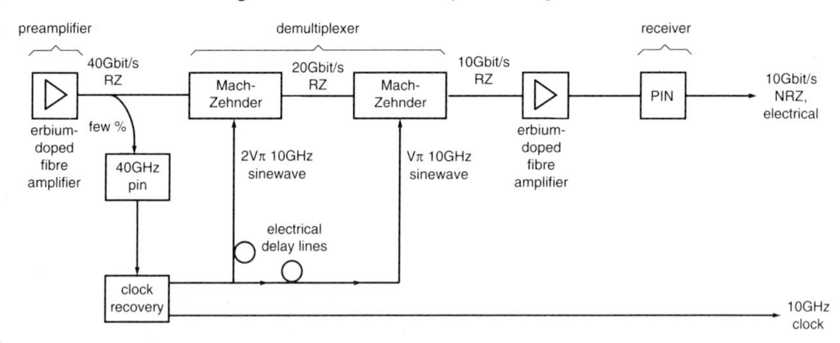

Fig. 4.6 40 Gbit/s to 10 Gbit/s electrooptic demultiplexer and receiver.

may be detected at a time, but integrated optic devices (or a 1×4 passive split followed by duplicates of the existing demultiplexer) would permit simultaneous detection of the complete 40 Gbit/s data.

All-optical

While extensions of OTDM transmitter technology to higher data rates should be relatively straightforward, implementation of an electrooptic demultiplexer to operate under the same conditions becomes increasingly difficult. The main constraint is the electrical power required to drive the first stage of the demultiplexer at twice the normal switching voltage and at these higher frequencies. As an alternative to electrooptic devices, all-optical demultiplexers have been demonstrated, operating with nonlinear effects in conventional optical fibres (see Chapter 10). The geometry of a fibre demultiplexer may take many forms, but is often based around a fibre loop mirror. Incoming data is split by a 3 dB coupler between both arms of the loop mirror. After propagation around several kilometres of fibre in the loop, the two pulse trains interfere, recombine and are reflected from the loop mirror under appropriate conditions. The operation here is essentially passive and linear. However, if a train of high-power clock pulses is introduced into the loop, co-incident with the data signal but travelling in one direction only, then the clock pulses modify the refractive index of the fibre core through the Kerr effect. Given sufficient cross-phase modulation, the appropriate phase change may be introduced on the data pulses to cause the co-incident data pulses to be switched out of the opposite port of the loop mirror. The functionality of the demultiplexer may be changed simply by adding in more clock channels with the appropriate time delay, allowing simple reconfiguration of the device. A device of this type has recently been used at BT Laboratories to demultiplex 10 and 20 Gbit/s RZ data from a 40 Gbit/s data stream [13], clearly demonstrating operation at speeds greatly in excess of those currently achievable with electrooptics. The ultimate speed of devices of this type is limited only by the speed of the nonlinearity in silica (≈ 1 fs). Similar functionality should be available from alternative nonlinear media, such as semiconductor and polymer materials, which offer the prospect of much more compact and robust devices (a few millimetres).

4.3.3 Clock recovery

One major part of any transmission system is clock recovery to enable correct synchronization at the receiver. The major advantage of RZ coding as used in OTDM systems is that the electrical spectrum of the data has a strong component at the clock frequency $n \times x$ GHz — unlike NRZ systems. Of additional importance in OTDM systems is the fact that the clock recovery is also used to drive the demultiplexer. Whether electrical or optical clock recovery methods are implemented, it must be decided if the clock recovery unit should be placed before or after the demultiplexer. In the former case, narrowband components must be used which are capable of operation at the clock frequency corresponding to the highest optical multiplexed rate ($n \times x$ GHz) in the transmission link, as used in the 4×10 Gbit/s testbed [10]. This is necessary since, for a perfectly interleaved system, no clock component exists at x GHz. However, the demultiplexer will obviously not function correctly until the clock has been recovered, leading to start-up ambiguities. It will be necessary to initiate an ordered start-up sequence, where only one channel is turned on until the clock recovery unit begins to function — then all the remaining channels may also be switched on.

Electronic clock recovery circuits for OTDM systems are no different from those for any other transmission system with RZ data as the input, except for the fact that, in some cases, x GHz is extracted from $n \times x$ Gbit/s data. However, there are at least two methods of active clock extraction using optical means which could be implemented in an OTDM system. In both cases, the optical clock is generated at a wavelength different from that of the signal. This may well be advantageous, as an amplified clock may then be combined almost without loss in an optical processing element (e.g. demultiplexer). A passive method of all-optical clock extraction is outlined in section 4.5 in the discussion on OTDM networks.

4.3.3.1 Optical clock recovery using self-pulsating laser diode

Under certain d.c. bias conditions of operation of a split contact semiconductor laser, one section of the laser can act as a passive Q-switch, i.e. the laser emits pulses of light with a well-defined repetition frequency (see Chapter 3). By injecting RZ format optical data with a similar clock frequency into such a laser diode, the self-pulsations may be locked on to the incoming data, providing a high-quality optical clock at the output of the device. In a two-contact self-pulsating laser diode (SPLD), the frequency of the self-pulsations may be varied by changing the bias current to each section and the temperature of the device as a whole. For example, the self-pulsation frequency of a single device has been varied between 3 GHz and 5 GHz by changing these parameters. Further details of the operation and applications of SPLDs are given in Chapter 3.

One of these devices has been evaluated in the 4×5 Gbit/s OTDM test-bed as a method of extracting a clock through optical means [14]. The output from the demultiplexer was split so that part of the optical data entered an SPLD biased to give a 5 GHz self-pulsation frequency — the remainder was incident on the optical receiver as usual. The BER error detector was driven using the electrical clock from the data transmitter and also from the all-optical clock derived from the SPLD output. There was no measurable difference in the BER data as a function of optical power incident on the receiver, indicating that the all-optical clock was at least of as good quality as the transmitter clock. Again, further detail is given in Chapter 3.

4.3.3.2 All-optical clock recovery using mode-locked fibre laser

In the discussion of the mode-locked fibre ring laser above, the intra-cavity modulation was provided by an electrooptic device provided with an RF drive. However, a recent development [15] has been to demonstrate a fibre laser containing a fibre modulator. Further details are given in Chapter 10. Incoming data enters the fibre modulator via a WDM coupler and modulates the fibre laser by cross-phase modulation. In this first proof-of-principle demonstration, the data rate was 1 Gbit/s, but, more recently, clock recovery at 40 Gbit/s has been demonstrated [16].

4.4 POINT-TO-POINT TRANSMISSION

This section raises issues which are important in the implementation of OTDM transmission systems and is illustrated with experimental and theoretical examples based on the OTDM demonstrators.

4.4.1 Linear transmission

To enable linear transmission over moderate distances, the signal wavelength of a multi-gigabit/s OTDM system must be within a few nanometres of the wavelength of zero dispersion of the transmission fibre. As an example, consider transmission of the 4×5 Gbit/s data over an optically amplified 205 km link [17]. The transmission section of the system consisted of $4 \times$ 51.2 km lengths of Corning dispersion-shifted fibre with an average wavelength of zero dispersion at 1554.2 nm. The mean group delay dispersion of the fibre spans was 0.085 ps/nm/km at the signal wavelength of 1555.4 nm, causing the broadening of the 14 ps pulses to 14.8 ps after transmission through the fibre. However, the system could be operated with sufficient optical power to introduce pulse compression to a minimum of 11 ps after weakly nonlinear transmission through the fibre. Three in-line optical amplifiers provided loss compensation, one after each 51 km of fibre. Back to back, the receiver sensitivity was -20.5 dBm for 20 Gbit/s data at the demultiplexer input — after transmission through the fibre, the sensitivity was degraded to -19.7 dBm, a penalty of only 0.8 dB. It has been established that this penalty is due to the accumulation of spontaneous emission from the chain of optical amplifiers in the system.

4.4.2 Nonlinear transmission

For signals further away from the wavelength of zero chromatic dispersion, linear dispersion of the optical pulses in an OTDM system become a major system constraint, e.g. a 15 ps transform limited pulse at 1550 nm will double in width in less than 10 km of step-index fibre. However, if the signal power is high enough and the signal wavelength lies on the long wavelength side of the fibre dispersion zero, nonlinear optical pulse compression can be used to maintain the pulsewidth as the light propagates along the transmission fibre. This is the mechanism responsible for soliton pulses, which has been used for many impressive transmission demonstrations over transoceanic distances [3, 7, 18, 19].

4.4.2.1 Nonlinear transmission demonstrations

For trunk applications of fixed maximum transmission distance, it is not necessary to create perfect soliton pulses — the prime requirement is that the optical pulsewidth at the receiver is still short enough to avoid inter-symbol interference between neighbouring channels at the receiver, even at the expense of some nonlinear distortion of the optical signal spectrum, i.e. with overall propagation conditions insufficient for ideal soliton propagation. For example, nonlinear transmission at 4×5 Gbit/s has been demonstrated using the 205 km system described in section 4.1 by increasing both the signal wavelength (6 and 11 nm above the wavelength of zero dispersion) and power [20]. These results illustrate that this system may be operated over at least a 10 nm range of signal wavelengths. Despite the variation in fibre dispersion over this range, successful operation of the system may be achieved by adjusting the output power of the repeater and power amplifiers. Figure 4.7 shows the BER results for each channel obtained following nonlinear transmission of 4×10 Gbit/s over the same 205 km link, 2.5 nm above the wavelength of zero chromatic dispersion. The performance on each of the four channels is identical. This amplfied link has been upgraded from

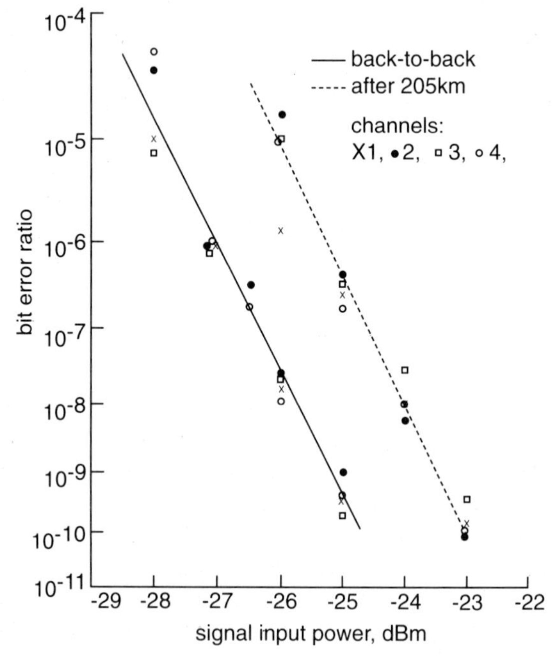

Fig. 4.7 4×10 Gbit/s, 205 km optically amplified BER results.

2.5 Gbit/s to 20 and 40 Gbit/s, the only change to the link being the selection of appropriate output powers from the booster and in-line amplifiers.

4.4.2.2 Modelling of limits to nonlinear system performance

Computer simulations have been performed to establish the feasibility of ultra-high bit rate OTDM transmission over representative distances for a speculative optically amplified network throughout the UK [21]. The distances of up to 2000 km were chosen primarily as a result of considering the length of an ultra-high-capacity fibre ring which would encompass all of the major centres of population in the UK. The model is identical to that used for studies of soliton transmission over transoceanic distances for optically amplified submarine systems, and includes many nonlinear optical processes. The major advance in this work has been to include the non-ideal chirp present on the transmitter pulses (semiconductor mode-locked laser), which can form a significant limit to system performance. The model has proved to be a powerful tool in establishing the tolerance of the system to, for example, transmitter optical pulsewidth and duty cycle. As an example, it has been established that transmission at a line rate of 20 Gbit/s should be possible over 2000 km with marginal eye closure, given an appropriate combination of fibre dispersion (0.5 ps/nm/km), amplifier output power (+4 dBm), soliton pulsewidth (10 ps), etc, even with a practical semiconductor mode-locked laser. Further constraints on the system configuration indicate that OTDM transmission at 40 Gbit/s should be possible, again with marginal eye closure over a similar distance if distributed amplification is used throughout the transmission medium [21].

4.4.3 Distributed amplification

For nonlinear transmission of pulses of pulsewidth τ in fibre with group delay dispersion D, the path average first-order soliton peak power $P_{n=1} \propto D/\tau^2$ and the soliton period $z_0 \propto \tau^2/D$. The practical significance of the soliton period is that it defines a characteristic length of nonlinear transmission — the soliton period should be several times the amplifier spacing. One consequence of this for ultra-high-speed nonlinear systems is that the repeater spacing may well be very short as a result of the squared dependency on the optical pulsewidth (e.g. soliton period ≈ 38 km for 10 ps pulse in fibre with 1 ps/nm/km dispersion). The highest-capacity systems could conceivably need amplifiers spaced at less than 10 km, an equipment-intensive configuration for a 2000 km system. An alternative is to use amplifying transmission fibre, providing low gain with low signal power excursion, distributed along the length of the fibre. The active connections (i.e. pumping locations) can then be further apart than in a system of discrete repeater amplifiers. This distributed amplification may be achieved either by using standard (step-index or dispersion-shifted) fibre using Raman amplification [22] or by lightly doping all the transmission fibre with erbium ions [23]. In either case, semiconductor pumping at wavelengths in the region 1460-1480 nm at similar powers is required for system operation in the 1550 nm window.

The potential of distributed amplification has been investigated by demonstrating linear and nonlinear transmission of a 10 GHz train of 18 ps pulses over an optically pumped 28.5 km fibre, doped with approximately 50 ppb of erbium ions throughout its length [24]. In the low signal power limit, the pulses broadened through linear chromatic dispersion to around 45 ps. When the optical signal power level was increased to 7.5 mW, the pulses compressed back to 18 ps. Here, the soliton period is approximately 25 km compared with the fibre length of 28.5 km.

However, distributed amplification may also be of benefit in linear transmission. Using such fibre, 4×5 Gbit/s OTDM unrepeatered transmission over 137 km has been demonstrated [25]. A schematic of the experiment is shown in Fig. 4.8. A 35 km fibre, lightly doped (≈ 50 ppb) with erbium ions was used in front of the optical receiver to provide both dispersion equalization over the system fibre as a whole and also to sustain the signal power above the receiver sensitivity level for the final 35 km of the system

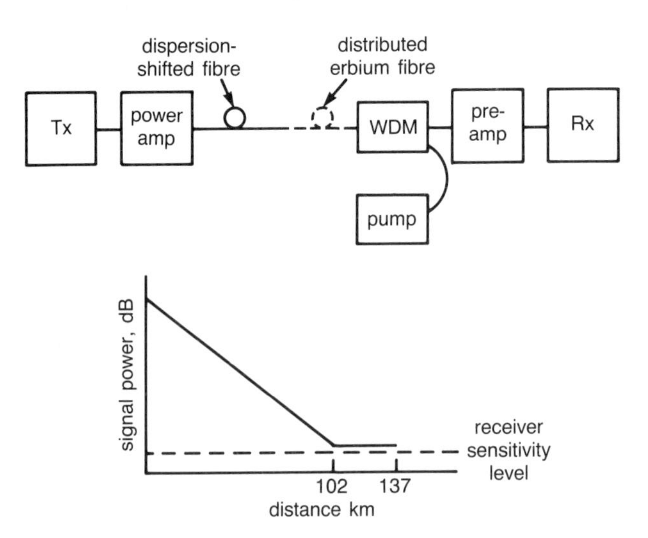

Fig. 4.8 Schematics showing configuration and principle of operation of a 137 km 4×5 Gbit/s OTDM link including 35 km distributed preamplifier.

(i.e. distributed preamplification). The first 102 km of the transmission path consisted of conventional dispersion-shifted fibre. The operating wavelength of 1536 nm was chosen to maximize the gain of the germano-silicate distributed erbium preamplifier, which had a group delay dispersion sufficient to cancel the linear dispersion of the 102 km of undoped fibre — the width of the pulses after transmission was identical to that at the transmitter output. System BER results are given in Spirit et al [25]. The lack of spectral broadening indicated the absence of any measurable nonlinear processes at any point in the system. These results demonstrate how OTDM may be implemented in a system where linear dispersion compensation techniques may be employed in preference to nonlinear transmission.

4.5 NETWORKS

The evolution of the core network into an optically amplified, bit-rate-transparent transport layer will give a technological base to enable the rapid provision of new, bandwidth-on-demand services. This will require optical multiplexing to provide sufficient capacity to future-proof the network, as well as increased flexibility for capacity allocation (slow switching) or even high-speed switching at the transport layer level. Besides management and control, one of the major issues facing the designers of such networks is synchronization, and the implementation of OTDM networks is no exception. In any optically multiplexed network, the synchronization of tributary channels must take place in both the time and optical frequency domains. For OTDM networks, this implies that the pulse trains arriving at any given node through different paths (including any laser source local to the node) must be synchronized in time to enable accurate demultiplexing and to have exactly the same spectral characteristics to enable zero walk-off between data streams as they are transmitted on to the network. For example, Fig. 4.9 shows a simple network comprising two physically separate RZ-format transmitters connected to a single receiver. In principle, the two lasers do not have to possess identical optical characteristics, since by programming in appropriate optical delays at the transmitters the two optical data streams may arrive suitably phased and can be successfully demultiplexed [26]. If, however, it is required to then pass these data streams on to another node for subsequent processing, control of the network may require demultiplexing of all channels at each node and a knowledge of the appropriate delays between each channel on each span of the network. In this case the network has become a set of individual OTDM channels, electronically linked at nodes, i.e. the main advantages of bit rate transparency have been lost.

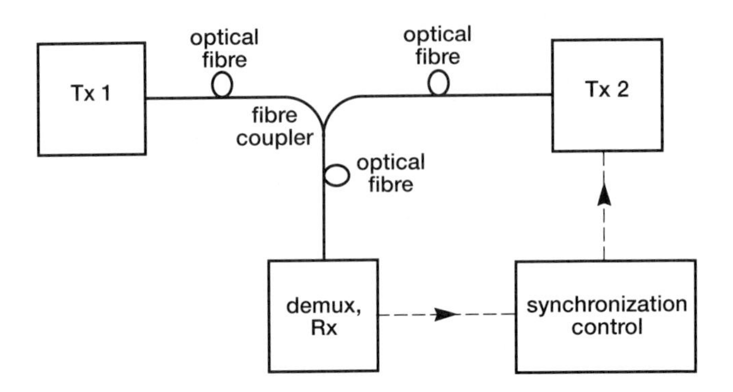

Fig. 4.9 Simple OTDM network — two sources transmitting to a single receiver.

One of the major issues in the implementation of large-scale OTDM networks is the matching of the optical characteristics of any inserted data channels to those already on the line. This is necessary to ensure minimal pulse walk-off between channels to minimize intersymbol interference and timeslot errors wherever processing functions such as demultiplexing or routeing are carried out. There are at least two approaches which may provide a solution to the matching of wavelengths — use of standardized wavelength sources at each transmitter site or, alternatively, a single transmitter laser to distribute a data signal around a network. In both cases, the optical pulse trains produced would be externally modulated to provide the added channel at the node. Optical procesing at a node can be used to generate an optical clock from the data (section 4.3.3). In the first case, the clock pulses can be used to provide an optical pulse stream at a fixed wavelength from a 'standard' CW source via, for example, nonlinear Kerr polarization rotation in a length of optical fibre [27]. The alternative approach of distributing the pulses from a master pulse source may be enabled through optical equivalents of electronic passive clock recovery techniques such as tank circuits. This idea is summarized in Fig 4.10, showing an schematic of a node in an OTDM network which uses passive clock extraction. This has the added advantage that the recovered clock may possess identical spectral characteristics (not just centre wavelength) to the incoming data, hence further reducing system impairments arising from the differences between the spectra of local laser beams and those which have been modified by nonlinear transmission.

If the line terminal equipment is co-located with the node, electrooptic modulation of the derived optical clock may well be most convenient. Alternatively, terminal equipment may be some distance away, linked in via conventional step-index fibre. In this case, linear dispersion of the pulses in

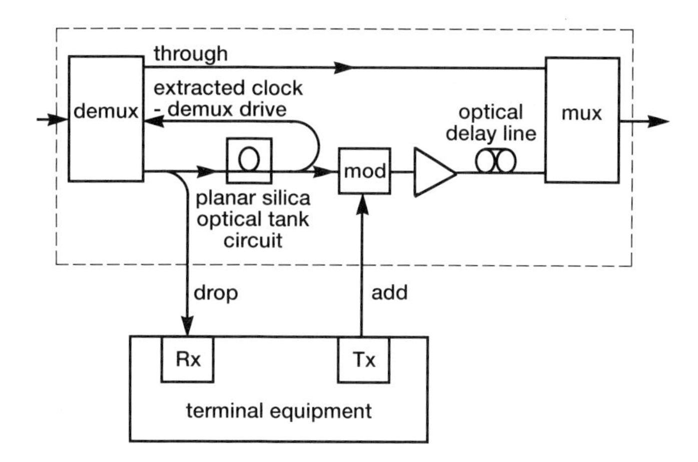

Fig. 4.10 Conceptual node with passive all-optical clock recovery.

the fibre between the node and the receiver on the 'drop' path provides some degree of passive conversion between the RZ picosecond pulse data towards NRZ format. Optical pulse shaping on the 'drop' path by either electrooptic or all-optical methods could significantly extend the maximum achievable length of this tributary, particularly in step-index fibre [28]. On the 'add' path, the format of the optical data is less critical, the main consideration being the interface to the unit which modulates the optical clock. Two options are possible — electrooptical modulation, where the interface to the modulator is a receiver followed by electronic amplification or, alternatively, an all-optical modulator (e.g. based on a nonlinear fibre device), where the interface may well maintain some bit-rate transparency as well as providing format conversion.

Building on these ideas, optically multiplexed networks may be designed which are compatible with the basic SDH ring structure, a simple example of which is given in Fig. 4.11 (many of the arguments which follow would apply equally to WDM or OTDM networks). To enable maximum network size by minimizing linear chromatic dispersion effects, the core of the ring should be dispersion-shifted fibre (to enable nonlinear transmission), or step-index fibre with appropriate dispersion-compensating elements (linear transmission). Erbium-doped fibre amplifiers would be used throughout the fibre ring to compensate for fibre attenuation and to supply sufficient optical power to provide nonlinear dispersion compensation if required. Existing conventional step-index fibre could be used for short (few kilometres) tributaries. Maximum network flexibility would be maintained if SDH management is applied end to end to individual channels rather than to the multiplexed data. In this manner, the network nodes may be controlled

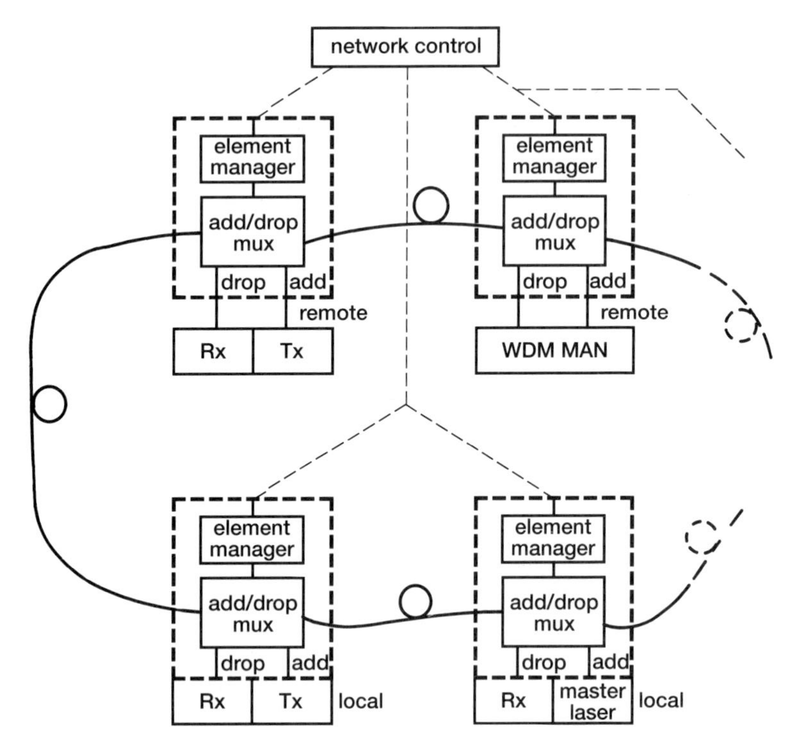

Fig. 4.11 Conceptual ultra-high-speed SDH-compatible self-healing trunk network. The 'add/drop multiplexer' corresponds to the hatched-line box in Fig. 4.10.

centrally but are essentially transparent to the SDH overhead bits. This implies that the nodes are incapable of modifying these bits (similar to the behaviour of optical amplifiers). The alternative would require complete demultiplexing at each node, again removing the bit-rate flexibility of the network. For the OTDM network, one approach is that there need be only a single source of optical pulses, which is distributed using passive all-optical clock recovery at each network node. This would ensure satisfactory system performance over each part of the network by removing channel walk-off effects. Other solutions to optimal wavelength synchronization may be developed in the future [27].

The eventual role of OTDM is most likely to be in the provision of a super-trunk inland ring network (see Chapter 3). As this is likely to evolve following the implementation of WDM in high-capacity trunk systems, interconnection with single- and multiple-wavelength SDH channels is a necessary part in the potential roll-out of OTDM. This may be achieved using electrooptic converters, or, alternatively, bit-rate-flexible interconnection between WDM and OTDM networks may be implemented using all-optical modulators.

4.6 CONCLUSIONS

The concepts involved in the implementation of OTDM to provide ultra-high (>>10 Gbit/s) bit-rate transport have been surveyed, and, currently, most of the components and subsystems are available in telecommunications laboratories. A number of demonstrations have been performed at 20 and 40 Gbit/s over optically amplified fibre to evaluate system performance. Ideas have been developed to consider a future all-optical super-trunk core network within the framework of SDH-managed self-healing ring topologies.

REFERENCES

1. Lynch T et al: 'Experimental field demonstration of a managed, multinoded, reconfigurable, wavelength-routed optical network', Proc of European Conf on Opt Comms (ECOC'92), Berlin (September 1992).
2. Walker G R, Spirit D M, Chidgey P J, Bryant E G and Batchellor C R: 'Effect of fibre dispersion on four-wave mixing in a multichannel coherent optical transmission system', Electron Lett, 28, pp 989-991 (1992).
3. Mollenauer L F, Lichtman E, Harvey G T, Neubelt M J and Nyman B N: 'Demonstration of error-free soliton transmission over more than 15 000 km at 5 Gbit/s, single channel, and over more than 11 000 km at 10 Gbit/s in two-channel WDM', Electron Lett, 28, pp 792-794 (1992).
4. Nakazawa M, Suzuki K and Yamada E: '20 Gbit/s 1020 km penalty-free soliton data transmission using erbium-doped fibre amplifiers', Electron Lett, 28, pp 1046-1047 (1992).
5. Hansen P B, Raybon G, Koren U, Miller G I, Young M G, Chien M, Burrus C A and Alferness R C: '5.5 mm long InGaAsP monolithic extended-cavity laser with an integrated Bragg reflector for active mode-locking', IEEE Photonics Technol, 4, pp 215-217 (1992).
6. Shan X, Cleland D and Ellis A D: 'Stabilizing erbium fibre soliton laser with pulse phase locking', Electron Lett, 28, pp 182-184 (1992).
7. Malyon D J, Widdowson T and Lord A: 'Assessment of the polarization loss dependence of transoceanic systems using a recirculating loop', Electron Lett, 29, pp 207-208 (1993).
8. Harvey G T and Mollenauer L F: 'Harmonically mode-locked fiber ring laser with an internal Fabry-Perot stabilizer for soliton transmission', Opt Lett, 18, pp 107-109 (1993).

9. Kawai S, Iwatsuki K, Suzuki K, Nishi S, Saruwatari M, Sato K and Wakita K: '10 Gbit/s optical soliton transmission over 6700 km using monolithically integrated MQW-DFB-LD/MQW-EA modulator light source and conventional EDFA repeaters', in 'Technical Digest of Optical Amplifiers and their Applications', Optical Society of America, Washington DC, paper PD3 (1993).

10. Ellis A D, Widdowson T, Shan X, Wickens G E and Spirit D M: 'Transmission of a true single polarization 40 Gbit/s soliton data signal over 205 km using a stabilized erbium fibre ring laser and 40 GHz electronic timing recovery', Electron Lett, 29, pp 990-2 (1993).

11. Kawanishi S, Takara H, Uchiyama K, Kitoh T and Saruwatari M: '100 Gbit/s, 50 km optical transmission employing all-optical multi/demultiplexing and PLL timing extraction', Technical Digest of OFC'93, postdeadline paper PD2 (1993).

12. Blank L C: 'Multi-Gbit/s optical time division multiplexing employing $LiNbO_3$ switches with low frequency sinewave drive', Electron Lett, 24, pp 1543-1544 (1988).

13. Patrick D M, Ellis A D and Spirit D M: 'Bit rate flexible all-optical demultiplexing using a nonlinear optical loop mirror', Electron Lett, 29, 702-703 (1993).

14. Barnsley P E, Wickens G E, Wickes H J and Spirit D M: 'A 4×5 Gbit/s transmission system with all-optical clock recovery', IEEE Photonics Technol, Lett, 4, pp 83-86 (1992).

15. Smith K and Lucek J: 'All-optical clock recovery using a mode-locked laser', Electron Lett, 28, pp 1814-1816 (1992).

16. Ellis A D, Smith K and Patrick D M: '10 GHz pulse train derived from a CW DFB laser using crossphase modulation in an optical fibre', Electron Lett, 29, pp 1323-1324 (1993).

17. Wickens G E, Spirit D M and Blank L C: '20 Gbit/s, 205 km optical time division multiplexed transmission system', Electron Letts, 27, pp 973-974 (1991).

18. Taga H, Suzuki M, Tanaka H, Yoshida Y, Edagawawa N, Yamamoto S and Wakabayashi H: 'Bit error rate measurement of 2.5 Gbit/s data modulated solitons generated by InGaAsP EA modulator using a circulating loop', Electron Lett, 28, pp 1280-1281 (1992).

19. Nakazawa M, Yamada E, Kubota H and Suzuki K: '10 Gbit/s soliton data transmission over one million kilometres', Electron Lett, 27, pp 1270-1272 (1992).

20. Wickens G E, Spirit D M and Blank L C: 'Nonlinear transmission of 20 Gbit/s optical time-division-multiplexed data over 205 km of dispersion-shifted fibre', Electron Lett, 28, pp 117-118 (1992).

21. Brown G N, Marshall I W and Spirit D M: 'Computer simulation of ultra-high capacity nonlinear lightwave transmission systems', Proceedings of Third IEE conference on Telecommunications, Edinburgh (1991).

22. Spirit D M, Blank L C, Davey S T and Williams D L: 'Systems aspect of Raman fibre amplifiers', IEE Proceedings Part J, 137, pp 221-224 (1990).

23. Davey S T, Williams D L and Ainslie B J: 'Distributed erbium-doped fibre for lossless link applications', in Digest of Conference on Optical Fibre Communication, OSA Technical Digest Series, 4, (Optical Society of America), Washington, DC, pp 200-202 (1991).

24. Spirit D M, Marshall I W, Constantine P D, Williams D L, Davey S T and Ainslie B J: 'Nonlinear, dispersion-free 10 GHz optical pulse train transmission in distributed erbium-doped fibre', Electron Lett, 27, pp 222-223 (1991).

25. Spirit D M, Wickens G E, Widdowson T, Walker G R and Williams D L: '137 km, 4×5 Gbit/s optical time division multiplexed unrepeatered system with distributed erbium fibre preamplifier', Electron Lett, 28, pp 1218-1220 (1992).

26. Blank L C, Bryant E G, Lord A, Boggis J M and Stallard W A: '150 km optical fibre transmission network experiment with 2 Gbit/s throughout', Electron Lett, 23, pp 977-978 (1987).

27. Patrick D M and Ellis A D: '10 GHz pulse train derived from a CW DFB laser using crossphase modulation in an optical fibre', Electron Lett, 29 pp 1323-1324 (1993).

28. Blank L C: 'High capacity optical fibre transmission systems', PhD thesis, University of Wales (1993).

5

IMPACT OF NEW OPTICAL TECHNOLOGY ON SPECTRALLY SLICED ACCESS AND DATA NETWORKS

L T Blair and S A Cassidy

5.1 INTRODUCTION

The aim of this chapter is to present new technological solutions to provide services ranging from simple telephony, to access to the home, to high-density data transfer. This is done with the aim of simplicity and low cost. An improved spectral-slicing technique is presented which uses new optical technology to impact on the areas of local access and computer data networks.

In section 5.2 the status of the world interest in WDM networks for data and access networking is briefly reviewed. Section 5.3 describes the basic principle of operation of the new spectral-slicing technique and section 5.4 considers some practical limitations. Two new networks, which may be applied to data transfer and local access, are then proposed in section 5.5.

5.2 BACKGROUND

5.2.1 World interest in networking

It is becoming evident that future telecommunications networks will demand flexibility and require multi-user and multifunctional capabilities. At the moment, proposals exist for a vast range of network topologies for a gamut of applications using a wide range of technologies [1]. Future networks will be required to provide simple low-bandwidth telephone links and, at the same time, deal with high-bandwidth services to the home or the workplace [2]. They will be needed to access sensor information at remote locations or to send uncompressed images for special applications, such as between hospitals. Current local network topologies including Ethernet and fibre distributed data interface (FDDI), which have total network capacities of 10 Mbit/s and 100 Mbit/s respectively, will become inadequate as improved connectivity is required. Also, new networks will be required to link hundreds or thousands of computers for data sharing or distributed processing applications. In particular, the drive for supercomputer visualization and the rapid increase in sophisticated computer users in our increasingly information-orientated society [3] is leading to the convergence of computer networks with telecommunications networks [4]. Such high-performance networks will be ideal for real-time interaction or applications where a large quantity of data is required to be transferred with low latency, such as virtual reality. The interest generated in the development of networks with gigabit data rates between many nodes has been so great that, in 1992 alone, special issues of IEEE Network [5], IEEE LTS [6] and IEEE Communications Magazine [7] have all been published. As the demand for increased data capacity increases, trends show that switched data services will cross the gigabits per second threshold by the year 2005 [8]. To keep pace with this, a consortium of American companies, universities and government laboratories (including GE, AT&T, IBM and Honeywell) has recently been funded by the US Defense Department's Research Projects Agency (DARPA) to speed up the development of all-optical networks within a five-year time span [9]. In a similar timescale, IBM predict the possibility of practical networks consisting of as many as 1000 nodes at a data capacity of 1 Gbit/s per node [4]. Also, within the next decade, a goal of America's National Research and Education Network (NREN) is to create a 3 Gbit/s backbone network across the USA.

5.2.2 Technologies

A way to achieve the above capabilities is through the use of optics and optical fibre systems [10]. BT's TPON network [11] demonstrates a single wavelength approach to the broadcasting of telephony over a passive optical network. In order to deliver large quantities of information (e.g. broadband services) complex protocols are required. One simple approach, to which much attention has been devoted, is to use wavelength-division multiplexing (WDM) [12] or, more specifically, high-density wavelength-division multiplexing (HDWDM) [13]. HDWDM may be said to occur when concurrent wavelength channels are separated by distances in the region of 1 nm. An example of such a scheme, described in Forrester et al [14], demonstrates its potential by distributing 2.5 Gbit/s signals to 43.8 million customers using 16 wavelength channels and erbium-doped fibre amplifiers. Excellent reviews of networking topologies using WDM technology may be found elsewhere [1, 13, 15, 16]. Bellcore's LAMBDANET [17], FOX [3] and HYPASS [18], IBM's RAINBOW [19], AT&T's SHUFFLENET [20] and BNR's SYMFONET [21] all describe examples of systems currently using HDWDM to investigate networking.

In the above networks, the most common source for transmission is a narrowband laser diode, such as a DFB semiconductor laser, which is capable of linewidths a fraction of a nanometre wide. To offer greater flexibility and capacity, there is a great deal of interest in making these devices tuneable over a wide range of wavelengths. At the same time, there is much interest in the design of tuneable receivers or tuneable passband filters of high finesse. Unfortunately, there exists a trade-off between the speed, linewidth and tuning range of these transceivers, with the main problem being the achievement of high-speed tuning over a wide spectral range. This increased complexity leads to an increase in cost and a search for a cheaper alternative.

5.3 BASIC PRINCIPLE OF NEW NETWORKS

In light of the above points and with the availability of new technologies, our aim is to design uncomplicated and low-cost networks. The use of laser diodes is avoided and studies are restricted to cheap broadband illumination, simple filtering and optical amplification. In particular, the concept of spectral slicing [22, 23] as the method of wavelength multiplexing is revisited. To date, this method of signal generation has used conventional diffraction-grating-based WDM devices to slice the broad bandwidth of LEDs into separate narrow bandwidth channels. Unfortunately, due to the low power available from LEDs and temperature instabilities, the wavelength slices have to be several nanometres wide, trading off the power within each slice to a number of possible wavelength channels. By operating at 1.5 μm, power levels may be improved by using optical amplifiers [24]. Now, with access to Bragg gratings recorded within the core of single-mode fibre [25] and optical amplification, improved and flexible network designs can be offered.

Figure 5.1 illustrates the use of a wide-bandwidth source, such as an ELED or a superfluorescent-doped fibre, to illuminate a single-mode optical fibre. The broadband signal (represented by a black arrow) passes through a modulator and is subsequently filtered by a Bragg fibre grating. Wavelength-selective reflection takes place at a specific wavelength defined by the grating period, and a narrowband signal of typically 0.1-0.5 nm bandwidth (represented by a grey arrow) is reflected back along the fibre. The modulated signal may be decoupled from the fibre at some later point. The light which is not reflected passes straight through the grating and is absorbed at the fibre output using index-matching termination. Many wavelengths may be multiplexed employing this technique [26] by using a single source and distributing it to many different gratings, each with a different fringe spacing and hence reflection wavelengths. The gratings are relatively insensitive to temperature fluctuations and have been measured as 0.0125 nm/°C [27],

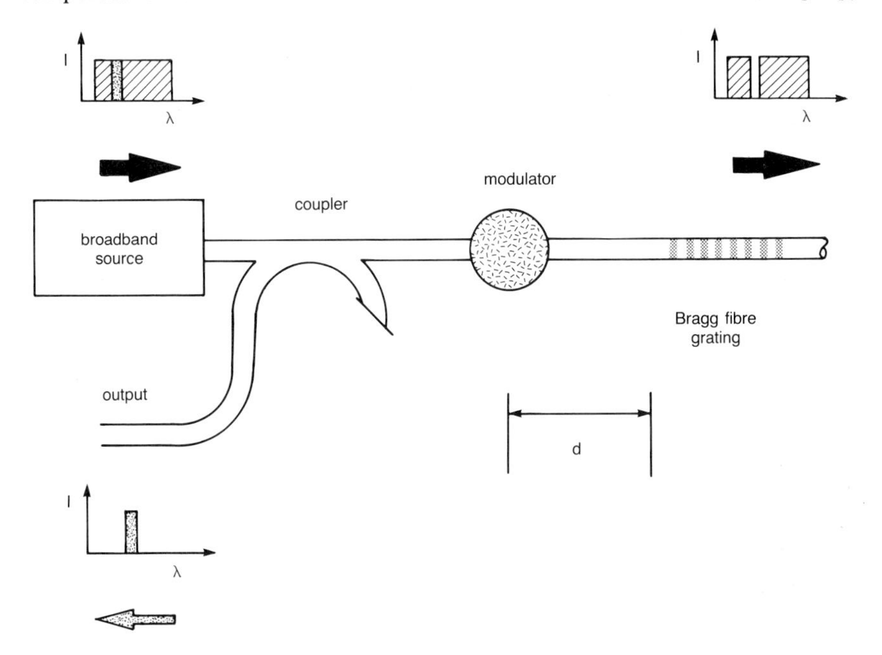

Fig. 5.1 Spectral slicing of broadband source using Bragg fibre grating.

which corresponds to about only a 0.5 nm wavelength shift over temperature fluctuations of 40 °C.

5.4 PRACTICAL CONSIDERATIONS OF SPECTRAL SLICING WITH NEW TECHNOLOGY

For such a multiplexing scheme it is necessary to investigate its information-carrying capability. This is dependent upon the maximum number of wavelength channels which may be squeezed into the passband of a broadband source and the maximum bit rate which may be supported by each channel. It is therefore necessary to take a preliminary look at:

- the method of modulation;
- the power carried within each wavelength channel;
- noise;
- the loss budget for the specific network in question.

Since the loss budget is dependent upon the specific network design, this will be discussed later in conjunction with some network examples. For the moment, the first three points only will be briefly considered.

5.4.1 Modulation

In order to encode information on each of these separate wavelength channels, it is necessary to modulate the signal wavelength. Since the source cannot be modulated (as is conventional with a laser diode) an external modulator must be used. A number of devices may be used such as lithium niobate modulators or gated amplifiers, which are capable of gigabits per second modulation rates, or other low-bit rate devices such as electro-mechanical switches. An interesting device is the fibre overlay modulator (FOM) (for reasons such as fibre compatibility and potential low cost) [28] shown schematically in Fig. 5.2. Essentially, the FOM consists of a D-fibre or polished half-coupler block which has been coated with an electro-polymer overlay. By applying a voltage across this polymer, it is possible to induce optical amplitude modulation through evanescent coupling. At present such work is at an early stage but promising results have been presented [29].

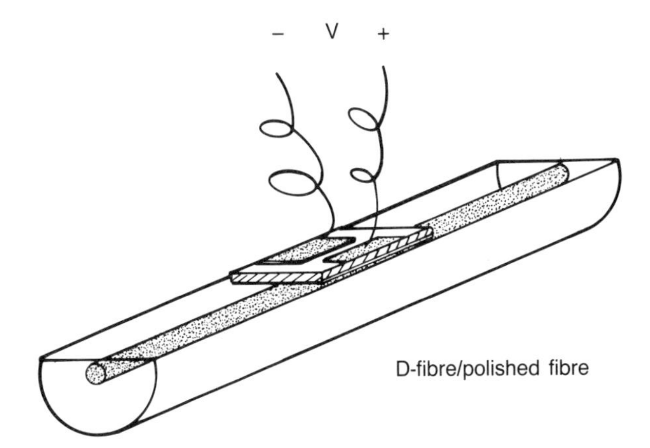

Fig. 5.2 Diagram of possible fibre overlay modulator (FOM).

An important consequence of using an external modulator in most of the above reflection-type topologies is that the separation between the reflecting element and the modulator (distance *d* in Fig. 5.1) determines an upper limit to the practical modulation frequency. In fact, the best achievable bit rate may be calculated as the ratio of the speed of light to twice the separation distance which, in the case of a separation in the order of 1 cm, corresponds to a maximum allowable bit rate of about 10 Gbit/s. This limit is of the same order as that which may be obtained elsewhere in the network by fundamental limits to electronic processing speeds.

Of course, if fibre gratings are used, other forms of modulation may be employed. If high powers are required then the modulator may be replaced by a semiconductor amplifier which, compared to a DFB or DBR laser, has a relatively broad linewidth. The device may then be modulated, with the wavelength selectivity being carried out by the fibre grating. Alternatively, it may be possible to modulate the grating directly by stretching it. The required shift of about 1 nm may be obtained with only a 0.1% strain [30], which is easily achieved and is equivalent to a temperature shift of approximately 80 °C over the grating length of a few millimetres. This shift is enough to move the central peak of a signal wavelength out of the filter passband at a receiver. It may then be possible to consider modulation speeds of megahertz [31] by placing the fibre grating on a piezoelectric transducer. A more ingenious way of modulating the grating is to form two gratings of the same wavelength very close together. It may then be possible to introduce a phase shift between the two reflected components to induce constructive or destructive interference. The filter design technology for this type of work already exists [32].

5.4.2 Power limitations

In this chapter we will assume the availability of optical amplification in the form of erbium-doped fibre amplifiers for both booster and power (source) applications. Assuming an ideal flat spectral response of 30 nm centred at 1.55 μm with a small-signal gain of 25 dB, it is important to note that the low-loss spectrum of optical fibre which may now be accessed is limited to the 'erbium window'. This sets a maximum limit to the modulation bandwidth available to approximately 4 THz and thus sets an upper limit to the total network capacity in terms of a bits $\times$ nodes product. For power budget calculations it is assumed that a broadband source exists which has a maximum spectral bandwidth equal to the erbium amplifier and has a total output power of +15 dBm. Such a device may be obtained through the use of a superfluorescent fibre source, or an ELED, or laser diode followed by an optical power amplifier. In the laboratory a diode-pumped erbium-doped fibre has been used which offers amplified spontaneous emission (ASE) output powers greater than 10 mW and optical bandwidths of 30 nm within the 1.5 μm window.

In order to increase the number of wavelength channels within a fixed-source bandwidth, it is necessary to know how the filtered optical power will depend on channel selectivity. In fact, as the filter becomes narrower, the reflected power decreases linearly. The availability of optical power and therefore the network power budget will place a lower limit on the filter width determining the maximum number of channels.

5.4.3 Sources of noise

In addition to the standard noise components which exist in a typical optical transmission system, further noise sources must be considered when a broadband source needs to be spectrally sliced. Firstly, section 5.4.3.1 describes how different wavelength channels may interfere with each other in terms of optical crosstalk if they are positioned too close to each other. The second noise source imposes a limit to signal-to-noise ratio at the receiver in addition to shot noise and is caused by the narrowband filtering of a broadband, incoherent source (section 5.4.3.2).

5.4.3.1 Crosstalk — wavelength channel separation

The closer each wavelength channel gets to the next one, the more likely interference will occur in the form of optical crosstalk [33]. To study this effect, it is assumed that the signal distribution is Gaussian. Figure 5.3 shows the interaction of three wavelength channels of equal amplitude separated by 0.5 nm and centred at wavelengths of 1549.5 nm, 1550 nm and 1550.5 nm.

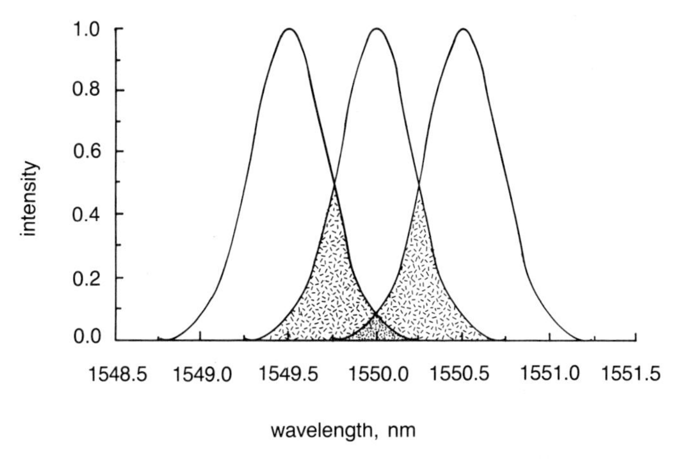

Fig. 5.3 Diagrammatic representation of crosstalk between three wavelength channels.

In order to measure the crosstalk at 1550 nm the contributions to the output at 1550 nm by other adjacent channels must be calculated. Using a theory developed by Senior et al [34], it is possible to obtain an estimate of the effects of crosstalk on a wavelength channel by its closest neighbours, in terms of channel selectivity and separation. Figure 5.4 shows a plot of crosstalk versus the ratio of wavelength channel separation to channel selectivity. It is observed that, in order to achieve crosstalk levels of less than −20 dB, a channel spacing of at least twice the selectivity is required. Thus, for a source bandwidth of 30 nm and a channel selectivity of 0.1 nm, no more than 150 channels can be achieved with crosstalk conditions better than −20 dB. For analogue systems (e.g. cable TV), which require high signal-to-noise ratios in excess of 50 dB, it is observed that a channel spacing of three times the selectivity may be needed.

5.4.3.2 Excess beat noise

In traditional transmission systems there exist noise terms which are based on the quality of the source and the receiver. Under ideal circumstances these

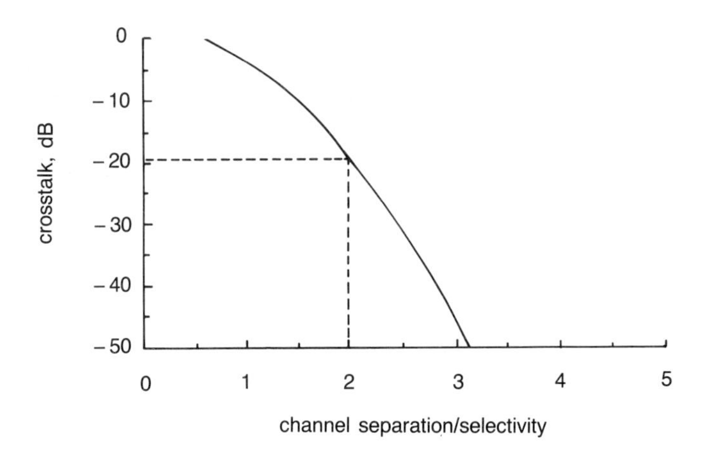

Fig. 5.4 Crosstalk versus channel separation/channel selectivity.

terms may be neglected and it can be assumed that this system is quantum limited with the only source of noise being caused by the quantum nature of photons (shot noise) incident at the receiver. For an ideal coherent source a quantum-limited receiver sensitivity may then be calculated which is dependent upon signal bit rate or receiver bandwidth. Based on experimental results [35], the assumption can then be made that a practical receiver will be about ten times less sensitive than this quantum limit, due to additional noise terms such as thermal noise in the receiver. This gives a rough indication of how a network will respond to different bit rates, and the conclusion that can be drawn is that the signal-to-noise ratio at the receiver may be improved by simply increasing the power incident on the photodetector.

When using signals derived from an incoherent broadband source as in this spectral-slicing technique, it is important to consider an additional noise component [36] which will occur due to the addition of spontaneous frequency components [37] — known in the study of optical fibre amplifiers as spontaneous-spontaneous beat noise [38]. This excess noise term is proportional to the square of the signal current and the receiver bandwidth, but is inversely proportional to the optical bandwidth of the signal channel. Because of the random nature of the signal, a penalty will be introduced at the detector which is not present for narrow linewidth sources. For good receiver sensitivities (low incident power), this excess beat noise is not very significant. For example, a 2 Mbit/s signal within a 0.5 nm filter at 1.5 μm which is detected at -50 dBm will exhibit an excess shot noise of less than 0.5 dB. This makes little difference to the signal-to-noise ratio and thus shows no noticeable degradation of BER characteristics. If, for the same conditions, the bit rate is increased to 1 Gbit/s, it is found that the optical power incident at the receiver must be increased. This leads to an increased dominance of the excess beat noise and causes the signal-to-noise ratio at the receiver to become depleted considerably, manifesting itself as a bit error floor of 10^{-8}. At such high bit rates a simple increase in optical power at the receiver will not improve the signal-to-noise ratio. By increasing the linewidth of the 1 Gbit/s signal to 1 nm, the problem is overcome and a bit error rate of 10^{-9} is easily achieved.

It can therefore be seen that there are conflicting aims. In order to squeeze many wavelength channels into the spectrum of a broadband source with low crosstalk, it is necessary to reduce the bandwidth of each channel. However, by reducing the filter bandwidth, the available power at the detector is reduced. Thus, there is a trade-off between the number of channels supported within the erbium window and the bit rate per channel. Fortunately, this problem is not so critical since high-power fibre sources or an optical amplifier placed within the network can now compensate for lost power. Therefore, because of the excess noise described above, the filter bandwidth is the determining factor for the maximum number of wavelength channels and bit rates which may be supported.

It should be noted that the above assumptions are based on Gaussian filtering characteristics. By optimizing grating filter design, using a technique developed at BT Laboratories [32] which is capable of designing filters of different shapes with customized frequency responses, the crosstalk limits may be further improved. Also, it has been noted that this spectrally sliced network is dependent upon excess beat noise under conditions of narrow filter bandwidths and high bit rates. If several filters of the same selectivity for the same wavelength channel were traversed in the network, the signal bandwidth would effectively be narrowed and the dependence on excess beat noise increased. By designing filters with rectangular characteristics this problem could be reduced.

5.5 EXAMPLES OF NEW NETWORKING TOPOLOGIES USING SPECTRAL-SLICING TECHNIQUES

5.5.1 Broadcast data network

This example takes a look at a network based on the passive broadcast star technique, which moves processing capabilities to distributed nodes. A number, N, of distributed nodes are connected together by means of an $N \times N$ star coupler, which acts as a hub. Each node is identified by a unique wavelength, which in most previous cases has been generated by a laser diode, but in this example is generated using spectral slicing. Figure 5.5 shows how such a scheme may be realized by using Bragg gratings for channel (or node) identification. In Fig. 5.5 a number of broadband sources are shown which illuminate an $N \times N$ power splitter which, in turn, distributes power through optical fibre to each of the N nodes. This fibre may be called the 'power supply fibre'. Each node is also connected by means of a second fibre to every other node using a second $N \times N$ star coupler. This second fibre is bidirectional and may be called the 'signal fibre'. Two examples of how each node may be constructed are shown. Node 1 incorporates the use of a circulator to control the direction of signalling traffic, whereas node 2 replaces the circulator with two simple fused-fibre couplers and an optical amplifier.

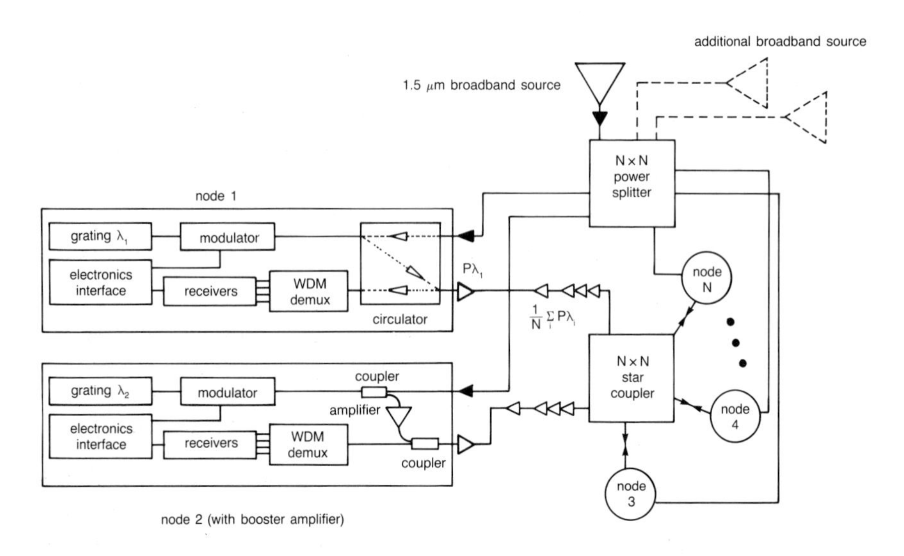

Fig. 5.5 Example of $N \times N$ data optical network using Bragg fibre gratings.

5.5.1.1 Network description

Broadband illumination passes from the $N \times N$ power splitter to each node via the power supply fibre, where it is directed, either by a circulator or a coupler, through a modulator to a grating. At the grating only the selective narrowband signal is diffracted, which is dependent upon the grating fringe spacing. The modulated narrowband signal is then returned along its inward path to the output of the node where it is diverted by the circulator or coupler to the signal fibre and passed through the broadcasting $N \times N$ star coupler to every other node. In the case of node 2, the signal may be amplified before leaving. On the signal fibre at the input to each node there now exist, in the opposite direction, N concurrent wavelength channels, which were generated at each of the N nodes. These enter the node and are directed, again by the circulator or coupler, to a WDM device which demultiplexes the wavelength channels to be detected independently by a receiver array. The information from the photodetectors is then transferred into the electronic domain where it is now suitable for processing. This processed information may now be used to control the operation of each respective modulator.

5.5.1.2 Features

This network design offers several features.

- Upgradability and redundancy — by adding extra broadband power sources at the input to the power distribution $N \times N$ splitter, the data bit rate or number of nodes may be increased. At the same time, by using more than one source, redundancy is included in the network, allowing it to continue functioning should one source fail.
- Bidirectional signal fibre — although two fibres enter each node, only one contains signalling information. This simplifies the network design.
- Multiple node update — each node is capable of receiving signals from any/all other nodes. This would be beneficial for distributed-memory applications in the area of parallel computing. Also, since each node is instantly updated with any changes, such a network could also be applied to a future signalling network, localizing customer information.
- Data speed and format transparency — since each node communicates over an independent wavelength channel the data speed and format from each node may be different without the need for complicated protocols.
- No tuneable elements — the network uses simple low-cost fibre Bragg gratings to define node wavelength selection. These may be programmed *in situ* or may be connected on location leading to flexible control over wavelength allocation.

5.5.1.3 Power budget

Table 5.1 gives an estimate of the power requirements for such a network assuming 32 nodes. In the simplest case, A, a node with a circulator is assumed and, in case B, a node with 3 dB couplers and a booster amplifier. Using a circulator at each node it can be seen that a very low power of -55.5 dBm is available at the receiver. This power level is only good enough to support

Table 5.1 Estimated power requirements for 32-node network.

A *Node 1* *(with circulator)*		*B* *Node 2* *(couplers and amplifier)*	
30 nm broadband source	+15	30 nm broadband source	+15
32 × 32 power splitter	−15	32 × 32 power splitter	−15
Circulator × 3	−3	50% coupler × 2	−6
Modulator × 2	−4	Modulator × 2	−4
Spectral slice loss	−18.5	Spectral slice loss	−18.5
32 × 32 star coupler	−15	Optical amplifier	+25
1 × 32 WDM insertion loss	−5	50% coupler × 2	−6
Excess system loss	−10	32 × 32 star coupler	−15
		1 × 32 WDM insertion loss	−5
		Excess system loss	−10
Required sensitivity	−55.5	Required sensitivity	−39.5

Note: All values in dB/dBm
Excess system loss includes — fibre loss
— splice loss
— component excess loss
— system margin

low bit rates less than 2 Mbit/s. This may be improved, however, by adding additional broadband sources to the input. If the circulator is removed and replaced by simple 50% couplers, then the network power budget becomes too large for practical applications. This can be overcome by adding an optical amplifier at each node. The required receiver sensitivity improves to −39.5 dBm, which should allow bit rates as high as 1 Gbit/s to be measured giving the overall network a possible data-handling capacity of 32 Gbit/s.

5.5.2 Application to the access network

Figure 5.6 shows one of many possible designs of a network which uses simple optical technologies and which may be applied to the area of local access. It avoids ranging protocols by identifying each customer by a unique signal wavelength and, in its simplest form, relies on time multiplexing to send information from the exchange (but this could easily be upgraded by offering WDM services in both directions). Its dimensions are modelled on a typical passive optical network for the distribution of telephony or broadband services to the customer's premises [11], which is currently under field trial. Each exchange line serves 32 customers by using a four-way passive splitter at the cabinet and an eight-way splitter at the distribution point.

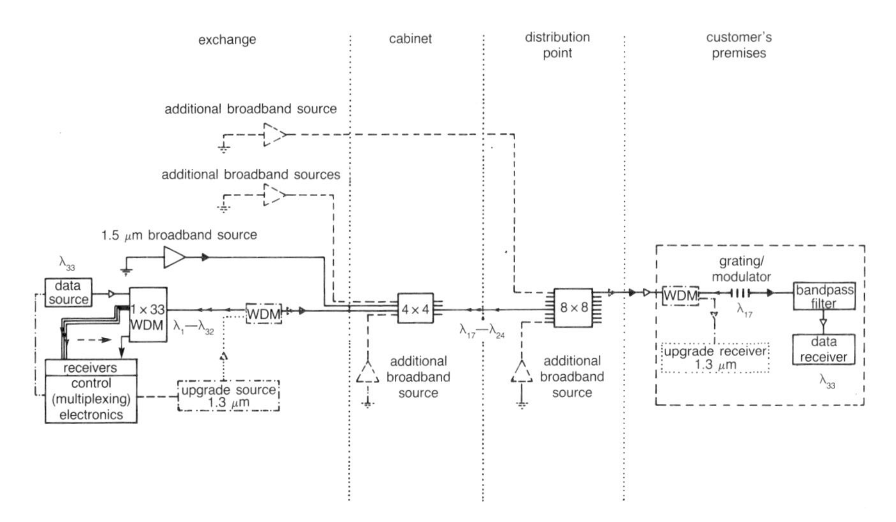

Fig. 5.6 Example of optical access network using Bragg fibre gratings.

5.5.2.1 Network description

For simple low-bandwidth operation it is proposed, for example, that only the 1.5 μm communications window is employed for data transmission. A signal (λ_{33}) outside the erbium-gain spectrum (e.g. at 1.52 μm) is used to transmit data to each of the 32 customers by the conventional means of time division multiplexing. At the customer's premises the signal passes through the grating, is filtered and then detected. The erbium-gain window is used to provide a return signal from each customer using wavelength-division multiplexing and high-density spectral slicing. A broadband source illuminates the grating/modulators present at each customer's location. Each signal is returned to the exchange to be demultiplexed using a 1×33 channel WDM, which is used to demultiplex the 32 individual wavelength channels returned from the customers into 32 serial outputs, each of which can be detected using its own photodetector. The 33rd channel is used to transmit time-multiplexed data to the customers. Full advantage may therefore be taken of the large available bandwidth within optical fibre to identify the signal from each customer by a unique wavelength. Additional sources may be added at the exchange to increase the number of services offered. Thus, one way of upgrading to a broadband network may be to install an optional 1.3 μm upgrade source at the exchange and provide an upgrade receiver at the upgrading customer's premises.

5.5.2.2 Features

Upgradability and redundancy

Additional broadband sources are optional and may be used to build in redundancy or upgradability to the network. Separate fibre links may be used to supply additional power of the same characteristics as the main source to the spare input ports at each splitter, thus ensuring an adequate power supply in the case of the main source failing. Similarly, by adding more sources, the power supplied to the network is increased, allowing the possibility of upgrading the network to higher speeds or a greater number of customers in the future.

It should be noted that additional customers may be added to the network by providing additional broadband sources with power spectra over different wavelength regions. An example of this would be the 1.6 μm fibre amplifier, which offers an ASE spectrum of 30 nm [39].

A 1.3 μm laser device (in conjunction with a 1.3 μm receiver) is optional and may be used to upgrade the network so that it can support high-data-rate broadband services. The 1.3 μm/1.5 μm WDM couplers required are shown in Fig. 5.6. The high-speed laser diode, which may be directly modulated at high data rates, should be capable of supplying 32 customers with required services within the 1.3 μm communications window.

Of course, there is still additional bandwidth available within optical fibre to further improve the network at some later date and use WDM to supply additional services from the exchange to the customer.

Customer identification

Each customer is identified in the return direction by a unique wavelength, avoiding the need for complex identification protocols.

Potential low cost at customer's premises

All the optical components shown at the customer's premises in Fig. 5.6 are potentially cheap. This would allow the possibility of a low initial installation cost.

5.5.2.3 Power budget

Taking into account the above constraints some simple power-budget calculations have been carried out. These are shown in Table 5.2 A for a signal passed from the exchange to the customer at 1.5 μm (no upgrade) and in

Table 5.2 C for signals passed from the customer to the exchange within the erbium ASE spectrum. The reflection bandwidth of the gratings has been assumed as 0.45 nm, the erbium-source bandwidth as 30 nm and the channel separation as 0.9 nm. Table 5.2 B shows a power budget calculation for an upgraded network with a laser source operating at 1.3 μm. No additional sources or amplification have been used.

From the calculations in Table 5.2 A, it can be seen that a receiver sensitivity of −35 dBm is required at each customer. Such a receiver is capable of receiving data rates greater than 150 Mbit/s and therefore may easily support broadband video services. In the opposite direction (Table 5.2 C), it is noted that the receivers at the exchange should be capable of receiving −51.5 dBm, making them able to detect signals of 2 Mbit/s. In Table 5.2 B, the required receiver sensitivity is −27 dBm. Using modern receiver technology, this should support bit rates up to at least 2.4 Gbit/s.

Table 5.2 Estimated power requirements for proposed access network.

	A *Exchange to customer*		*B* *Exchange to customer (upgrade)*		*C* *Customer to exchange*	
Exchange	1.5 μm laser diode source	0	1.3 μm laser diode source	0	30 nm broadband source	+15
	1×33 WDM insertion loss	−5	1.3 μm/1.5 μm WDM (upgrade)	−1	1×33 WDM insertion loss	−5
	1.3 μm/1.5 μm WDM (upgrade)	−1			1.3 μm/1.5 μm WDM (upgrade)	−1
	Sub-total	−6	Sub-total	−1	Sub-total	+9
Cabinet/ distribution point	4-way splitter	−6	4-way splitter	−6	4-way splitter × 2	−12
	8-way splitter	−9	8-way splitter	−9	8-way splitter × 2	−18
	Sub-total	−15	Sub-total	−15	Sub-total	−30
Customer's premises	1.3 μm/1.5 μm WDM (upgrade)	−1	1.3 μm/1.5 μm WDM (upgrade)	−1	1.3 μm/1.5μm WDM (upgrade)	−1
	Grating/modulator	−1			Grating/modulator	−1
	Bandpass filter	−2			Spectral slice loss	−18.5
	Sub-total	−4	Sub-total	−1	Sub-total	−20.5
Others	Excess system loss	−10	Excess system loss	−10	Excess system loss	−10
Required receiver sensitivity		−35		−27		−51.5

Note: All values in dB/dBm
Excess system loss includes — fibre loss
— splice loss
— component excess loss
— system margin

It can be seen therefore that this network is capable of passing a single wavelength with a high data rate to every customer, as is required by a low-bandwidth-type network. In the opposite direction, the network is capable of passing 32 different wavelength channels of moderate bit rate, each identifying a customer, which may be used for lower-bandwidth services such as telephony. In order to upgrade the downstream data capacity of this network, it is only necessary to insert a second high-speed source at the exchange and a high-speed receiver at the customer's premises. For small upgrades additional 1.5 μm sources may be inserted at the splitting points with no great effort. To upgrade the data rate from the customer's end in the far future, the grating module may simply be replaced by a laser module of the same wavelength. It can be seen that this network keeps all the complexity within the exchange end and thus allows this cost to be distributed between all the customers.

5.6 CONCLUSIONS

Having established a need for future multifunctional and high data-capacity networks, a possible low-cost and practical alternative to present laser-based network designs is offered. Assuming the availability of a single source, optical amplification, external modulation and spectral-slicing techniques, the design of two spectrally-sliced networks capable of supporting computer data or local access services has been described. These offer flexibility in terms of node location, data format and data speed. The use of fibre gratings offers the advantage that wavelength filtering/multiplexing may take place at distributed locations, thus not limiting any wavelength-demultiplexing function to any one location. Gratings may be preprogrammed and simply installed where needed. Costly multiwavelength WDM devices can be avoided because of the advantages of new high-powered fibre sources. It has been demonstrated through power-budget calculations that the proposed network topologies are capable of large data rates between nodes; however, excess noise due to the narrowband filtering of a broadband source may limit the maximum number of nodes which may be supported.

By applying new optical fibre technologies such as doped-fibre sources/amplifiers and gratings to the technique of spectral slicing, the possibility exists to reduce the cost of WDM networks to a practical level and at the same time increase network functionality.

REFERENCES

1. Green P E: 'Fiber optic networks', Prentice-Hall, Inc (1993).
2. Special issue of IEEE Commun (April 1990).
3. Goodman M S: 'Multiwavelength networks and new approaches to packet switching', IEEE Commun, 27 , No 10, pp 27-35 (October 1989).
4. Green P E: 'The future of fibre-optic computer networks', Computer, pp 78-87 (September 1991).
5. Special issues of IEEE Network, 6 , Nos 2 and 3 (1992).
6. Special issue of IEEE LTS, 3 , No 2 (May 1992).
7. Special issue of IEEE Commun, 30 , No 4 (April 1992).
8. Ransom M N and Spears D R: 'Applications of public gigabit networks', IEEE Network, pp 30-40 (March 1992).
9. Shimazu M: 'DARPA funds $8 million consortium to pursue optical backplane interconnect', Photonics Spectra, pp 28-30 (September 1992).
10. Hartman D H: 'Digital high speed interconnects: a study of the optical alternative', Opt Eng, 25 , No 10, pp 1086-1102 (1986).
11. Hill A M et al: 'An experimental broadband and telephony passive optical network', IEEE Global Telecommun Conf (Globecom '90), pp 1856-1860, San Diego (December 1990).
12. Senior J M and Cusworth S D: 'Devices for wavelength division multiplexing techniques for high capacity and multiple access communication systems', IEEE J Sel Areas in Commun 8 , No 6 (August 1990).
13. 'Dense Wavelength Division Multiplexing Techniques for High Capacity and Multiple Access Communication Systems', IEEE J Sel Areas in Commun (Special Issue), 8 , No 6 (August 1990).
14. Forrester D S et al: '39.81 Gbit/s, 43.8 million-way WDM broadcast network with 527 km range', Electron Lett, 27 , No 22, pp 2051-2053 (1991).
15. Mukherjee B: 'WDM-based local lightwave networks Part I: single-hop systems', IEEE Network, pp 12-27 (May 1992).
16. Mukherjee B: 'WDM-based local lightwave networks Part II: multi-hop systems', IEEE Network, pp 20-32 (July 1992).
17. Goodman M S et al: 'The LAMBDANET multiwavelength network: architecture, applications, and demonstrations', IEEE J Sel Areas in Commun, 8 , No 6, pp 995-1004 (1990).
18. Arthurs E et al: 'HYPASS: an optoelectronic hybrid packet switching system', IEEE J Select Areas in Commun, 6 , No 9, pp 1500-1510 (1988).

19. Green P E: 'An all-optical computer network: lessons learned', IEEE Network, pp 56-60 (May 1992).

20. Hluchyj M G and Karol M J: 'Shuffle net: an application of generalized perfect shuffles to multihop lightwave networks', IEEE Infocom '88, pp 379-390 (March 1988).

21 Kirkby P A: 'SYMFONET: Ultra-high-capacity distributed packet switching network for telecoms and multiprocessor computer applications', Electron Lett, 26 , No 1, pp 19-21 (1990).

22. Reeve M H et al: 'LED spectral slicing for single-mode local loop applications', Electron Lett, 24 , No 7, pp 389-390 (1988).

23. Hunwicks A R et al: 'A spectrally sliced, single-mode, optical transmission system installed in the UK local loop network', IEEE Globecom '89, pp 1303-1307 (1989).

24. Kilkelly P D D et al: 'Experimental demonstration of a three channel WDM system over 110 km using superluminescent diodes', Electron Lett, 26 , No 20, pp 1671-1673 (1990).

25. Kashyap R et al: 'All-fibre narrowband reflection gratings at 1500 nm', Electron Lett, 26 , pp 730-732 (1990).

26. Blair L T and Cassidy S A: 'Wavelength division multiplexed sensor network using Bragg fibre reflection gratings', Electron Lett, 28 , No 18, pp 1734-1735 (1992).

27. Morey W W et al: 'Multiplexing fibre Bragg grating sensors', Proc SPIE, 1586 , pp 216-224 (1991).

28. Wilkinson M et al: 'Optical fibre modulator using electro-optic polymer overlay', Electron Lett, 27 , No 11, pp 979-980 (1991).

29. Fawcet G et al: 'In-line fibre-optic intensity modulator using electro-optic polymer', Electron Lett, 28 , No 11, pp 985-986 (1992).

30. Campbell R J et al: 'Narrow-band optical fibre grating sensors', OFS'90, pp 237-240 (December 1990).

31. Imai M et al: 'Wide-frequency fiber-optic phase modulator using piezoelectric polymer coating', IEEE Photon Technol Lett, 2 , No 10, pp 727-729 (1990).

32. Wilkinson M et al: 'Novel computer designed waveguide grating structures with optimised reflection characteristics', Electron Lett, 28 , No 17, pp 1660-1661 (1992).

33. Rosher P A and Hunwicks A R: 'The analysis of multichannel wavelength division multiplexed optical transmission systems and its impact on multiplexer design', IEEE J Sel Areas in Commun, 8 , No 6, pp 1108-1114 (1990).

34. Senior J M et al: 'Wavelength division multiplexed multiple sensor network', OSCA Report (1991).

35. Smyth P P et al: '152 photons per bit detection at 622 Mbit/s to 2.5 Gbit/s using erbium fibre preamplifier', Electron Lett, 26 , pp 1604-1605 (1990).

36. Liu K: 'Noise limits of spectral slicing in wavelength-multiplexing applications', Proc Optical Fiber Communications Conference, p 174 (February 1992).

37. Morkel P R et al: 'Noise characteristics of high-power doped-fibre superluminescent sources', Electron Lett, 26 , No 2, pp 96-98 (1990).

38. Yamamoto Y: 'Noise and error rate performance of semiconductor laser amplifiers in PCM-IM optical transmission systems', IEEE J Quantum Electron, QE-16 , No 10 (1980).

39. Massicott J F et al: 'Low noise operation of Er^{3+} doped silica fibre amplifier around 1.6 μm', Electron Lett, 28 , p 1924 (1992).

6

OPTICAL WIRELESS LOCAL AREA NETWORKS — ENABLING TECHNOLOGIES

P P Smyth, M McCullagh, D Wiseley, D Wood, S Ritchie, P Eardley and S A Cassidy

6.1 INTRODUCTION

An exciting prospect for the 1990s is the development of the new generation of laptop and pen computers. People will be able to use these as truly personal computers with high-resolution graphics and telecommunications services such as videophones, electronic messaging and access to multimedia databases over the network. They could, by the end of the century, emerge as the computer equivalent to the mobile phone.

Future networks will allow people arriving at the office with their mobile computer to 'log on' at their desk. They will then receive and transmit information as they work. As they move around the building with their computers, electronic messages or videophone calls will be routed to them automatically. Away from work, at public buildings such as railway stations or libraries, they will use their personal mobile computer to obtain information, study timetables and make videophone calls. Both offices and public buildings will have wireless local area networks (LANs) supporting these computers with connection to the public switched network with sophisticated network intelligence.

Until recently wireless communication has been restricted to radio or microwave techniques. However, the radio spectrum is a scarce resource and there are increasing pressures for its economic use. Therefore allocated bands are usually narrow, typically tens of megahertz, and usually sub-divided into many channels. This is adequate for present networks, which might provide each user with around 100 kbit/s, but future networks, requiring perhaps 10 to 100 Mbit/s per user, would be almost impossible to implement.

Optical free-space transmission is the new alternative to radio and microwave communications for future mobile computer networks — it is capable of offering far greater bandwidth, it does not interfere with other systems since optical radiation is confined to the room in which it is used, and it is globally compatible since it is outside the radio frequency planning controls.

Low cost is fundamental to mass exploitation of the mobile computer for telecommunications purposes. The projected cost of optical free-space wireless systems is much lower than microwave-based systems since they could be based on very low-cost consumer optoelectronic components such as 850 nm sources and silicon detectors.

This chapter details work on broadband optical wireless LANs. Results are presented for a 50 Mbit/s system based on a new 'optical cellular' topology. It also focuses on the various enabling technologies developed at BT Laboratories. These include new forms of transmitter technology to allow the safe distribution of relatively large amounts of optical power, and highly efficient optical collection systems to increase system range and performance. The chapter also discusses the important issue of optical safety, which is fundamental to optical wireless, and includes proposed changes to the safety standards.

6.2 OPTICAL WIRELESS SYSTEMS

6.2.1 Propagation techniques

There are two basic propagation techniques for optical wireless systems — diffuse (multidirectional) or line of sight (LOS). Optical wireless LANs can be designed which use either or both of these techniques.

Diffuse systems have the major advantage of supporting complete mobility, as the optical transceivers can be positioned almost anywhere in the system area. The infra-red light from the source is spread through a wide angle and is reflected from objects, creating a diffusely scattered background illumination. Optical receivers used in a diffuse system receive signals from more than one path from the transmitter and the ultimate system speed is limited by the multipath delay differences. It has been shown [1] that multipath effects limit the bandwidth distance product to a maximum of 260 Mbit m/s for baseband modulation. This suggests that for a typical room size of 10 m $\times$ 10 m $\times$ 3 m a diffuse system could typically support a data rate of 16 Mbit/s.

LOS propagation supports much higher bit rates than diffuse propagation (100s Mbit/s) as there are no multipath effects. However, this system is susceptible to path blockage. The extent to which the system is susceptible is directly dependent on the field of view (FOV) of the receiver and the beam size of the transmitter. Spatial diversity techniques, as in radio systems, can be used to overcome path blockage problems. An adaptation of the LOS technique is reflective propagation, where all transceiver units are aimed at a common spot on the ceiling, which partially alleviates the problem of path blockage. LOS-based LANs are available as commercial products for bit rates up to 16 Mbit/s with future developments including links for FDDI (fibre distributed data interface), which has a data rate of 100 Mbit/s.

6.2.2 Realization of broadband optical wireless LANs

An optical wireless LAN which supports broadband communications to mobile computers could be created by fixing a satellite unit to the ceiling of each office and connecting them together with a high-speed optical fibre backbone. The satellite units would act as base stations, networking the optical wireless terminals. Optical wireless LANs will need to offer similar high-speed networking as cabled systems, including FDDI. Therefore optical wireless LANs will need to use semiconductor lasers rather than LEDs. These also have the advantage of much greater electrical-to-optical conversion efficiency. Unfortunately, to operate an optical wireless LAN at high data rates ($>$10 Mbit/s) requires transmit powers much higher than the eye safe limit ($<$1 mW).

The use of holographic transmitters developed at BT Laboratories (BTL) allows for the safe transmission of these high optical powers (see section 6.4) and for the area of operation to be divided into ‘optical cells’ as depicted in Fig. 6.1. The confinement of optical transmit power within a cell allows for a much higher data rate operation than in a diffuse transmission system. The cellular structure also provides spatial reuse of bandwidth.

Similarly, the use of concentrating optics in the satellite allows the room area to be broken up into a number of collection cells, each supporting a narrower FOV than in a diffuse system. The narrower FOV reduces the transmit power requirement from mobile terminals. This, in turn, increases the battery life, which is very important for mobile terminals. In order to maximize the power margin it is essential that the optical receiver has good sensitivity while still providing the broad bandwidth required. It is concluded that the key enabling technologies for this optical cellular system include holographic-based transmitters, high gain collection optics and broadband, low-noise receivers.

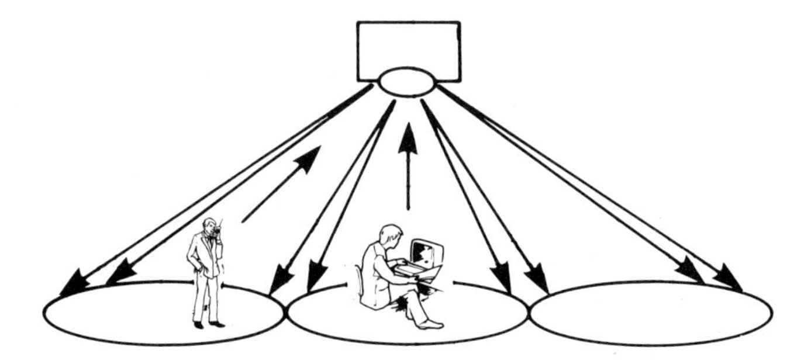

Fig. 6.1 Mobile communications in an optical cellular system.

A major issue in the design of an optical wireless LAN is to ensure that the ambient light level, which generates noise in the optical receiver, does not significantly impair the overall system performance. With the optical cellular approach, the restricted receiver field of view allows the incorporation of narrowband optical filtering, which minimizes the noise generated by the ambient light. However, a further problem is the high level of noise generated by fluorescent lighting when switching on, with noise frequencies up to 500 kHz. To minimize the effects of this noise source, the system can use a data coding scheme with a suppressed spectrum at lower frequencies. Alternatively, subcarrier modulation can be used. Although this is more complex, it results in maximum suppression of fluorescent-light-generated noise.

The office is divided into a number of optical cells produced by a hologram (see section 6.4). The media access control protocol used for an optical cellular LAN should allow the networked terminals to access the large bandwidth available whenever they require it. It should also allow for roaming of the mobile computers between cells. High-speed wired LANs use time-based multiple access protocols such as Token Ring and FDDI. However, these networks suffer from the problem that, as the number of users on the network increases, so the share of bandwidth per user decreases. To avoid this reduction in bandwidth per user, future optical wireless LANs could employ channel-based, as opposed to time-based, multiple access techniques. Two particularly applicable channel-based techniques are subcarrier multiple access and wavelength-division multiple access. The separate subcarriers or optical wavelengths could be dynamically assigned to terminals communicating across the network, and/or could be used for accessing different network services.

6.3 OPTICAL SAFETY FOR OPTICAL WIRELESS SYSTEMS

Optical radiation can present a hazard to the eye (both the cornea and the retina) and to the skin if the exposure is high enough. The degree of hazard depends on a number of factors, including the exposure level (total energy or power density), exposure time and wavelength. The greatest hazard presented by near-infra-red sources is retinal damage. This is because they are focused to a point by the eye in the same way as visible light. Moreover, since the light is invisible there is no tendency for the eye to blink. Another important factor is the size of the optical source — the lens of the eye may focus the radiation from a point source to a small intense spot on the retina, while the same optical energy from a large-area source will give a much larger spot on the retina and will thus result in less of a hazard.

Various national and international standards on the safety of lasers (such as IEC825 [2]) specify what optical emission levels from lasers should be considered as safe. In these standards, products are classified according to the hazard, with class 1 being defined as inherently safe (even when viewed with optical instruments such as binoculars). It will be necessary for optical wireless applications to fall into this category.

It is important to understand how the performance of optical wireless LANs may be affected by safety standards. In a typical optical wireless system, the satellite on the ceiling illuminates the floor of a room and also receives signals from terminals located within the room. For a typical office, with a floor area of 10 m by 10 m, the transmitter power required can be simply estimated assuming a uniform distribution on the floor, and a receiver collection area of 1 cm^2. The sensitivity of the receiver at 10 and 100 Mbit/s, determined by ambient light shot noise, is estimated at 0.1 μW and 1 μW respectively. Therefore transmitter launch powers of 100 mW and 1 W would be required. However, significant reductions are possible through the use of optical concentrators (see section 6.6).

The relatively low safe-power level (class 1) for a point source at 850 nm wavelength (0.24 mW) means that a large number of spatially separated devices would be needed to give adequate illumination for optical wireless applications. Four hundred devices would be required for 10 Mbit/s and 4000 devices for 100 Mbit/s operation! However, semiconductor lasers are capable of producing much greater powers than the safety limits, up to 100 mW. To use these devices effectively and safely it is necessary to change the laser from a point source to a large-area source that cannot be focused back to a single spot on the retina. Both holograms and 'optical leaky feeders' can be used to solve this problem (see sections 6.4 and 6.5).

6.3.1 LED sources

Some LEDs now have output power that is similar to semiconductor lasers (>40 mW) and pose similar hazards. Therefore it is planned that they will be included in the next amendment to the safety standard IEC825 (Amendment 2). Several commercially available optical wireless products using LEDs give exposures well over the internationally accepted maximum permissible exposure (MPE) limits and will be formally classified as hazardous when LEDs are included in IEC825.

6.3.2 Wavelengths of operation

The increasing exploitation of wavelengths in the near-infra-red range for communications and other applications has stimulated a more detailed examination of the biological damage levels for these wavelengths. The wavelengths relevant to optical wireless systems will be around 850 nm, 1300 nm and 1550 nm where suitable emitter and detector components are available. In general, the maximum safe-power levels are lower, nearer the visible part of the spectrum, because the eye is particularly efficient at focusing and absorbing the radiation. At wavelengths greater than 1400 nm the retina is not at risk, but skin damage is possible. Proposed changes, in Amendment 2 of IEC825, will incorporate relaxations to the safe-power levels. These changes will result in increasing the maximum power for point sources for class 1 at 850 nm from 0.24 mW to 0.4 mW, at 1300 nm from 0.6 mW to 8 mW, and at 1480 nm and 1550 nm from 0.8 mW to 10 mW. In addition, further analysis of the effect of radiation from large-area sources has resulted in a new, more accurate, method of assessing the hazard, which is also planned to be included in Amendment 2.

6.3.3 Large-area optical sources for safe deployment

A simple method of reducing the danger of retinal damage is to use a large-area source. A simple example is a diffusing screen placed some distance in front of a laser. The light is now scattered and appears as diffuse light coming from the extended area of the screen. For larger sources, the power density on the retina, and thus the hazard, will depend on the radiance or brightness ($\mathrm{Wm^{-2}ster^{-1}}$) of the source. As it is not possible for the radiance to be increased using an optical instrument, this value is a fundamental measure for assessing safety. For the wavelength 850 nm, the maximum safe radiance for continuous viewing (IEC825) is 12.8 $\mathrm{kWm^{-2}ster^{-1}}$. Assuming a large area Lambertian source (the distribution expected from a perfect diffuser), the radiance, L, is related to the power density on the emitting surface, H (Wm^{-2}), by $L = H/\pi$. Therefore the maximum safe-power density will be 41 kWm^{-2}. For sources which give distributions of power that are different from Lambertian, this power density will be scaled according to the total solid angle covered.

A value of 41 kWm^{-2} would appear to be very large to be considered as eye-safe, especially as it exceeds the MPE level for skin damage, 2 kWm^{-2}. Recently there have been suggestions that the conditions specified in IEC825 underestimate the hazard for large sources. New proposals for dealing with large sources will appear in Amendment 2, and are likely to reduce this value considerably. Initial proposals indicate that the maximum safe power density from a Lambertian source for continuous viewing at a wavelength of 850 nm could be as low as 0.37 kWm^{-2} (or 37 $mWcm^{-2}$), and BT's systems are now being designed based on this lower figure.

6.4 COMPUTER-GENERATED HOLOGRAMS

Diffusing screens and arrays of lasers are practical ways of realizing large-area sources with safe distributions of high optical powers. However, they give the system designer very little control over the shape of the optical beams. Consequently, the range of design options for tailoring the cell size, room coverage or shape of the area covered by the wireless system would be severely restricted.

In the optical wireless systems developed at BTL a controlled, safe distribution of high amounts of optical power can be achieved with a computer-generated hologram (CGH). The hologram used as the 'antenna' is not of the conventional display type, which converts a coherent input beam into a different coherent beam representing the image of some scene, but instead is designed on a computer to produce light in many beams with equal intensity that overlap in the far field to produce the desired distribution in the target area specified by the designer. The much shorter wavelength of light means that there is much greater control of the beam shape than is possible with radio or microwave antennas.

A schematic diagram of the optical arrangement in which the hologram is used is shown in Fig. 6.2. Here the light is emitted from the laser source and expands with its natural beam divergence until it encounters the hologram that determines the pattern of light emerging from the source package. To achieve a good diffraction efficiency the hologram is etched as a surface relief phase pattern on a transparent quartz substrate [3]. For near-infra-red wavelength sources, the feature size of these hologram patterns is submicron and,

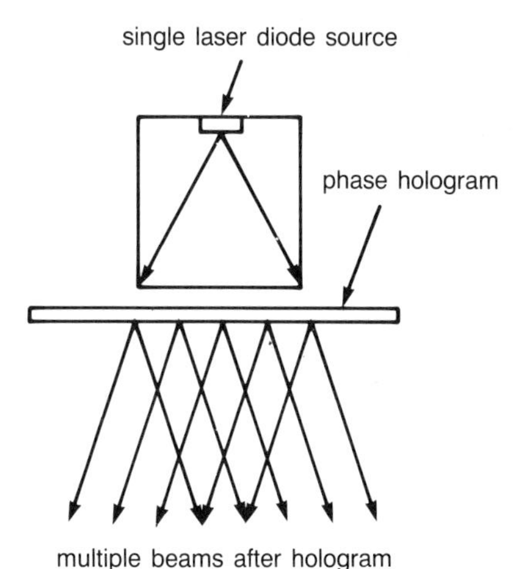

Fig. 6.2 Schematic diagram of a hologramic large-area source.

electron-beam lithography is needed to write the resist mark directly on to the quartz. As the holograms are relatively large in size, the electron beam machine is in use for several hours, if not days. It is therefore very expensive to produce a master hologram. However, in conjunction with a credit card embossing company, it has been possible to produce a nickel copy of the master which allows unlimited copies to be pressed into plastic. These holograms combine the advantages of low cost and high diffraction efficiency. The performance of these is very close to the quartz masters.

The phase pattern on the hologram determines the angular distribution of light emitted from its surface. A single small cell is repeated in a lattice to cover the area of the top surface of the emitter package. The repeated cells are equivalent to the lines on a conventional diffraction grating and produce an infinite series of beams or orders. The phase pattern of each cell determines the distribution of intensity between these orders, just as the blaze on a grating does. It is arranged that the intensity from the orders merges to create a uniform distribution in the far field.

Figure 6.3 shows the design of a single-cell phase pattern for three holograms, with the far-field distribution of the light emerging from them. Here the black areas denote etched regions that produce a half-wavelength phase shift in the light, and the white areas denote unetched regions. Figure 6.4 shows a scanning electron microscope measurement of the surface of one of BTL's plastic holograms. From this it is possible to distinguish the surface binary phase pattern with submicron features. If a lens (for example a binocular lens or the eye pupil) is placed in the emerging radiation from the hologram then the image in the back focal plane is not a single

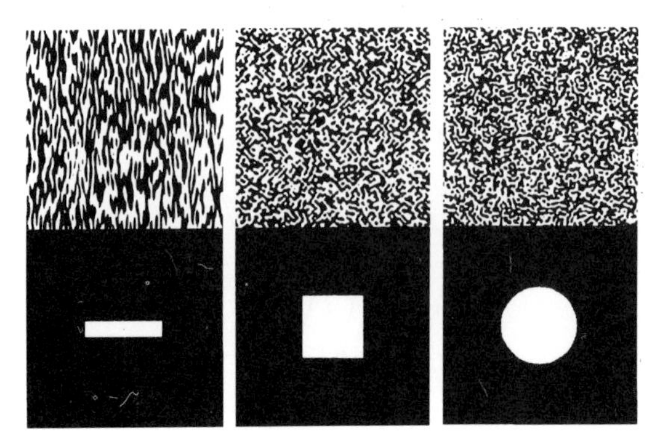

Fig. 6.3 Single hologram cells with their far-field distribution below.

Fig. 6.4 Scanning electron micrograph of surface of plastic hologram.

focused spot but a two-dimensional array of spots or spatial frequencies. If the ratio of etched and unetched regions in the hologram blaze is exactly 50% then the power contained in the zero-order spatial frequency is zero. In practice, the limitations associated with etch control mean that there is always a small discrepancy, and consequently there is power in the zero-order spatial frequency. Since this 'd.c. spot' cannot exceed the MPE for point sources, this limits the total amount of output power from a holographic transmitter. The power present in this spot is minimized at the correct wavelength of operation where it is typically between 15 dB and 20 dB down on the total power contained within the optical cell. Figure 6.5 shows a measurement for a plastic hologram of the percentage of the light in the 'd.c. spot' (plotted on a log scale) versus wavelength; for this hologram, the 'd.c. spot' contains only about 1% of the light (i.e. nearly 20 dB down) at the optimum wavelength (about 700 nm).

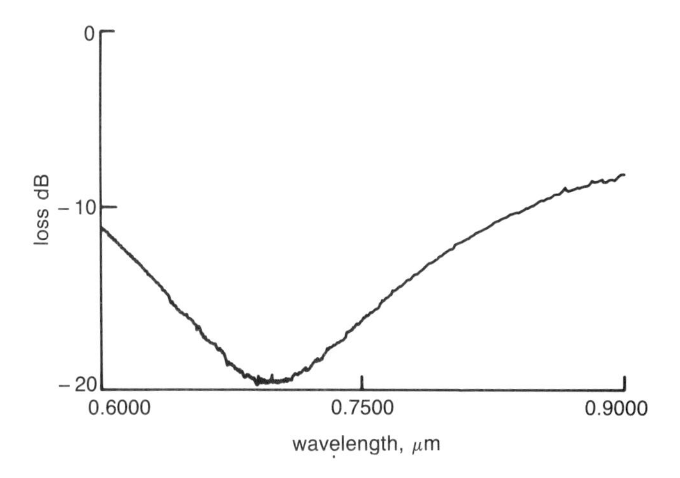

Fig. 6.5 Percentage of light in 'd.c. spot' (log scale) versus wavelength for plastic hologram.

The holograms shown in Fig. 6.3 act as super diffusers. Each part of the hologram contributes to each spot in the far field and so the hologram is truly equivalent to a large-area source. In addition, the cone of radiation from the hologram can be accuractely tailored to meet the application, in marked contrast to the Lambertian emission from a ground-glass screen. Consequently, this effective large-area source is obtained without losing any of the advantages of a single laser diode, such as, for example, light efficiency, ease of packaging and modulation.

6.5 OPTICAL LEAKY FEEDERS — D-FIBRE

The variety of applications of optical wireless distribution will lead to a diversity of physical layouts that will need to be serviced. These may well include corridors, connected rooms, galleries, open-plan offices, etc. In many of these applications it would be an advantage to use an optical leaky feeder where light, from a single source, is coupled out of the fibre selectively at 'taps' or distributively. This gives accurate control of the amount and distribution of the optical signal. It is not possible to efficiently couple light back from terminals into a single-mode fibre and therefore this system is particularly suitable for broadcast use within buildings. The feeder would be installed along a ceiling or wall running the length of the area to be covered. By this means, information can be relayed to a large number of recipients, arranged in a convoluted layout, in a way that is simple to install and distributes power in a safe and uniform manner.

BT Laboratories have implemented the optical leaky feeder using a D-fibre. This fibre is produced by removing a portion of the circular cladding at the fibre preform stage to give a D-shaped cross-section with the core close to the flat of the 'D'. When the fibre is drawn the D-shape is maintained, and long lengths of fibre can be fabricated with the core only a matter of microns below the flat surface [4]. Optical power is guided by this structure with low loss, providing the material in contact with the flat of the D-fibre has a lower refractive index than that of the fibre core. If a material with higher index meets the flat surface, light will couple out from the core into the material overlaid. The rate at which the light is coupled out is determined primarily by the distance from the core to the flat of the D-fibre. The refractive index of the overlay material determines the angle at which the light emerges, and, to a lesser extent, the coupling rate.

The 'optical taps' can be formed by microlens or scattering surfaces on the D-fibre flat (see Fig. 6.6). The emergent optical power can be made to cover wide or narrow angles with different inclinations to the fibre surface according to the application. In principle, the lensing or scattering features could be incorporated into the fibre or overlay materials themselves. These overlay materials might simply form the fibre coatings. The overlay material could carry holographic features to take advantage of the techniques described above.

In these experiments, a length of D-fibre was fusion-spliced to the tail of a semiconductor communication laser of wavelength 1500 nm. A silicon microlens array measuring 1 cm^2 and containing 6400 lenses on a 250 μm grid was placed, in optical contact, on to part of the D-fibre flat. The light emitted by the panel of microlenses is symmetrically distributed normal to the panel and spatially incoherent. The power thus distributed can therefore be higher, before safety considerations impose a limit, than is the case for direct laser or LED distribution. The power emission lobe has a 38°, 3 dB half angle. In fact, 80% of the input power from the laser diode was detected in the output lobe. It is calculated that this would illuminate a 4 m width at desk level, from a ceiling height of 2.5 m.

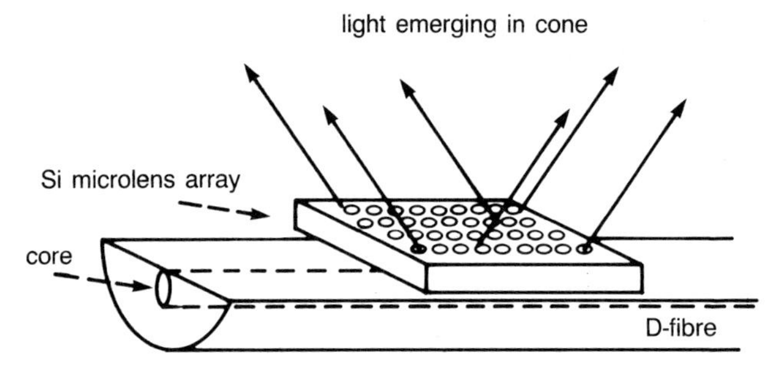

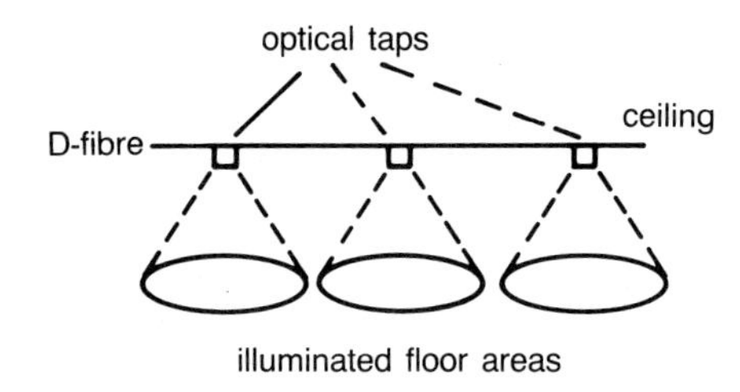

Fig. 6.6 Schematic diagram of an 'optical leaky feeder' and its use.

Quasi-continuous illumination could therefore be obtained by placing a microlens array every 4 m along a cable run, spliced or connected into the transmission path. It is easy to imagine continuous distribution from suitably coated fibre, installed as a continuous length into a transparent conduit. This installation might be conveniently effected by the blown-fibre technique for installing fibre into pre-installed conduit [5].

6.6 OPTICAL CONCENTRATORS FOR WIRELESS SYSTEMS

In an optical communications system the power budget is increased by concentrating as much of the incident radiation as possible on to the detector, with the resultant advantages described in section 6.2.2. In general, this is achieved by optical components such as lenses and mirrors, the whole optical system having a large entrance aperture to collect radiation over a wider area than simply the photodiode size. The concentration ratio of such a collector can be defined as the ratio of the photocurrent in the photodiode with the concentrator in place, compared to that generated when the photodiode alone is exposed to the same radiation source (Fig. 6.7). This ratio cannot be increased indefinitely and is ultimately limited by the fact that no optical system, imaging or otherwise, can increase optical radiance (brightness for visible light). Specifically the maximum concentration ratio achievable, with typical silicon large area diodes, is:

$$C = 0.82 \left\{ \frac{n'}{\sin\phi} \right\}^2$$

where ϕ is the (semi-) field of view of the concentrator (i.e. the maximum input angle which is accepted by the optics), and n' is the index of the material in contact with the detector surface. This material is typically air ($n' = 1$) or glue ($n' = 1.5$). The above equation is valid up to the maximum acceptance angle of the photodiode ($\phi = 65°$), while above this angle the ratio is simply n'^2.

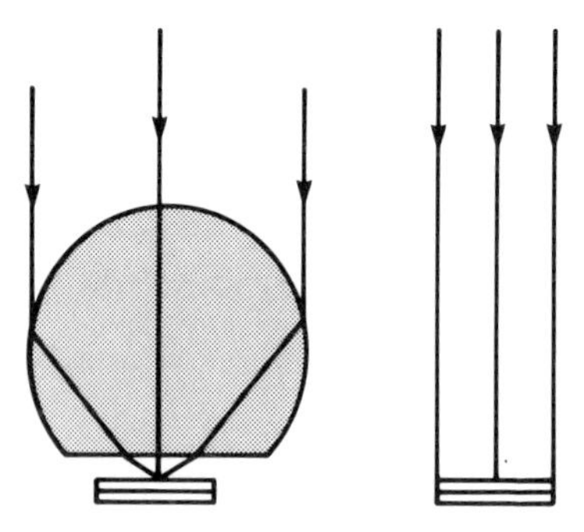

Fig. 6.7 Definition of concentration.

This equation contains the counter-intuitive result that, in a fully diffuse system, which receives light from an entire hemisphere ($\phi = 90°$), the maximum concentration ratio is n'^2, and, if air separates the photodiode from the concentrator, that no increase in received power is possible, whatever scheme of lenses and reflectors is devised. Only when radiation is incident over a very narrow range of angles are large concentrations possible.

In the case of the receiver in this 50 Mbit/s optical cellular system, a half-field of view of 20° is required to collect radiation from any terminal within the cell. This translates to a maximum concentration ratio of 7, since air separated the photodiode from the concentrator. If rays were incident on the concentrator from only one angle, then a perfect concentrator would have a concentration ratio of 7 at all angles up to 20° and zero concentration for radiation incident at higher angles (Fig. 6.8). Any practical concentrator, however, would never reach this level of concentration and would always show some fall-off in concentration towards the edge of the field, owing to optical aberrations. Edge performance is important in the optical cellular system because radiation is only ever incident at more or less one angle at a time, depending on the position of the ground-based transmitter within the cell. This contrasts with the case of a conventional concentrator, which receives radiation from a broad range of angles simultaneously.

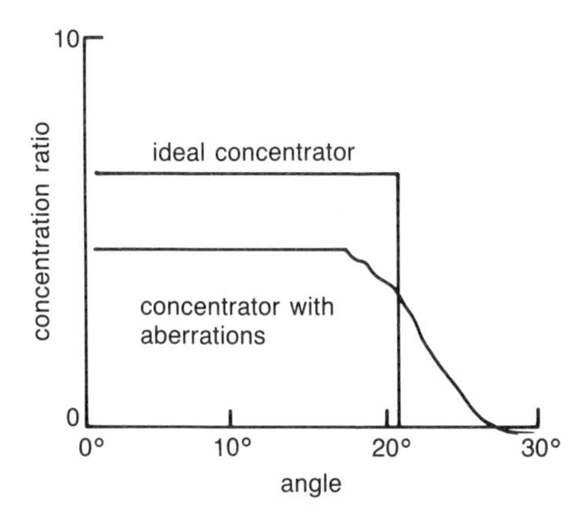

Fig. 6.8 Effect of aberrations on ratio concentrator performance.

Non-imaging reflective and refractive concentrators have been designed for use in the optical receiver of a 50 Mbit/s link which used 50 mm^2 area photodiodes (Fig. 6.9). These were designed using modified commercial ray-tracing packages, and are non-imaging because the optics design was optimized to achieve a maximum number of rays striking the detector. This contrasts with a typical imaging system, in which blur-spot size or the modulation transfer function is optimized. Figure 6.10 shows the performance of the best aspheric refractive concentrator as a function of angle. The refractor had a nominal focal length of 12 mm, a diameter of 15 mm and, consequently, an aperture with f number of 0.8 (f#). The effect of aberrations can clearly be seen in the non-ideal profile, maximum concentration ratio of only 2.5 and effective half field of view of 15°. The large levels of aberration seen in this concentrator were largely due to the use of only a

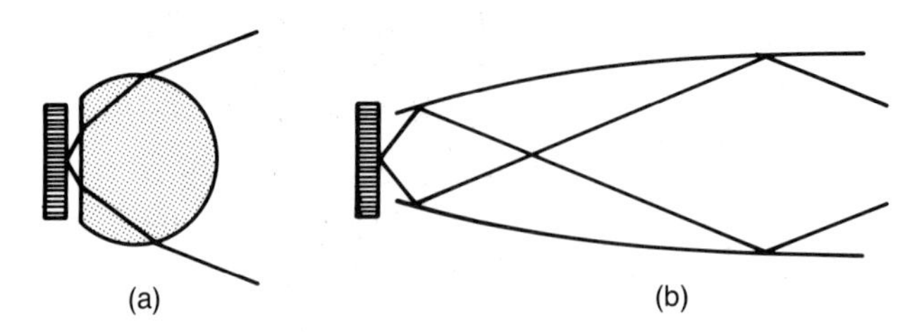

Fig. 6.9 Refractive (a) and reflective (b) non-imaging concentrators for the wireless office system.

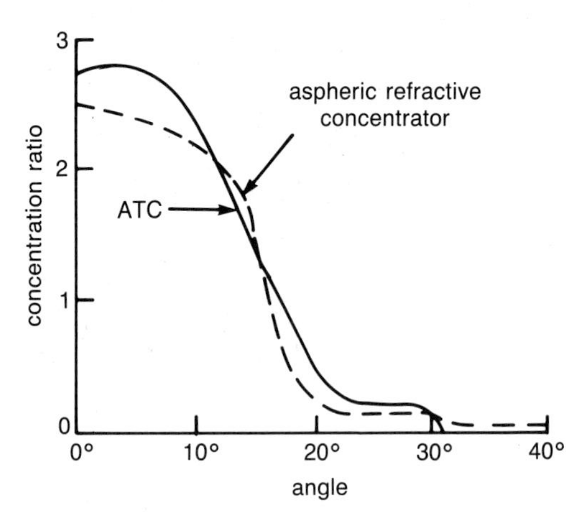

Fig. 6.10 Performance of an ATC and aspheric refractive concentrator.

single-element lens for reasons of low cost. It is impossible to correct off-axis aberrations, such as astigmatism and distortion, in a single element for fields of view exceeding a few degrees.

Figure 6.11 shows a reflective, non-imaging, concentrator known as an angle-transforming concentrator (ATC). ATCs were originally used as solar concentrators but it has been possible to adapt the design principles to fabricate a version with a 20° (semi-) field of view, and its performance is shown in Fig. 6.10. The shape of the curve, maximum concentration and effective field of view are all a slight improvement on the aspheric lens performance but still fall well short of the ideal performance. The low concentration ratio is attributed to absorption losses from the many reflections the rays undergo before reaching the receiver and from rays reflected back out of the concentrator.

Fig. 6.11 Angle-transforming concentrator (ATC).

6.7 RECEIVER DESIGN FOR OPTICAL WIRELESS LANS

The receiver for use in a high-speed optical wireless LAN must have a wide bandwidth to allow for high system data rates (perhaps 100 Mbit/s), and good sensitivity to maximize the system power margin. It also needs a large dynamic range to allow for the changing receive power levels in a wireless environment. These requirements suggest the use of a transimpedance preamplifier, as this topology combines the characteristics of good sensitivity, high dynamic range and high gain, and does not require equalization. The two most popular photodetectors for the 850 nm window of operation are the silicon PIN photodiode and the silicon avalanche photodiode (APD). The PIN photodiode offers the advantages of lower cost, simpler biasing circuitry and a wider range of freely available large-area devices. However, the gain associated with the APD can offer substantial improvements in receiver sensitivity in cases where narrowband optical filtering can be employed.

For a diffuse-based optical wireless system, it is usually necessary for the receiver's FOV to be maximized, hence maximizing the number of transmission paths available. This also means that the input optical bandwidth will be relatively large (typically 300 nm), as only simple absorptive optical filtering techniques can be used, e.g. coloured glass. Narrowband interference-type optical filters only operate correctly over a small range of incident angles, typically $\pm 10°$. Under broadband optical filtering conditions, the shot noise generated by the ambient light level can be much larger than the noise level resulting from the preamplifier electronics. The receiver's signal-to-noise ratio is then maximized by maximizing the active area of the photodiode, as the received signal power increases linearly with the area while the noise power increases with its square root.

With the proposed optical cellular system, the restricted receiver FOV allows for the use of narrowband optical filters, which minimizes the noise contribution of ambient light-generated shot noise. Although a refractive non-imaging concentrator has been used to increase the effective area of the photodiode, the receiver still has a much larger-area photodiode than that of an optical-fibre-based receiver. This is required to maximise the amount of light collected by the photodiode and hence maximize the system's power margin. For the optical cellular half-angle FOV of 20°, the photodiode diameter was 8 mm compared with 50 μm for a typical optical-fibre-based receiver. This larger photodiode area means that the photodiode capacitance at the input to an optical wireless receiver is much larger than in the optical fibre case. The value for this example was close to 50 pF. To accommodate this very large input capacitance, while ensuring adequate bandwidth and minimum sensitivity degradation, the circuit technique known as 'bootstrapping' can be used. A schematic of a bootstrapped common collector preamplifier is contained in Fig. 6.12.

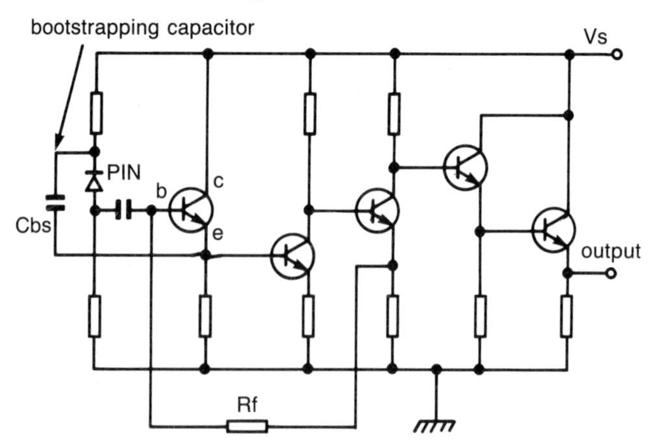

Fig. 6.12 Schematic of bootstrapped common collector optical preamplifier.

The action of bootstrapping can easily be understood by observing that the a.c. voltage swing generated at the input transistor base, by the photocurrent, appears slightly attenuated at the emitter and at a much lower impedance level. A bootstrapping capacitor C_{bs} positively feeds back the emitter voltage signal to the cathode of the photodiode. The effective photodiode capacitance observed by the signal source has now been reduced by a factor of $(1—A_v)$, where A_v is the attenuation of the common collector stage, which is close to unity. Reduction of the effective photodiode input capacitance allows a much larger value of receiver transimpedance resistance (R_f) to be used than would be the case without bootstrapping, typically an order of magnitude increase. This reduces the thermal noise power generated at the receiver input by the same amount, and hence improves the receiver sensitivity. It also means that the transimpedance (the amount of output voltage generated for a given input photocurrent) of the preamplifier is increased, and reduces the need for more than one postamplification stage.

The measured sensitivity for a bootstrapped optical receiver operating at 50 Mbit/s was −32.8 dBm compared with a value of −33.3 dBm predicted by a PSPICE simulation and optical receiver noise theory. The receiver's small signal bandwidth was almost 30 MHz and the measured optical dynamic range in excess of 23 dB.

6.8 50 Mbit/s LAN LINK

A single optical cellular down-link was established in the laboratory with a ceiling height of 3 m. The transmitter used an 850 nm semiconductor laser, an f#1.5 collection lens and a hologram which defined a transmit half angle of 20°. The resultant cell size was 2.2 m × 2.2 m. The ground-based receiver contained an aspheric refractive non-imaging concentrator that defined a half angle FOV of 10°, an interference-type optical filter with 50 nm bandwidth and the bootstrapped common collector preamplifier described above. The laser source was modulated with a 50 Mbit/s pseudo-random bit stream at various pattern lengths greater than 2^7-1, and the receiver performance was observed throughout the cell. Orientation of the receiver with respect to the transmitter was not very critical as a result of the large receive optics FOV. The power budget for the link is detailed in Table 6.1.

The power margin of 5 dB can be increased for this cell size either by using a higher-power laser or having more than one laser within the transmitting unit. A substantial improvement (5 to 10 dB) could be achieved

Table 6.1 Power budget for 50 Mbit/s optical cellular LAN link.

Laser output power	+16 dBm	mean power level, transmit half angle of 20°
Tx optics loss	−2 dB	single collection lens
Spreading loss	−46 dB	geometrical loss (Rx at 1 m off floor)
Optical concentrator gain	+7 dB	refractive concentrator
Optical filter insertion loss	−3 dB	interference type, 50 nm bandwidth
Rx power	−28 dBm	
Rx sensitivity	−33 dBm	for BER = 10^{-9}
Power margin	+5 dB	

in receiver sensitivity by using a silicon APD as the receiver sensitivity is limited by preamplifier noise. Ultimately this improvement is limited by the amplified shot noise associated with the ambient light after narrowband filtering. The improvement in sensitivity increases the available power margin, albeit at greater expense and complexity of the receiver unit.

This work is being extended to show 155 Mbit/s operation with an optical cellular system and hence demonstrate the potential of optical wireless technology to provide the capacity required by future ATM-compatible wireless LAN links.

6.9 CONCLUSIONS

Optical wireless LANs offer the possibility of satisfying the dual need for both mobility and broadband networking for present-day and future indoor communications systems. They have the advantages of operation within an unregulated part of the electromagnetic spectrum and a lower-cost technology base than their microwave counterparts. These systems also offer global compatibility. The success of optical wireless systems will in part depend on them being classified as inherently safe, Class 1. The design of the system must anticipate changes to the safety standards, such as the inclusion of LEDs and the new power limits.

The enabling technologies to achieve safe operation have been developed, and work on computer-generated holograms (CGH) has shown that very efficient large-area sources can be implemented which enable eye-safe distribution of large amounts of optical power. These holograms also allow the system designer to control accurately the optical far-field distribution. This not only improves the power efficiency and the receiver dynamic range but will enable new system concepts for optical wireless, such as the optical cellular network. The optical leaky feeder, implemented with D-fibre, has the ability to provide optical distribution in halls and corridors for broadcast use such as optical paging. Like the holograms, this technology is inherently eye-safe. Allowing for the likely new safety regulations, the total emissions through the hologram and D-fibre devices are several orders of magnitude greater than would be allowed for single-point sources and are sufficient to enable the safe deployment of future broadband optical wireless systems.

For mobile applications with moderate speeds (up to $\approx$16 Mbit/s), diffusely scattered transmission is likely to be preferred. This will require transmitter powers much greater than that of the safety limits for point sources, i.e. <1 mW. CGHs can be used to achieve the safe distribution of optical power.

For mobile applications requiring high speed, an optical cellular system can be implemented with a 50 Mbit/s down-link between a ceiling-mounted optical wireless transmitter and a ground-based receiver. The single cell covers almost 5 m^2, and this can be replicated to cover a whole room. The key optical and electronic technologies required to implement this optical wireless link safely and efficiently include holographic transmitters, high-efficiency collection optics, and high-sensitivity (bootstrapped) optical receivers. These three technologies have been successfully developed, with good agreement between measured and computer simulated results.

By dividing the future office into optical cells there are great gains to be made in collection efficiency and speed. Novel optical concentrator design may increase power margins by up to 10 dB, while there are likely to be further improvements in receiver sensitivity since receiver design for optical wireless is still in its infancy. The reduction in received power required will allow the systems designer to reduce transmitter power in both the satellite and mobile terminal. Reduced transmitter power will enhance the battery life and reduce the weight of the mobile computer. These are very important features for the mass exploitation of laptop and pen computers to include telecommunications services.

REFERENCES

1. Gfeller G and Bapst U: 'Wireless in-house data communications via diffused infra-red radiation', Proc IEEE, 6, No 11 (November 1979).

2. IEC825: 'Radiation safety of laser products', equipment classification, requirements and user's guide (1984) plus Amendment 1 (1990).

3. Dames M P, Dowling R J, McKee P and Wood D: 'Efficient optical elements to generate intensity weighted spot arrays — design and fabrication', Appl Opt 30 , pp 2685-2691 (1991).

4. Millar C A et al: 'Fabrication and characterization of D-fibres with a range of accurately controlled core-flat distances', Electron Lett, 22 , No 6 (March 1986).

5. Cassidy S A et al: 'A radically new approach to the installation of optical fibre using viscous flow of air', 32nd International wire and cable symposium, New Jersey, USA (1983).

7

QUANTUM CRYPTOGRAPHY AND SECURE OPTICAL COMMUNICATIONS

S J D Phoenix and P D Townsend

7.1 INTRODUCTION

With the globalization of telecommunications and the introduction of integrated voice, data and image services it is becoming increasingly necessary to consider the security of information over highly connected broadband optical networks. Clearly, if, for example, important meetings will in future be conducted over video links, the integrity of those links to unwarranted interception is of vital importance. The ease with which mobile communications have been intercepted recently is a telling reminder of the need for information security. The problem is likely to be exacerbated with the widespread introduction of broadband ISDN and virtual network services. Information security is an important service which any global telecommunications provider must be able to offer its customers. The need for such services is likely to increase sharply as the global information infrastructure becomes ubiquitous.

There are various ways in which security can be provided. The most obvious, and the most difficult to achieve reliably, is to prevent access to the channel over which information is carried. The most conventional method is to scramble the message in some way such that only the authorized users of the channel can easily recover the message. The various techniques which achieve this fall within the domain of cryptography. A cryptographic technique works because the users of the channel possess a secret random sequence of bits which can be used with a specified algorithm to scramble the content of a message. Such a sequence of bits is known as a key. The basic idea is that anyone with access to the key can easily reconstruct the message, whereas this process is extremely difficult if the key is not known. The security depends crucially on the secrecy of the key. The key can take the form of shared secret information between the legitimate users of the channel, or each user can possess an individual secret key. With each reuse of a secret key the security of the channel becomes weaker as any interceptor has a larger amount of encrypted data to analyse. As the volume of traffic carried on telecommunications networks becomes larger, the central issue therefore becomes one of key management and distribution. Two users must be able to guarantee the secrecy of their shared key or their individual keys, and these keys must be regularly updated if the system is to remain secure.

Figure 7.1 shows the basic processes involved in the protection of a communications channel by a cryptographic technique. Alice wishes to send a message **m** (also known as the 'plaintext') to Bob. She possesses a random secret sequence of bits $k(E)$ which is used in conjunction with an encryption algorithm to produce a cryptogram **c** (also known as the 'ciphertext'). Bob receives the ciphertext **c** and performs an 'inverse' transformation with his secret key $k(D)$ and a decryption algorithm. There are two types of cipher system — a conventional or symmetric system in which $k(D)$ is easily obtainable from $k(E)$, and public or asymmetric systems in which it is computationally unfeasible to obtain $k(D)$ from $k(E)$. The eavesdropper, Eve, is assumed to have access to the encryption and decryption algorithms and the ciphertext **c**[1]. The secrecy of the channel is assumed to depend solely on the secrecy of the keys. Asymmetric or public-key cryptosystems are particularly important because it is thought that they might provide a method whereby two parties, who initially possess no shared secret information, can communicate securely. However, the security of these systems depends on the supposed difficulty of particular mathematical operations, such as factoring large numbers into the product of two primes, for example. While such operations are currently undoubtedly difficult, they have not been proven to be 'difficult' in a strict mathematical sense. The security of these public-key cryptosystems is in a kind of mathematical limbo.

---

[1] Clearly, the encryption and decryption algorithms may also be secret and not known by the eavesdropper. There are examples of commercial encryption and decryption chips for which the algorithms are a closely guarded secret. However, some of the main encryption algorithms, such as DES and RSA, do have publicly known algorithms [1].

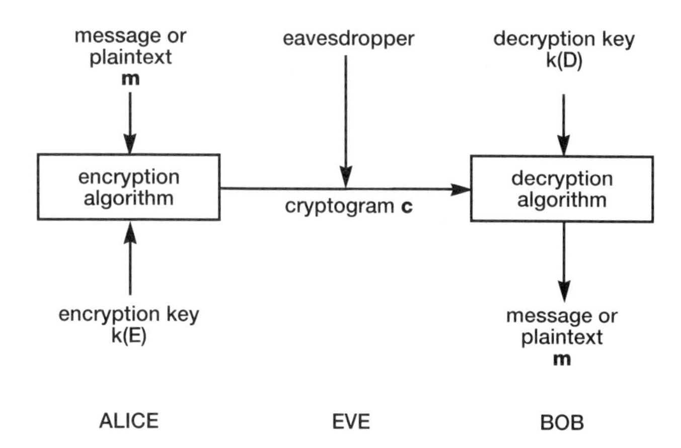

Fig. 7.1 The basic processes involved in the protection of a communication channel by a cryptographic technique.

The only cipher system known to have guaranteed security is a symmetric cipher system known as the Vernam cipher or the one-time pad. In this system a bitwise exclusive-or of the message **m** with the key $k(E)$ is performed and the resulting cryptogram can be deciphered by performing a bitwise exclusive-or of the cryptogram **c** with the key $k(D)$, which must be equal to $k(E)$. For the system to be unbreakable the key must be as long as the message and must be used only once. The key management and distribution problems quickly become terrifying with this kind of cipher system, although for important messages (the Moscow—Washington hotline, for example) this technique might be employed. For most symmetric systems, however, the key is much shorter than the message and in this case the ciphertext is given a 'minimal cover time', i.e. the length of time for which it is believed the system can withstand a cryptanalytic attack. Central to this supposed security is the secrecy of the key data.

Quantum cryptography is a radical new technique which allows the distribution of key data in such a way as to guarantee its secrecy. The secrecy is achieved by exploitation of the laws of physics and is immune to attack from more powerful analytic or computational tools. In this sense the term 'quantum cryptography' is a misnomer and we are really concerned with 'quantum key-distribution' techniques. However, the accepted usage appears to favour the former term. On the one hand we have public-key cryptosystems which do not require the sharing of secret information, but which also do not have provable security as yet. On the other we have symmetric cryptosystems for which the sharing of secret information is required but which also have a given degree of security as specified by their cover time.

Quantum cryptography is a method whereby key secrecy can be guaranteed, for use in symmetric cryptosystems. If the quantum techniques can be made reliable and practical the implications for the protection of communications are very significant.

It is important, at this stage, to stress two things. Firstly, the security of a quantum channel for key distribution is guaranteed by a physical law. As we shall show, it is impossible, even in principle, to eavesdrop on such channels. Secondly, although these quantum techniques appear at first sight to be somewhat esoteric, quantum key-distribution has been achieved in the laboratory using readily available components [2]. In this scheme the key transmission took place over a distance of less than 1 m. However, the technique has been shown to be effective, and developments at BT Laboratories have shown that secure key transmission over several kilometres of optical fibre is possible using readily available components. The issue is not one of whether these quantum techniques work but whether they can be made robust and practical enough for use over an optical network. Although it is evident that quantum cryptography is not a panacea for all of the security issues that face a global telecommunications company, the provision of a secure key-distribution facility would be an important service to be able to offer customers who require an added guarantee of security.

Having established the relevance of quantum cryptography to real issues of concern, we shall now outline the basic physics of such schemes and indicate how quantum optical methods have been crucial in the development of these techniques for implementation in optical fibre. Central to the success of any quantum cryptography scheme is the probabilistic nature of quantum mechanics. This is not something that can be switched on and off at will — the world behaves according to the laws of quantum mechanics whether we like it or not. For the most part, the probabilistic nature of individual quantum events is not evident on the macroscopic scale of our everyday experience. For events at a single photon level, however, the laws of quantum mechanics are all-important and cannot be ignored. It is events at a single photon level that are exploited to give the security over an optical quantum cryptography link. The description of these events and their exploitation in a quantum cryptography scheme requires some familiarity with the peculiar laws which govern the quantum world and their consequences. One of the consequences with which we shall be concerned is the Heisenberg uncertainty principle which states that certain pairs of quantities, such as position and momentum for example, cannot be simultaneously measured with an arbitrary accuracy. In effect the principle enunciates the reasonable assertion that an object cannot be measured without disturbing that object. The very act of measurement necessarily requires the interaction of the measured system with the measurement device. This interaction unavoidably perturbs the measured system unless that system is prepared in a special state with respect to the measurement device. (Technically we say that a measurement projects the system into a new quantum state unless the system has previously been prepared in an eigenstate of the measured operator.) The Heisenberg principle, which we have already mentioned, is in itself a consequence of deeper quantum laws which we shall later need to examine rather carefully. Let us suppose that we wish to determine the position and momentum of an electron. As we increase the resolution of the measurement to determine, for example, the momentum, the Heisenberg uncertainty principle ensures that we become more and more unsure of where the electron is. In the limit of perfect measurement resolution, where the momentum is determined precisely, we know absolutely nothing about the whereabouts of the electron. Furthermore, we cannot even design 'better' measurements to overcome this limitation. In effect, the act of measurement here prepares the electron in a precise momentum state at the expense of its position state being indeterminate (we can loosely think of the electron as 'being' in an infinite number of precise position states). This situation also obtains for the polarization states of single photons, where there exists a Heisenberg uncertainty relationship between measurements of circular and linear polarizations, for example.

Suppose that we prepare a photon in a precise, or definite, polarization state. As we have mentioned above, the act of state preparation is equivalent to a precise measurement with perfect resolution. We shall suppose, for the sake of argument, that the photon is prepared in a precise circularly polarized state (there are two possible precise states of circular polarization — 'right' and 'left'). Because of the uncertainty relation between circular and linear polarization states, we know, by analogy with the position and momentum problem discussed above, that precise knowledge of the circular polarization implies imprecise knowledge of the linear polarization. In the same way that an electron in a precise momentum state is equally likely to be found in any precise position state (the uncertainty principle prevents knowledge of this position state), then a photon in a precise circular polarization state is equally likely to be found in any of the linear polarization states. In this case there are just two states to choose from — 'vertical' linear polarization and 'horizontal'. We now meet another of these strange quantum laws. Suppose that our photon, prepared in a circular polarization state (either right or left circular polarization), enters an apparatus designed to measure, with perfect resolution, the linear polarization (remember that precise determination of both linear and circular polarizations is not possible so that a precise measurement of circular polarization, for example, precludes any precise knowledge of the linear polarization). If a photon were divisible, half a photon would yield a result of vertical linear polarization, the other half yielding the result of horizontal linear polarization. However, a photon is not divisible and this measurement will give a vertical linear polarization with a 50% probability and a horizontal linear polarization with 50% probability. This is a consequence of the superposition principle of quantum mechanics, a discussion of which is contained in the next section. The photon is projected into a precise state of linear polarization, but the actual state, vertical or horizontal, is a matter of chance. There is no way to predict a unique outcome of this experiment. If it were possible to predict the outcome this could constitute a violation of the Heisenberg uncertainty principle in that precise knowledge about both kinds of polarization, circular and linear, would have been obtained. It is precisely this probabilistic outcome of the experiment which is crucial in the implementation of a quantum cryptography scheme. We have that a photon prepared in a precise state of circular polarization will give a probabilistic outcome to an experiment designed to measure a precise state of linear polarization (and vice versa). Furthermore, the measurement projects the photon into the precise state of polarization which has been measured. **All information about the original circularly polarized state is lost.** All that the single experiment can determine is that the photon is now in a definite state of linear polarization. If the experiment is set up to measure a precise state of circular polarization the state of the photon (which has been prepared in a circular polarization state) is unaffected by the measurement and there is a unique outcome to the measurement.

We are now in a position to describe, in broad terms, how a quantum key distribution scheme works. Alice and Bob agree that right circular and vertical linear polarization states are to be read as a logical '1'. The left circular and horizontal linear polarization states are to be read as a logical '0'. Alice sends a right circular polarization state to Bob, i.e. she sends a '1'. Bob, who doesn't know that Alice has chosen this state, orients his polarizer to measure a circular polarization. All things being well, he will determine the state of the photon to be right circularly polarized, i.e. he should read a '1' from Alice's transmission. However, let us suppose that Eve wishes to intercept the communication. Eve also does not know which state Alice has transmitted. In order to determine the information encoded in the photon's polarization, she must choose to measure a particular polarization, either circular or linear. This is an either/or choice — she cannot measure both. Let us suppose that Eve chooses incorrectly and decides to measure a linear polarization. This gives her a 50% chance of reading the correct bit, i.e. she will obtain a vertical linear polarization state with 50% probability (and a horizontal state with 50% probability). Eve has no way of knowing that she has misaligned her measurement apparatus so she faithfully retransmits her result, a photon with a precise linear polarization, on to Bob. Bob has chosen a correct orientation for his measurement and would, in the absence of Eve, obtain the result '1' with unit probability from Alice's transmission. However, Eve has corrupted the channel and has sent a state of linear polarization on to Bob. Bob now has only a 50% chance of reading the correct bit. If Alice and Bob subsequently communicate about their quantum transmission there is a finite probability, due to Eve's intervention, that they will disagree about what bit was sent. This situation is depicted in Fig. 7.2. Eve will eventually cause discrepancies between Alice and Bob where none were expected. There is no strategy, measurement or calculation which will allow Eve to escape this eventuality. The purpose of a quantum cryptography scheme is for Alice and Bob to communicate in such a way as to make these disagreements unavoidable if there is an eavesdropping attempt.

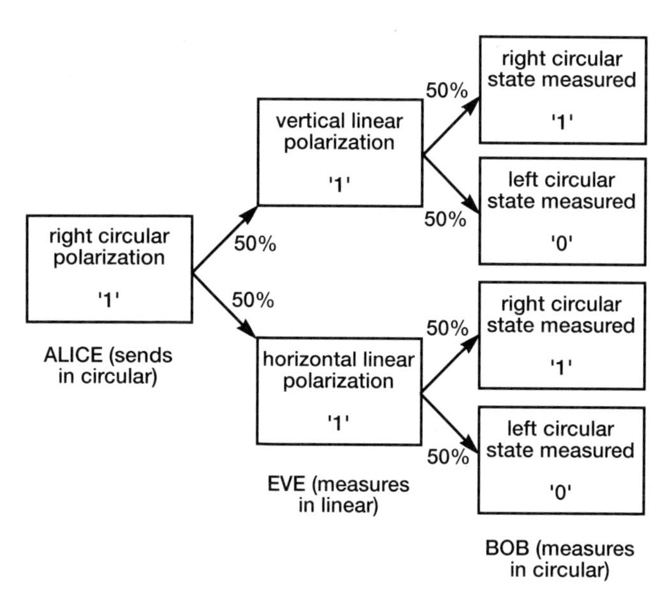

Fig. 7.2 The probability tree for a particular realization of the quantum communication between Alice and Bob. Alice, Bob and Eve have chosen particular alphabets and the probabilities of their various results are shown.

It should now be obvious how physics conspires to foil an eavesdropper, provided Alice and Bob use a protocol which exploits these physical laws. If Alice and Bob randomly and independently choose which polarizations to transmit and measure, respectively, then Eve cannot avoid situations in which she is likely to cause disagreements between Alice and Bob. There are still a few more details to discuss concerning exactly how a key is established, how a protocol is designed and other such questions. We have, however, described the fundamental physics which gives the channel its security. In the next sections we shall discuss in detail the theoretical and experimental aspects involved in the actual realization of a quantum cryptography scheme. Any practical realization of these secure key distribution schemes will necessarily involve a considerable amount of expertise in the understanding, production, manipulation and propagation of single photons.

7.2 QUANTUM CRYPTOGRAPHY — THEORETICAL ASPECTS

As with many problems in communications there are several approaches necessary to obtain a complete understanding. Quantum cryptography is no exception. At BT Laboratories we have developed the most general theoretical approach so far to quantum cryptography by extending Shannon's pioneering work on information theory into the quantum regime. This information-theoretic approach allows us to set the fundamental limits on the quantum transmission and its security. The gain in generality, however, is compensated by a loss in detail. While we are able to set limits, the formalism does nothing to indicate how those limits might be achieved. We shall, accordingly, take two approaches — the general approach based on information theory, and a specific approach using the polarization states of single photons. The latter will be useful in highlighting features of the former. By an understanding of the information-theoretic limits we are led directly to the development of new protocols for quantum cryptography and the discovery of a simple topological method whereby the properties of these quantum channels can be easily evaluated by drawing simple diagrams. The discussion will, however, necessarily require some technical facility with the concepts of information theory and quantum mechanics but we shall arrive at a fundamental formalism which can be applied to any of the quantum cryptography schemes so far developed.

Let us suppose that Alice and Bob merely wish to communicate over some quantum channel. The properties of that quantum channel are described by a set of operators, $\hat{A}$, $\hat{B}$, $\hat{C}$,... and the available alphabets described by the eigenstates $|\{\alpha_j\}>$, $|\{\beta_k\}>$, $|\{\gamma_m\}>$,... associated with these operators, respectively. The notation $\{...\}$ refers to a complete set of eigenstates. It is not, in general, necessary to map each eigenstate on to an alphabet symbol, but, for convenience, we adopt this convention in our present treatment. The operators which characterize the channel describe physically measurable quantities, for example, $\hat{A}$ could be the operator describing circular polarization. This would give rise to an alphabet of just two symbols, the right and left circular polarization states. In general, then, Alice has access to several alphabets formed from the eigenstates of the operators which describe the channel, as does Bob. If, for example, Alice chooses $\hat{A}$ then the eigenstates $|\{\alpha_j\}>$ form the symbols of an alphabet with which she can communicate with Bob. If Bob chooses to measure some different operator, say $\hat{B}$, then the outcome of his measurement will be one of the eigenstates of $\hat{B}$ which constitute the symbols of another alphabet. For any communication channel, quantum or classical, to be maximally effective, Alice and Bob must be able to uniquely associate a given output symbol with a given input symbol, thereby giving perfect error-free communication. In general, however, we can suppose that there is noise on the channel which causes error so that given input and output symbols can only be associated with one another with a finite probability of less than unity. The quality of the channel is therefore characterized by the channel transition probabilities which give the probability that a particular symbol is received, given that a particular symbol was transmitted. This is depicted in Fig. 7.3.

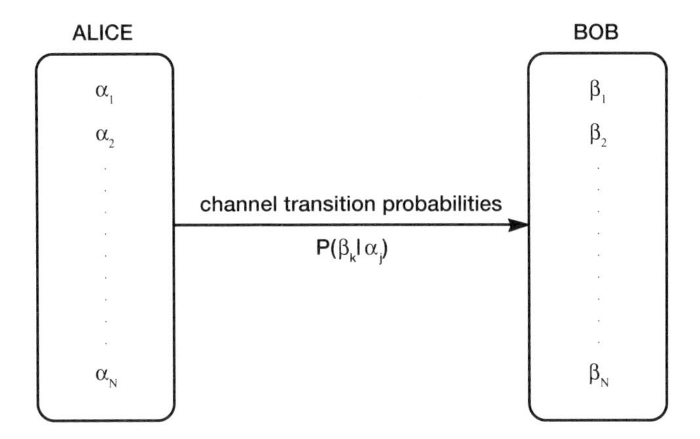

Fig. 7.3 A schematic representation of the alphabets used by Alice and Bob.

The central reason why these channels are quantum mechanical is the possibility that transmission and reception may occur using different alphabets where the transition probabilities for the symbols of these alphabets are entirely determined by the laws of quantum mechanics. It is a consequence of the Heisenberg uncertainty principle that by measurement of the wrong alphabet Bob can irretrievably scramble the information that Alice is attempting to send. Here, then, the transition probabilities are determined by the choice of transmission and reception operator or basis. (We have used the word 'alphabet' to emphasize that the eigenstates of a particular operator can be thought of as constituting the symbols of an alphabet. These eigenstates also form a 'basis' and this word can be used interchangeably with 'alphabet'.) Even on noiseless channels, i.e. channels with no external source of noise, the choice of measurement basis can affect the transition probabilities, thereby acting as an effective noice source. Even on perfect quantum channels, therefore, 'noise' can occur as a result of the Heisenberg uncertainty principle.

Let us suppose now that Alice sends to Bob the symbol $|\alpha_j>$. Bob, however, for reasons best known to himself, has set his measurement apparatus to read the $\hat{B}$ alphabet. The superposition principle of quantum mechanics tells us that any state can be expanded as a linear superposition of a complete set of states. Alice's state as it arrives at Bob's measurement apparatus is given by:

$$|\alpha_j> = \sum_{k=1}^{N} |\beta_k><\beta_k|\alpha_j> \qquad \ldots (7.1)$$

where we have assumed that the basis is N-dimensional and $<\beta_k|\alpha_j>$ is the quantum mechanical overlap between the alphabet symbols $|\alpha_j>$ and $|\beta_k>$. The interpretation of this result is straightforward — Bob has a probability of $|<\alpha_j|\beta_k>|^2$ of 'reading' the symbol $|\beta_k>$. Furthermore, Bob's measurement projects the system into this state so that some knowledge of the transmitted state has been lost and a subsequent measurement by Bob will give no additional information about $\hat{A}$. The channel transition probability is thus given by:

$$P(\beta_k|\alpha_j) = |<\alpha_j|\beta_k>|^2 \qquad \ldots (7.2)$$

where we have used an obvious notation. If we restrict ourselves to a discussion of finite N-dimensional state spaces and assume that Alice makes an equal *a priori* choice of input symbols, we find that the channel mutual information [3-5] is given by:

$$J(\hat{A},\hat{B}) = \ln N + \frac{1}{N} \sum_{j=1}^{N} \sum_{k=1}^{N} |<\alpha_j|\beta_k>|^2 \ln |<\alpha_j|\beta_k>|^2 \qquad \ldots (7.3)$$

This quantity measures the correlation between the input and output symbols and therefore defines the information flow rate between Alice and Bob. Maximization of $J(\hat{A},\hat{B})$ over the input alphabet gives the channel capacity which in this case is just $\ln N$. It should be noted that this is also the channel capacity for a perfect classical channel with finite input and output alphabets of equal size. Minimization of $J(\hat{A},\hat{B})$ occurs when Alice and Bob use conjugate alphabets, i.e. when $|<\alpha_j|\beta_k>|^2 = 1/N$, so that any input symbol is equally likely to cause any output symbol. This is just Wiesner's definition of conjugacy [6] and allows us to restate the notion of operator conjugacy in information theory terms [3, 5]. If Alice and Bob use conjugate alphabets it is seen from equation (7.3) that $J(\hat{A},\hat{B})=0$ and no information can be transmitted on the channel. The circular and linear polarization operators are conjugate in the above sense. This 'cancellation' of information, if Bob uses an operator conjugate to that of Alice, is an information-theoretic expression of the uncertainty principle.

The above discussion has shown how Alice and Bob can reduce the information flow rate on their channel by inappropriate choices of alphabets. Indeed if Alice and Bob use conjugate bases they will reduce this information flow to zero. Let us suppose that Eve wishes to intercept the communication between Alice and Bob. We shall further assume that Eve does not know which bases Alice and Bob are using, and, furthermore, that Bob has no knowledge of which basis Alice will choose. Eve aligns her apparatus to measure some operator $\hat{E}$ with an associated alphabet of eigenstates $|\{\epsilon_m\}>$. We then have three operators which characterize the transmission — $\hat{A}$, Alice's operator, $\hat{B}$, Bob's operator and $\hat{E}$, Eve's operator. These could, of course, all be the same operator if Eve and Bob guess Alice's basis correctly, but we shall consider the general situation first. Eve has the same problem as Bob; she must guess Alice's basis correctly if she is to reliably retrieve the information in Alice's transmission. If she doesn't want to be detected she must then send on to Bob the result of her measurement, which, if she has guessed incorrectly, may well be different to Alice's original transmission. There is a likelihood, therefore, that Eve will affect the channel transition probabilities, thereby causing a disturbance in the flow of information between Alice and Bob. It is this change in information flow which can be exploited by Alice and Bob to reveal the presence of Eve. In the presence of Eve the channel transition probability between Alice and Bob, previously given by equation (7.2), is now given by:

$$P_E(\beta_k|\alpha_j)=\sum_{m=1}^{N} |<\alpha_j|\epsilon_m>|^2|<\epsilon_m|\beta_k>|^2 \qquad \text{... (7.4)}$$

The channel mutual information in the presence of Eve, $J_E(\hat{A},\hat{B})$, can now be calculated from equation (7.3) by substitution of P_E for the overlaps $|<\alpha_j|\beta_k>|^2$. We are now in a position to define a 'channel disturbance parameter' which gives a measure of by how much the eavesdropper affects the flow of information on the channel. This parameter, which we denote by ξ, is the normalized difference between the information flow rates in the presence and absence of Eve and is given by:

$$\xi = \frac{J_E(\hat{A},\hat{B})-J(\hat{A},\hat{B})}{J^{max}(\hat{A},\hat{B})} \qquad \text{... (7.5)}$$

where $J^{\max}$ defines the maximum possible information flow between Alice and Bob given their initial choice of input symbols (this is just the channel capacity of $\ln N$ in the above case). We find that $-1 \leq \xi \leq 1$, so that a negative value implies that the eavesdropper has caused a reduction in the flow of information. However, we note that Eve can also cause a flow of information on the channel if $\xi > 0$. Both positive and negative changes in the information flow can be detected and the presence of Eve revealed. It is only if $\xi = 0$ precisely that Eve can escape detection. In order to escape detection Eve must choose the same basis as either Alice or Bob. Alice and Bob must set up their quantum cryptography protocol so that it is nearly impossible for Eve to do this, i.e. Alice and Bob must set up the channel to exploit the situation where $\xi \neq 0$. This can indeed be achieved and, after discussing a topological method whereby these properties can be understood, we shall describe a protocol that can be used.

Consider the situation depicted in Fig. 7.4. Alice and Bob choose to use the alphabets $\hat{A}$ and $\hat{B}$. Eve, we shall assume, chooses to use three alphabets $\hat{A}$, $\hat{B}$, and $\hat{C}$. (It has been suggested that Eve can derive some benefit from choosing to measure in some basis intermediate to those chosen by Alice and Bob. This is not the case as we shall now demonstrate.) We define a pathway to be a particular realization of the communication between Alice and Bob. In the absence of Eve there would be four distinct pathways.

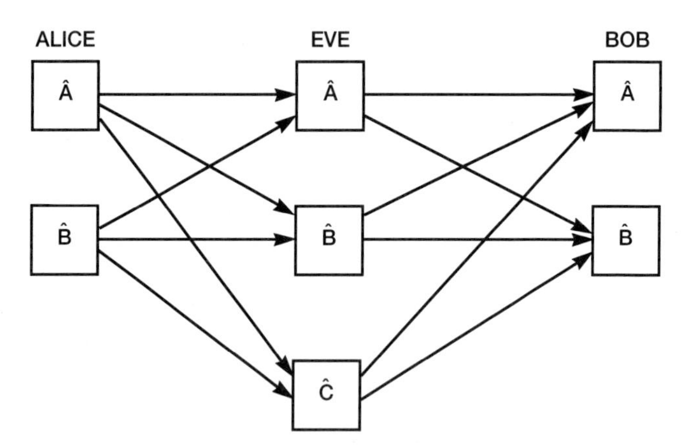

Fig. 7.4 A schematic illustration of the possible pathways between Alice and Bob in the presence of Eve. Alice, Bob and Eve use the alphabets generated by the operators $\hat{A}$ and $\hat{B}$ and Eve also uses the additional alphabet generated by the operator C.

However, in the presence of Eve there are 12 distinct pathways, i.e. there are 12 different combinations of operator in the chain Alice-Eve-Bob. For example, one particular realization is the choice $\hat{A}$—$\hat{B}$—$\hat{A}$. Another might be $\hat{A}$—$\hat{C}$—$\hat{B}$. As we have discussed above, Eve can be detected on any pathway, for which $\xi \neq 0$. Consider now the pathway represented by the chain $\hat{A}$—$\hat{A}$—$\hat{B}$. As far as Bob is concerned, Eve cannot be distinguished from Alice, and indeed we find that $\xi = 0$. Eve cannot be detected on this pathway and it is, in fact, equivalent to the pathway $\hat{A}$—$\hat{B}$ between Alice and Bob in the absence of Eve. Clearly, any pathway for which Eve chooses an alphabet used by either Alice or Bob can be reduced in this fashion. This leads us to a statement of the fundamental theorem of quantum cryptography [7] — '*An eavesdropper can only be detected on an irreducible pathway*'.

An equivalent statement of this theorem is that Eve can only be detected on a pathway for which she has chosen a different basis from both Alice and Bob. This follows quite naturally from the fact that it is only for these pathways that $\xi \neq 0$. If Alice and Bob use the same basis, then Eve's intervention on an irreducible pathway (e.g. $\hat{A}$—$\hat{B}$—$\hat{A}$) causes a reduction in the flow of information between Alice and Bob, i.e. $\xi < 0$. If Alice and Bob use different bases, and Eve chooses a different basis from both Alice and Bob (e.g. $\hat{A}$—$\hat{C}$—$\hat{B}$), then Eve's intervention can cause an increase in the flow of information between Alice and Bob, i.e. $\xi > 0$. Different protocols have to be developed for each of these situations [7]. A protocol for exploitation of single-channel quantum cryptography schemes where $\xi > 0$ has been developed jointly by BT and Strathclyde University [4], but we shall concentrate here on the Bennett-Brassard-Wiesner (BBW) protocol necessary to exploit the region $\xi < 0$ [8]. This is simply because our current experiments are set up to examine this region and the BBW protocol saturates the inequality for ξ so that $\xi = -1$. We have described here, however, a topological method, whereby the efficacy of a quantum cryptography scheme can be determined simply from a diagram. Alice and Bob must use a protocol which causes at least one irreducible pathway for each of Eve's choices of measurement basis. The BT-Strathclyde protocol can then be used in conjunction with the BBW protocol, if necessary, to yield a guaranteed security for the key distribution.

The BBW protocol is designed specifically to exploit the region $\xi < 0$ as mentioned above. We shall describe the protocol in terms of the polarization bases discussed earlier. The protocol can be described as follows.

- For each transmitted photon Alice randomly chooses the state of transmission to be one of the four polarization states (|*right circular*> = '1', |*left circular*> = '0', |*vertical linear*> = '1' or |*horizontal linear*> = '0'). She records her chosen basis (circular or linear) and the bit encoded ('1' or '0').

- Bob randomly and independently chooses his measurement basis (circular or linear) for each incoming photon and records his choice of basis and the result ('1' or '0').

- Alice and Bob then communicate, possibly over a public channel, and Bob tells Alice which measurement basis he chose but **not** the result of his measurement. Alice and Bob agree to discard all those results for which they chose different bases.

- They should, in the absence of an eavesdropper, now possess a shared random sequence of bits, which they could use as a secret key. However, they choose a random subset of this data and publicly compare results. The intervention of Eve will cause errors in this chosen data, thereby allowing Alice and Bob to infer her presence. Because Eve cannot know which bases Alice and Bob will choose there will almost certainly be times when Eve's choice results in an irreducible pathway.

- If no eavesdropper is detected Alice and Bob can then use their remaining untested bits as a secret key. The secrecy of the key is guaranteed to an extremely high level of confidence.

The exact experimental implementation of this protocol will be discussed in detail in the next section. We note here, without proof, that for the above protocol Eve has a 3/4 chance, per compared bit, of escaping detection. If M bits are compared Eve's chance of escaping detection is $(3/4)^{M}$ which rapidly becomes negligible as M increases. Clearly, the eavesdropper must be extremely cautious. Of course, any practical realization will deviate somewhat from the idealizations of the theory. These non-ideal situations will be discussed in detail in the subsequent section. However, we note that techniques have been developed to overcome system inefficiencies [2] and also to reduce an eavesdropper's knowledge still further. This latter technique is known as 'privacy amplification' and allows statistically secret information to be manipulated to yield a secret shared sequence of data [9].

Quantum cryptography is based deeply in fundamental principles of quantum mechanics. Any implementation of such schemes requires a significant understanding of the quantum properties of the channel, in this case the properties of single photons. We have shown how Shannon's information theory can be extended into the quantum regime allowing us to place fundamental limits on the security of quantum cryptography schemes. This theoretical development leads us to the discovery of new protocols and also to the discovery of a simple topological understanding of quantum cryptography channels. Quantum cryptography is a rapidly expanding field of investigation, with many groups world-wide involved in trying to develop new schemes and experimental demonstrations. It can only be a matter of time before this technology is routinely available. The work at BT is focused on the need to increase the distance over which these schemes are effective, and some experimental considerations will be discussed in the next section.

7.3 EXPERIMENTAL PHASE

Quantum cryptography entered the experimental phase in October 1989 with the pioneering work of Bennett et al, who performed the first ever secure key distribution scheme using a quantum transmission channel [2]. The prototype constructed at IBM was based on a coding scheme involving polarized photons, in which the linear and circular polarization states formed the required pair of conjugate bases. As discussed above, in quantum systems the act of measurement is no longer a totally passive process as it is in the classical case, and, as a consequence of the uncertainty principle, any eavesdropper on the channel cannot simultaneously measure both polarization directions of the same photon with arbitrary accuracy. Moreover, if the wrong choice of measurement basis is made, the bit of information carried by that photon is irretrievably randomized. The errors produced by this randomization can be detected by the legitimate users of the channel, thereby revealing the presence of the eavesdropper.

The IBM apparatus, which is shown schematically in Fig. 7.5, used very faint flashes of polarized light from a green LED to transmit the random sequence of key bits over a free-space link of approximately 30 cm in length. A computer containing software representations of Alice, Bob and Eve was used to control the transmission. Alice's light source produced a beam of incoherent pulses of 5 μs duration at a repetition rate of a few kilohertz, which was collimated and passed through a spectral filter and a polarizer. The mean intensity of the beam was very low, with an average of about 0.1 photons per pulse. Each pulse was encoded by Alice with one of the four polarization states as described above, i.e. |*right circular*> = '1', |*left circular*> = '0' |*vertical linear*> = '1' or |*horizontal linear*> = '0'. This was achieved by random switching of the voltage drive to her Pockel's cell between the appropriate levels. Bob's receiving apparatus consisted of another randomly (and independently) switched Pockel's cell followed by a calcite polarizer oriented so as to split the beam into horizontally and vertically polarized beams. These beams were directed on to a pair of photomultipliers which had sufficient sensitivity to detect single photons. Hence Bob's choice of basis (circular or linear polarization) was determined by the selected Pockel's cell voltage, and the bit type ('1' or '0') by the destination detector.

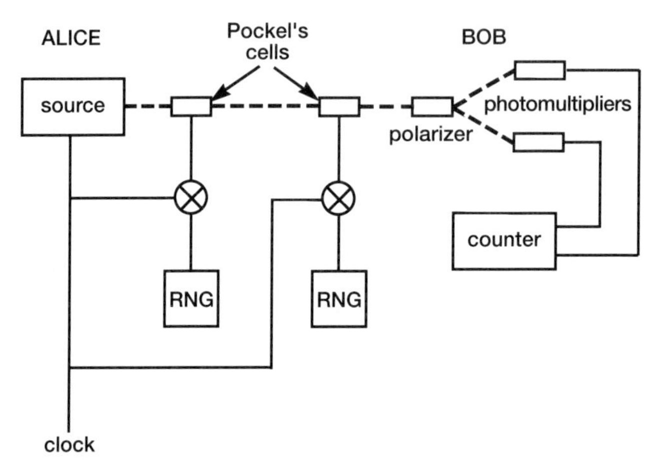

Fig. 7.5 IBM quantum key transmission apparatus (RNG = random number generator).

Firstly, let us consider the way in which a typical quantum transmission would proceed with an ideal version of the IBM apparatus. We shall then consider some of the problems that occur in the real system and how these were overcome. In the ideal case Alice sends a sequence of photons, each randomly encoded with one of the four polarization states, each occupying a well-defined time slot. Bob shares the same clock as Alice and in each of the time slots he randomly chooses to measure one of the two polarization types, i.e. circular or linear. He records the result of that measurement. Table 7.1 shows a possible outcome of this procedure. After the transmission Alice and Bob confer publicly; in the experiment, this is carried out via the software in the control computer, but in a real system this discussion could be performed using a standard telephony circuit which is potentially accessible to an eavesdropper. Bob tells Alice which type of polarization measurement he made in each time slot (linear or circular), **but not the result of that measurement** ('1' or '0'). They then agree to discard all the data from time slots in which they used a different measurement basis (recall that we are discussing the BBW protocol here and that the BT-Strathclyde protocol [4] would not reject any data at this stage prior to analysis). Alice and Bob now choose a random subset of the remaining data which they use to test for the presence of an eavesdropper. This test is again carried out over the public channel but now Alice and Bob perform a comparison of bits which were

Table 7.1 Ideal quantum communication between Alice and Bob using linear and circular polarization states. There are 14 time slots labelled in row (a). Alice sends a sequence of randomly encoded photons in these time slots [row (b)]. Bob detects these photons using a random choice of measurement basis shown in row (c), and deduces the data sequence shown in row (d). In step (e) Alice and Bob discard all data where they used a different basis. Data from time slots 3, 6 and 14 are randomly chosen for comparison and, since they find no errors, Alice and Bob deduce that no eavesdroppers are present on the channel. The remaining data [row (f)] can then be used to generate a secret binary key [row (g)].

(a)	1	2	3	4	5	6	7	8	9	10	11	12	13	14
(b)	↻	↺	↻	↑	→	↻	→	↑	↑	↺	↻	→	↺	↑
(c)	+	○	○	+	○	○	+	○	○	○	+	+	+	+
(d)	↑	↺	↻	↑	↻	↻	→	↺	↻	↺	↑	→	↑	↑
(e)		↺	↻	↑		↻	→			↺		→		↑
(f)		↺		↑			→			↺		→		
(g)		'1'		'1'			'0'			'1'		'0'		

sent and received in the selected time slots. If no errors are found Alice and Bob can be sure that the remaining data, which has not been subject to public scrutiny, is secure and therefore constitutes a useful secret key. As discussed above, if an eavesdropper is present on the channel her measurements will inevitably result in Alice and Bob detecting some errors in their bit comparison. As we have already stated, the probability that Eve escapes detection is 3/4 per compared bit. If Alice and Bob compare M bits the probability that Eve remains hidden is $(3/4)^{M}$. As M increases this probability becomes extremely small, so that for $M = 100$, for example, the probability that Eve evades detection is approximately 3.2×10^{-13}, i.e. Eve has about 3 chances in 10 trillion of not triggering the alarm.

In practice, experimental factors present in the real system significantly change the quantum transmission protocol from the ideal case shown in Table 7.1. One problem stems from the fact that Alice does not have a source of single photons. Instead, her source produces a temporal pulse in each clock period which contains, on average, μ photons. If $\mu \geq 2$, Eve can perform an undetectable beam-splitting attack in which she splits off one or more of the photons in the pulse, leaving at least one photon which is then detected by Bob. Consequently, a small value, $\mu \approx 0.1$, was chosen in the experiment so that the probability of obtaining a pulse containing two or more photons was also small ($\leq \mu^2$). A second important consideration was the relatively low quantum efficiency of the detectors employed, which was only about 9%. This means that Bob fails to register the arrival of most of Alice's photons. If there is any additional loss in the transmission channel itself the number of photons measured by Bob will be further reduced. However, loss is not, in itself, a severe problem. This is because the security of the channel and the secret key are only established from Bob's measured bits. Any loss will reduce the data rate, but **will not intrinsically compromise the security of the channel**. This robustness to loss is a very appealing feature of the quantum transmission and shows that genuine quantum effects can be exploited even over lossy channels. These non-ideal aspects of the experiment reduce the rate at which Bob receives photons, R_B, to:

$$R_B = \mu \eta T R_A \qquad \text{... (7.6)}$$

where η is the detector quantum efficiency, T is the channel transmission coefficient, and R_A is the rate at which Alice's source is pulsed, i.e. the clock rate.

In the IBM experiment, $R_A \approx 3$ bits/s, which may seem excessively low when compared with the gigabits per second data rate achievable with modern optical communications networks. It is, however, important to remember that the quantum system is only intended to transmit relatively short sequences of key data (a key size of 56 bits is required for standard implementations of DES, although for more important transmissions, such as financial transactions between banks, for example, a key size of 112 bits is employed). This consideration notwithstanding, however, it should be noted that the principle limitation on the data rate in the IBM scheme is the speed with which the Pockel's cells can be switched (a few kilohertz). This places a limitation on R_A because it is important to ensure that only one pulse is processed at a time by the Pockel's cell. With modern integrated optical modulations gigahertz switching speeds are achievable. It is, therefore, expected that the data rate, R_A, can be increased by many orders of magnitude, thus leading to a concomitant increase in R_B (the detector time resolution will limit R_B before the limit on R_A is achieved).

A potentially more serious problem than bit-rate reduction is the fact that Bob's detectors suffer from 'noise', i.e. they occasionally register a count even when no photon is incident from the transmission channel. They will also occasionally measure the polarization state of an incoming photon incorrectly, due to misalignment in the optics for example. This noise occurs randomly and leads to an error rate which, if sufficiently large, would mask the errors caused by the presence of an eavesdropper. Obviously some form of error detection and correction procedure must be adopted or the key data obtained from the quantum channel would be meaningless and of no use. The first part of the solution is for Bob to turn his detectors on only in the short time interval during each clock period when he knows a photon may arrive. This is a standard technique for dark count reduction in photon-counting experiments and it requires Alice's pulse duration to be significantly less than the clock period. Any residual error rate due to noise in the system must now only be a fraction of the total received bit rate that Bob registers. As we discuss below, however, protocols have been developed that allow the system to achieve arbitrarily high levels of secrecy (at the expense of key length) even when this noise-induced error fraction is relatively large. A typical quantum transmission over a noisy channel could proceed as illustrated in Table 7.2. Alice and Bob perform the transmission as if using an ideal channel as before. However, the received data rate is now reduced since both the average number of photons per pulse, μ, and the detector quantum efficiency, η, are less than unity (but, for simplicity, not taken here to be as small as in the IBM experiment), and Bob's detectors now suffer from noise. This noise manifests itself in timeslot 13, where a dark count has occurred in Bob's detectors, causing him to register a result even though Alice's source did not generate a photon. This would lead to an error which is indistinguishable from that caused by an eavesdropper. Consequently, as noted in the previous section, Alice and Bob adopt a protocol that allows them to distil a smaller body of shared secret key information from data containing a significant amount of errors [2, 9]. This protocol (which we do not discuss in detail here) is based on the random permutations of the bit positions in Alice and

Table 7.2 Quantum communication between Alice and Bob using an imperfect channel. Alice sets her Pockel's cell to encode the polarization states shown in row (b). However, $\mu < 1$ for her source so that some of the timeslots do not contain a photon (indicated by a dashed cross). Bob randomly chooses his measurement basis [row (c)] and obtains the data sequence shown in row (d). Bob's detectors have quantum efficiency $\eta < 1$ (so he sometimes fails to register incoming photons as in timeslots 1 and 12) and the system is noisy (so he sometimes registers a photon where none has been sent, as in timeslot 13). In step (e) Alice and Bob discard all data where they used a different basis. After performing an error detection and correction procedure, followed by 'privacy amplification', Alice and Bob end up with a smaller sequence of shared secret key data with an arbitrarily high degree of security [rows (f) and (g)].

(a)	1	2	3	4	5	6	7	8	9	10	11	12	13	14
(b)	↻	↻	↺ ×	↑	→	↺	→	↑ ×	↑	↻	↺	→	↑ ×	↑
(c)	+	○	○	+	○	○	+	○	○	○	+	+	+	+
(d)	×	↻	×	↑	↺	↺	→	×	↺	↻	↑	×	→	↑
(e)		↻		↑		↺	→			↻			→	↑
(f)		↻					→			↻				↑
(g)		'1'					'0'			'1'				'1'

Bob's data strings followed by division of the data into blocks and parity comparison of these blocks over the public channel. By performing this procedure iteratively with different block sizes, Alice and Bob can locate and correct errors in their data without actually revealing the results of individual measurements publicly. This procedure is followed by 'privacy amplification' [9] involving random subset hashing of the remaining data which reduces, to arbitrarily small amounts, any information that Eve may have obtained by monitoring both the quantum transmission and the public exchanges between Alice and Bob. For example, in one test of the IBM apparatus, where the combined error rate due to eavesdropping and noise was about 8% (that is, Eve only occasionally eavesdrops and does not, therefore, cause the 25% error rate that might be expected from an interception of every transmitted bit), the implementation of the protocol enabled a 105-bit key to be distilled from 2000 received bits [2]. In this case, Eve's information about the key was estimated to be infinitesimally small at about 6×10^{-171} bits, illustrating the extreme degree of secrecy obtained for the shared key.

Having established that quantum cryptography is now an experimental reality, at least over short distances in the laboratory, we discuss the technological challenge of implementing such schemes over optical fibre networks. As is usually the case, the optimum system design will likely be determined by a series of trade-offs involving individual component characteristics, such as detector efficiency, dark count rate, etc, and systems requirements, such as bit rate and transmission distance. In order to maximize the latter it will be necessary to choose an operating wavelength in one of the low fibre-loss telecommunications windows at ≈ 0.85, 1.3, or 1.5 μm. Silicon avalanche photodiodes (APDs) are an established detection technology for single-photon counting experiments in the visible and near-IR region of the optical spectrum and these would be an excellent choice for quantum cryptography systems operating in the first telecommunications window. However, the relatively high fibre loss in this window, ≈ 2 dB/km, would limit the ultimate transmission distance achievable. At BT Laboratories we are concentrating on the 1.3/1.5 μm windows where fibre loss can be as low as 0.2 dB/km, but detector technology (InGAs or germanium APDs) is less well developed. We expect that quantum cryptography systems should be able to tolerate transmission losses as high as ≈ 10 dB and still be able to achieve useful data rates (perhaps ≈ 1 MHz), and consequently system spans of $\approx$ 50 km may be envisaged. This points towards applications in secure local area networks, perhaps in dedicated networks linking dispersed sites of business customers, for example. An important consideration, however, is that as quantum cryptography is an optical technique it could only be implemented using a fibre-based, rather than copper-based, access network. There are other important considerations related to the compatibility of this optical technique with BT's existing network. As it stands above, quantum key distribution can only be implemented over an unrepeatered, unamplified optical link. The technology is, however, very much in its infancy, and the practical limits of the technique have not yet been firmly established. It is apparent that further work is necessary to address some of these important issues.

Finally, we note that standard sources such as semiconductor lasers and LEDs are available for all the telecommunications windows and could be used as cheap, compact sources of pulses for these systems. However, improvements in both security and bit rate can be made if true single-photon sources are available. Fortunately, we have some experience of such sources, called optical parametric amplifiers [10]. They are based on nonlinear crystals which, when excited with an intense stream of pump photons, perform the function of 'photon fission', i.e. splitting some of the pump photons into pairs of lower-energy photons which are correlated in both phase and time. One of these pair photons can be detected and used to open a shutter or trigger an electrical gate in order to isolate or identify its 'twin', which is used to transmit the key information. Unlike schemes which use conventional sources, such as the one discussed above, it is therefore possible, in principle, to obtain $\mu = 1$ such that there is guaranteed to be one and only one photon per pulse, and hence Eve has no chance of successfully performing a beamsplitting attack. Also, as indicated by equation (7.6), a higher data rate can be achieved. This type of source has recently been exploited in secure free-space communications schemes (not based on quantum cryptography) [11]. We also note that the twin photons produced by the parametric amplifier exhibit strong violations of the Bell inequality. These quantum correlations can also be exploited in a quantum cryptography scheme to give a guaranteed security for the key [12, 13]. These schemes also give the additional feature of secure key storage, provided certain technological improvements can be made.

In summary, although experimental quantum cryptography is still in its infancy, it seems likely that the application of optical telecommunications techniques will lead to rapid advances in this area. In particular, dramatic improvements in both bit rates and system span can be anticipated. Furthermore, quantum cryptography is a rapidly expanding field of investigation, and further improvements in the design of new protocols and experimental implementations are expected.

7.4 CONCLUSIONS

Quantum cryptography is an exciting new technique which represents a radical departure in both physics and cryptology. The implications of this technology for communications networks are potentially very significant indeed. The pioneering results of the IBM experiment show that secure key distribution has been achieved in the laboratory. Moreover, as already discussed, experiments at BT Laboratories have indicated that improvements of many orders of magnitude in both data rate and distance are achievable. Despite this promising beginning there remain unanswered questions concerning the applicability and compatibility of this technique over communications networks. Some of these issues have been touched upon in the previous section and it is evident that much further work needs to be done. This work needs to be focused quite clearly on two issues. Can we make the technique work over practical distances on BT's networks? How can BT best exploit this technology? The first issue concerns the basic technology required to achieve secure key distribution between two remote parties. The second issue concerns a more strategic viewpoint on how BT can best offer such a service if it were to become practical.

The new field of quantum cryptography is still very much in its infancy. Despite its immaturity, however, the technique shows considerable promise and has generated a great deal of interest world-wide, and further developments can be expected over the next few years. Quantum cryptography is a very exciting field both from the viewpoint of basic physics and the perspective of telecommunications. It is an example of how an unexpected technological discovery can lead to the possible provision of a new service — the guaranteed secrecy of a cryptographic key. Furthermore, the provision of other information security services may be possible over quantum channels including discrete decision making and authentication [14]. It is clear that the full potential of quantum communications is only just beginning to emerge.

REFERENCES

1. Beker H and Piper F: 'Cipher systems', Northwood, London (1982).

2. Bennett C H, Bessette F, Brassard G, Salvail L and Smolin J: 'Experimental quantum cryptography', J Cryptology, 5, p 3-9 (1992).

3. Phoenix S J D: 'Quantum cryptography without conjugate coding', Phys Rev A, 48, No 1, pp 96-102 (July 1993).

4. Barnett S M and Phoenix S J D: 'Information-theoretic limits to quantum cryptography', Phys Rev A, 48, No 1, pp R5-R8 (July 1993).

5. Barnett S M, Pegg D T and Phoenix S J D: 'Quantum communication channels, correlations and cryptography', manuscript unpublished.

6. Wiesner S: 'Conjugate coding', SIGACT News, 15, p 78 (1983).

7. Blow K J and Phoenix S J D: 'On a fundamental theorem of quantum cryptography', J Mod Opt, 40, p 33 (1993).

8. Bennett C H, Brassard G, Breidbart S and Wiesner S: 'Quantum cryptography, or unforgeable subway tokens', in Chaun D, Rivest R L and Sherman A T (Eds): 'Advances in Cryptology: Proceedings of Crypto'82', Plenum Press, New York, p 267 (1983).

9. Bennett C H, Brassard G and Robert J M: 'Privacy amplification by public discussion', SIAM J, Comput, 17, p 210 (1988).

10. Townsend P D and Loudon R: 'Quantum noise reduction at frequencies up to 0.5 GHz using pulsed parametric amplification', Phys Rev A, 45, p 458 (1992).

11. Seward S F, Tapster P R, Walker J G and Rarity J G: 'Daylight demonstration of a low-light-level communication system using correlated photon pairs', Quantum Opt, 3, p 201 (1992).

12. Ekert A K: 'Quantum cryptography based on Bell's Thoerem', Phys Rev Lett, 67, p 661 (1991).

13. Ekert A K, Rarity J G, Tapster P R and Palma G M: 'Practical quantum cryptography based on two-photon interferometry', Phys Rev Lett, 69, p 1293 (1992).

14. Bennett C H, Brassard G and Ekert A K: 'Quantum cryptography', Scientific American (October 1992).

8

MICROWAVE AND MILLIMETRE-WAVE RADIO FIBRE

D Wake, L D Westbrook, N G Walker and I C Smith

8.1 INTRODUCTION

Transmission systems incorporating both radio and optical fibre elements, so-called radio-fibre systems, are expected to find an increasing role in telecommunications networks over the next decade. These systems can give operational benefits in terms of diverse or mixed service provision in areas such as mobile/cellular, local/rural access, and remote feeding of radio stations. Furthermore, demand for broadband video and data is expected to increase substantially during this period; while implementation of fibre-to-the-home requires a good deal of time and investment, then radio-fibre systems can be expected to play a significant part in the roll-out of broadband local access networks. This approach is in line with the strategic objectives of the access network regarding dropwire replacement by radio at UHF, in the sense that broadband could take advantage of the same fibre infrastructure. Examples of some proposed radio-fibre systems include:

- antenna remoting at satellite earth stations;
- remote operation of satellite earth stations themselves;
- dropwire replacement by radio;
- rural access by radio;
- the RACE mobile broadband system (MBS).

For any transmission system, practical issues such as the size, weight, reliability, cost and power consumption of remote equipment are of critical importance. One of the principal advantages of radio-fibre technology is the freedom to concentrate some or all remote equipment at a centralized location where these factors are more easily controlled. Whether radio-fibre systems find widespread use in BT's networks will depend in part on the resolution of these practical issues, as well as technical aspects such as linearity and sensitivity of the radio and optical segments of the link. The first three examples of radio-fibre systems given above utilize frequencies in the UHF/microwave bands where suitable optoelectronic components are now available at reasonable cost.

In the satellite arena, the next decade is likely to see increasing use of optical links at earth stations world-wide for two main applications. The first involves short-range (0.1-1.0 km) remoting of antennas at SHF (between 1 GHz and 15 GHz), and allows cost savings via more efficient redundancy/spares holding by locating equipment centrally wherever possible. There are also significant operational advantages to be gained, e.g. quick and easy re-routeing of traffic to different antennas via centralized switching. Fibre systems must cost in against dry gas pressurized waveguide runs, although these are expensive at L and C bands[1] and impractical at Ku band because of high electrical losses. Another application is long-range (10-1000 km) remoting of earth stations at VHF (typically at 70 MHz). The main operational advantage is that users can site their antenna many kilometres from the control area, for example to improve satellite visibility or reduce interference with terrestrial systems. Cost savings are possible, for example, by removing control equipment from expensive metropolitan sites to allow more efficient use of high-cost premises. Multi-hop links of up to 500 km are also under consideration. Since satellite services generate large amounts of revenue they can bear the cost of installing and maintaining relatively expensive equipment. For this reason satellite earth station operators are among the first practical users of radio-fibre technology. There remains a need to improve the dynamic range of optical systems, mainly to increase the usable link range. There is also scope for further equipment rationalization by using some of the novel ideas arising out of this work, for example up-conversion over a fibre link.

---

[1]

Microwave band	Frequency range (GHz)
L	1.12-1.7
C	3.95-5.85
X	8.2-12.4
Ku	12.4-18
K	18-26.5

In the near future there will be opportunities for the implementation of radio-fibre systems in the local access network, where there is a need to provide telephony, possibly alongside other narrowband services such as ISDN, over radio. Such cordless access networks must cost in against copper-fed radio systems. These systems are expected to appear within four years, operating either at CT2 (866 MHz) or DECT (1890 MHz) frequencies. Again there is a high dynamic range requirement from the optical and radio components. Roll-out of these systems will help the case for asymmetric broadband delivery at millimetre-wave frequencies since existing duct and, possibly, fibre will be available.

In all the systems described so far, subcarrier multiplex (SCM) is used for the transmission of radio signals over the optical links. SCM is a technique in which a conventional frequency-division-multiplexed radio signal is applied directly to an intensity-modulated laser. This composite radio signal typically comprises a free mixture of digital and analogue signals. SCM, described more fully in section 8.2, is feasible using present-day technology and can be obtained from several commercial sources. The work described here is part of a fresh approach to the technology used to implement radio-fibre interfaces. Investigations are under way to find alternative subsystem topologies which can simplify the hardware needed at radio outstations. In this context a development of interest is the recent allocation, across the European Community, of spectrum between 40.5 GHz and 42.5 GHz for multipoint video distribution services (MVDS). MVDS is a transmit-only service which can be used to serve areas the size of a small town (8-9 km^2) with 30 or so channels of analogue FM TV (more if the signals are compressed, fewer if HDTV). The coverage area is served by a transmitter located on a mast or tall building. Currently this is done using an array of directly modulated Gunn oscillators, each with its own horn antenna and heat pipe for frequency stabilization. The rooftop hardware could be simpified considerably by feeding either a travelling wave tube (TWT) or a solid-state amplifier, over an optical fibre link, at the transmit frequency. All that would then be required is a detector, the amplifier and a single horn antenna. The weight and wind loading of the transmitter would be greatly reduced, and the unit could be fed by a single optical fibre from a distance of several hundred metres. Unfortunately the operating frequency is far above that currently achievable using a directly modulated laser, whereas an external modulator would be very lossy and extremely difficult to produce with adequate bandwidth. A new technological approach which makes this kind of system enhancement feasible is therefore extremely useful.

The RACE mobile broadband system (MBS) is a futuristic radio-fibre system aimed at satisfying demand for broadband services to pedestrians, trains and road vehicles. Millimetre-wave radio LANs could be implemented in a very similar manner. It would be useful to be able to feed such radio nodes via fibre. In this case, the high frequencies involved (62-66 GHz) are well beyond the capability of existing optoelectronic components, if SCM is to be used. From the radio perspective the 60 GHz bands are of special interest because of the oxygen absorption peak, which allows excellent frequency reuse for cellular operation. In addition, high-gain antennas are physically small at this frequency and can be manufactured cheaply in volume for consumer applications. The long-term frequency stability of millimetre-wave sources is questionable, however, and this is most likely to dictate the use of phase-locked loop or injection-locking techniques in any practical system. Efforts to realize such components, and at the same time take a fresh approach to the implementation of this type of system, are outlined in section 8.3.

8.2 MICROWAVE RADIO FIBRE

8.2.1 Introduction

Current interest in the distribution of microwave signals over optical fibre owes much to the pioneering work carried out in the late 1970s and early 1980s on GaAs laser diodes. At that time, experiments were mostly driven by military interest. Although small signal modulation of GaAs lasers up to X band was achieved (see, for example, Lau and Yariv [1]), nevertheless fibre attenuation and dispersion at 850 nm limited their application. The development of high-quality lasers at the low-loss fibre windows, $\lambda = 1.3\ \mu m$ and $\lambda = 1.55\ \mu m$, has permitted the transmission of microwave signals over useful distances with low loss [2]. Way et al [3] and others [4] demonstrated the use of fibre optics to transmit C band microwave signals to and from a remote satellite antenna using directly modulated 1.3 μm lasers. The development of lasers operating to Ku band by Olshansky et al [5] at GTE and K band (with cooling) by Bowers et al [6] at AT&T further spurred the interest in wideband techniques and made possible directly modulated fibre-optic links at X and Ku bands. At the same time, Olshansky [7] and others [8] have demonstrated the potential of the subcarrier technique as an alternative to time division multiplex (TDM) for the distribution of wideband FM and digital signals. Further improvement in laser performance has led to a rapidly growing market in fibre systems for upgrading amplitude-modulated cable TV (CATV) distribution links which have stringent signal-to-noise and intermodulation requirements using optical overlays to improve the signal quality [9, 10]. Currently much of the effort devoted to SCM systems is directed at these video services, which are seen to be key to introducing fibre to the home. The number of military applications has also grown during this period with microwave over fibre used to control beam steering in phased array and conformable antenna, and as a means of electromagnetically isolating microwave sub-systems.

The performance of a simple directly modulated SCM link, such as that shown in Fig. 8.1, is considered now. In the case where the laser diode

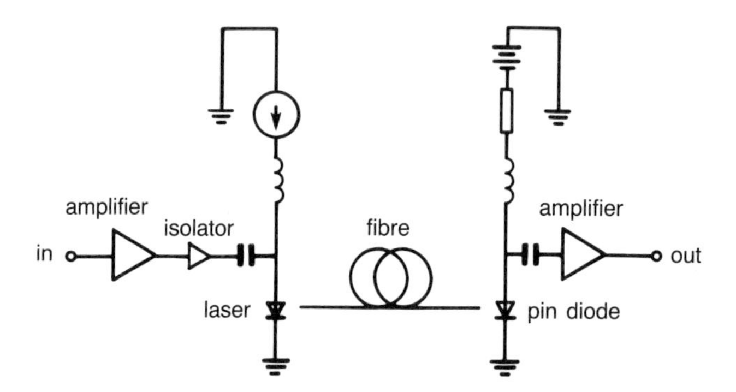

Fig. 8.1 Directly modulated subcarrier multiplexed link.

transmitter is resistively matched to a 50 Ω environment, Cox et al [11] give the end-to-end link loss of the fibre-optic part as:

$$G = C^2 \, (\partial P/\partial I)^2 \, t_{\text{opt}}{}^2 \, \eta_{\text{d}}{}^2$$

where $(\partial P/\partial I)$ is the laser ex-facet slope efficiency in W/A, C is the fraction of the laser power coupled into the fibre, t_{opt} is the optical system loss due to attenuation, splicing and coupling, and η_{d} is the detector responsivity in A/W.

Even using optimistic values for the laser efficiency (0.4 W/A), laser-fibre coupling (3 dB), fibre loss (3 dB) and detector responsivity (0.9 A/W) gives a link loss of ≈ -20 dB. It is assumed that in order to be useful a 0 dB link loss is required. Therefore additional electrical amplification is required, and in general, electrical amplification is required at both the transmit and receive ends, due to the relatively large noise figure for the fibre-optic part of the link. Again, Cox et al [11] gives this as:

$$NF = 10\log\left(2 + \frac{50}{kTG}\,(2\,e\,I_{\text{d}} + \text{RIN}\,I_{\text{d}}{}^2)\right)$$

where e is the electronic charge, and I_{d} is the d.c. photocurrent. The noise figure comprises contributions due to thermal noise of any matching resistor, shot noise ($2eI_d$) and relative intensity noise (RIN I_d^2). Figure 8.2 shows the dependence of this noise figure on the received photocurrent, assuming a constant laser RIN of −150 dB Hz. At a typical photocurrent of 1 mA the noise figure is ≈30 dB.

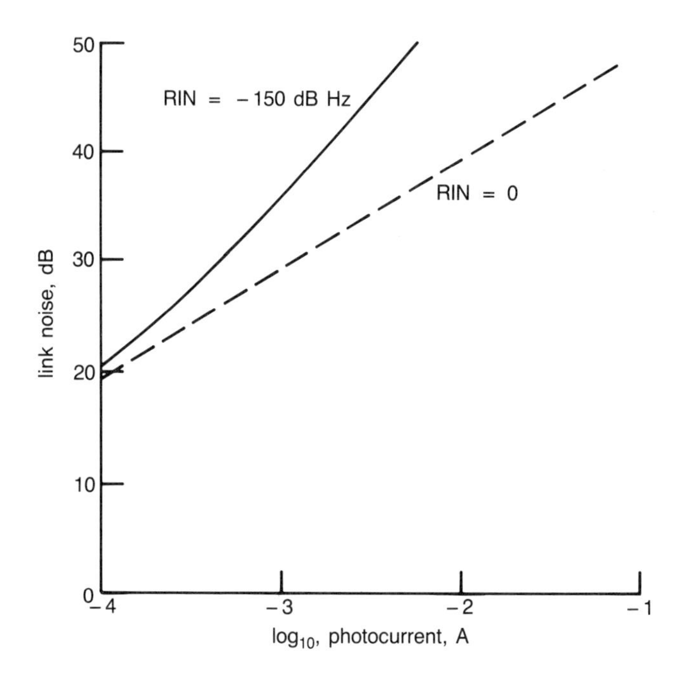

Fig. 8.2 Link noise figure (without electrical preamplification) against received photocurrent.

In general, microwave links are 'narrowband' and therefore 3rd order intermodulations are the dominant distortion products. The detected intermodulation product $IM = \beta S^3$, where S is the detected signal power and β is a constant of proportionality related to the 3rd order intercept. The specification on noise and distortion may be combined into one figure — the intermodulation-free dynamic range [11]. As shown in Fig. 8.3 the intermodulation-free dynamic range is given by the point at which the signal-to-noise and signal-to-intermodulation levels are equal. If NP is the detected noise power then the intermodulation-free dynamic range can be shown to be $1/(NP^2\beta)^{1/3}$. As shown in the next section, both NP and β are strong functions of the laser output power and this can make link optimization a more complex process than is suggested by Fig. 8.3.

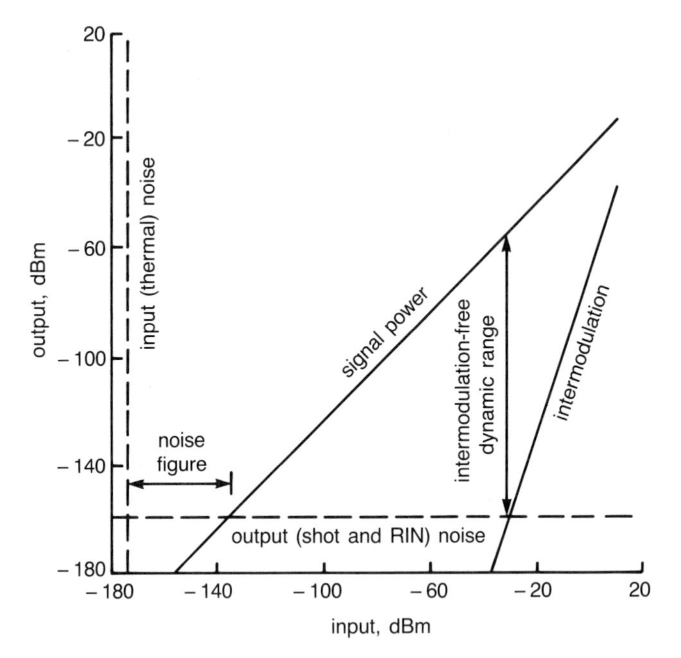

Fig. 8.3 Theoretical link performance showing intermodulation-free dynamic range.

8.2.2 Sources

The analogue performance of laser diodes may be separated into intrinsic effects — resulting from the dynamic interaction of electrons and photons — and extrinsic effects — parasitic capacitances, leakage currents, etc. Both effects strongly influence laser performance.

8.2.2.1 Intrinsic effects

The amplitude modulation (AM) response, relative intensity noise (RIN) and distortion in a laser are strongly influenced by the resonant interaction of electrons and photons [6]. An internal particle balance exists only for frequencies below an internal angular resonance frequency ω_0. The intrinsic 3 dB modulation bandwidth is given by $(\sqrt{(1+\sqrt{2})})\omega_0$. Figure 8.4 shows the dependence of the laser RIN, 3rd order intermodulation, and intermodulation-free dynamic range with the ratio $(\omega_0/\omega_m)^2$ — where ω_m is the modulation frequency. It is evident that laser performance benefits from having as high a resonance frequency as is possible. It is well known that the resonance frequency increases as the square root of the optical output

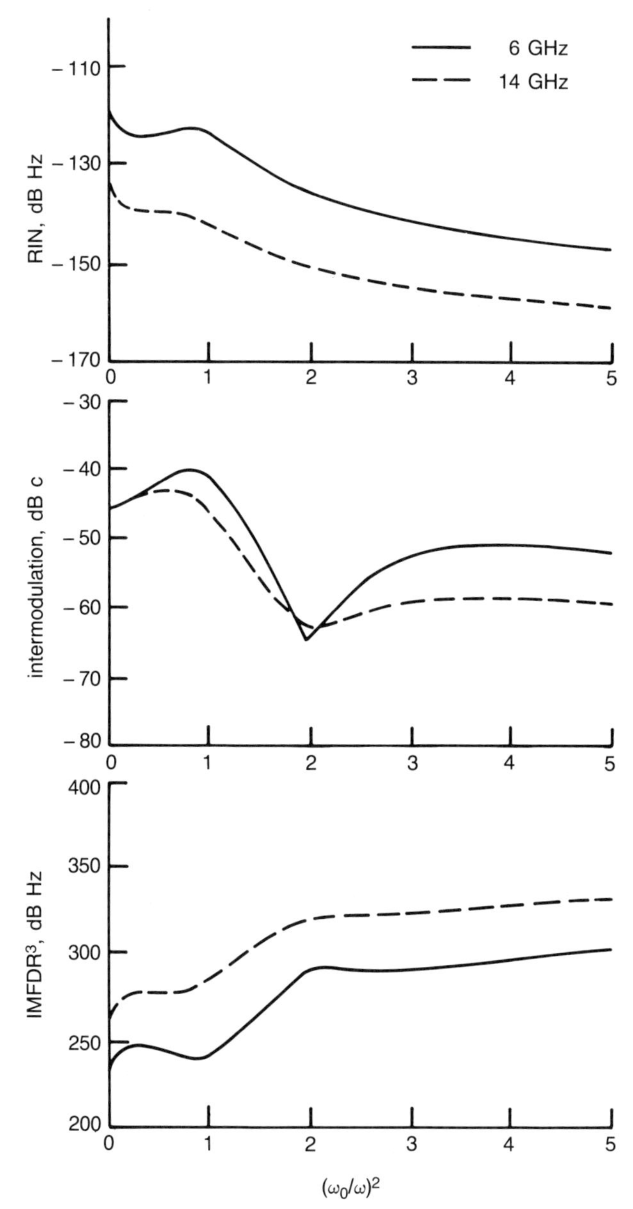

Fig. 8.4 Dependence of RIN, 3rd order intermodulation, and IMFDR with modulation frequency ratio.

power P, and therefore the horizontal axis of Fig. 8.4 may be considered as scaling with laser output power. Increasing the output power in order to improve microwave performance may be used up to a point; however, this demands an increase in the electrical preamplifier gain (to achieve the intermodulation-free dynamic range) and extrinsic linearity often suffers at higher powers. Bowers et al [6] have shown that the figure of merit ($\omega_0/(2\pi\sqrt{P})$) can be optimized by altering the laser dimensions, in particular by maximizing the ratio of the optical confinement factor to the cross-sectional area. The degree to which this may be employed is limited to some extent by the need to be able to manufacture reliable devices reproducibly.

A complementary strategy is to employ multiple quantum well (MQW) active regions. Size quantization in 8 nm thick semiconductor layers increases the gain constant (the derivative of the optical gain with electron number) which, in turn, increases ω_0^2 by a factor of up to 3. Experimentally [12, 13] the optimum number of quantum wells for microwave SCM has been found to be in the region of 16 to 20.

8.2.2.2 Extrinsic effects

Conventional buried heterostructure (BH) lasers use a low leakage p-n-p-n blocking structure to achieve low threshold current, high output power and high-temperature operation. Unfortunately these p-n-p-n blocking layers have parasitic capacitance of the order of 50 pF and this seriously limits the modulation bandwidth. Etching grooves (trenches) through the blocking structure on either side of the active mesa can reduce this capacitance to a few picofarads. However, to achieve reproducibly the ≈ 1 pF required for Ku band microwave operation requires a different strategy. To this end, Fe-doped semi-insulating blocking layers grown by metal-organic vapour phase epitaxy (MOVPE) have been developed [12]. Figure 8.5 shows a schematic cross-section of a 16 multi-quantum well laser diode with Fe-doped semi-insulating blocking layers. At present these low parasitic capacitance Fe-doped blocking layers are electrically inferior to p-n-p-n blocking structures and result in slightly lower output powers and less linear P-I characteristics. This leads to increased intermodulation via the transfer (P-I) characteristic and, for frequencies well below the resonance frequency ($\omega << \omega_0$) where intrinsic distortion is low (as in CATV transmission), this extrinsic distortion may be dominant.

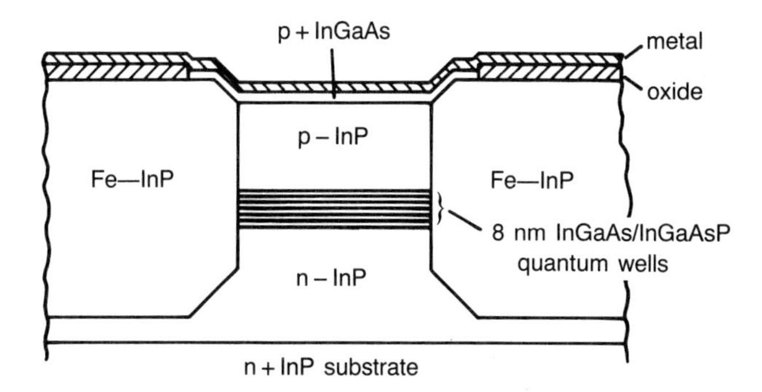

Fig. 8.5 Cross-section of 16 quantum-well laser with Fe-doped semi-insulating blocking layers.

Figure 8.6 shows the measured intermodulation-free dynamic range for our 16 multi-quantum well lasers at 14 GHz. These lasers were packaged in a modified DIL package with a Wiltron K microwave input connector and have a packaged 3 dB bandwidth of ≈ 17 GHz and a maximum intermodulation-free range of ≈ 100 dB.

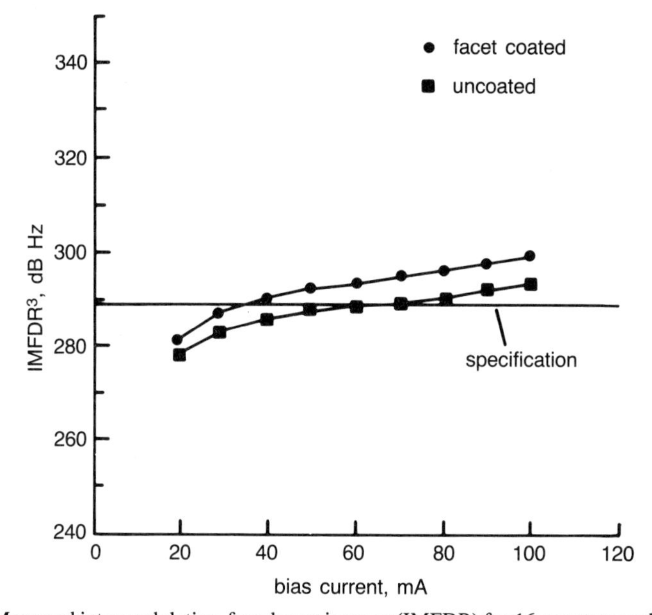

Fig. 8.6 Measured intermodulation-free dynamic range (IMFDR) for 16 quantum-well lasers.

Currently the widest 3 dB laser bandwidth in the research laboratory is 25 GHz. However, the reliability of some of the reported structures is a matter of concern. Assuming parasitic capacitance can be further reduced and resonance frequency increased without compromising the reliability, laser modulation bandwidth is ultimately limited by the intrinsic nonlinear gain process to around 30 GHz [13].

8.2.3 Photodetectors

For RF and microwave frequencies the surface-illuminated PIN photodetector offers the best compromise between responsivity and bandwidth [14]. Surface-illuminated detector design requires optimization of the depletion (absorbing) layer thickness and active area. Thick depletion layers result in high responsivity and low capacitance per unit area, but also increased transit times. Optimum performance occurs when the bandwidth due to capacitance and transmit time are made equal. Likewise, small surface areas result in reduced capacitance but make contacting difficult and ultimately limit the responsivity. Reliability is also an issue since microwave PIN detectors fabricated using mesa etching have exposed p-n junctions. Careful optimization using planar technology has resulted in reliable PIN detectors with 25 GHz bandwidth and a responsivity of 1 A/W at $\lambda = 1.55\ \mu m$ [15]. Figure 8.7 shows a photograph of the detector structure grown by MOVPE and shows an air bridge used to connect the 30 μm PIN to an insulated bond pad. The p-n junction is formed by Zn diffusion. The detector is mounted in a specially made fibre package with a Wiltron K output connector. Total capacitance is ~ 100 pF with 90% of devices having leakages < 50 pA.

Linearity is not considered to be a serious problem for detectors for received powers of 1 mW or less; however, space charge effects and heating are likely to cause problems for some externally modulated microwave-fibre systems where reduced figures result from high (~ 100 mW) received optical powers [11].

8.2.4 A Ku band SCM system

Microwave SCM links based on the laser transmitter and PIN receiver described above are currently on trial at BT's Goonhilly satellite earth station [16]. Figure 8.8 shows a photograph of the complete transmitter and receiver modules for one of the links, which operates at the 4, 6, 11 or 14 GHz satellite bands. These links will be used to carry mixed (i.e. analogue and digital)

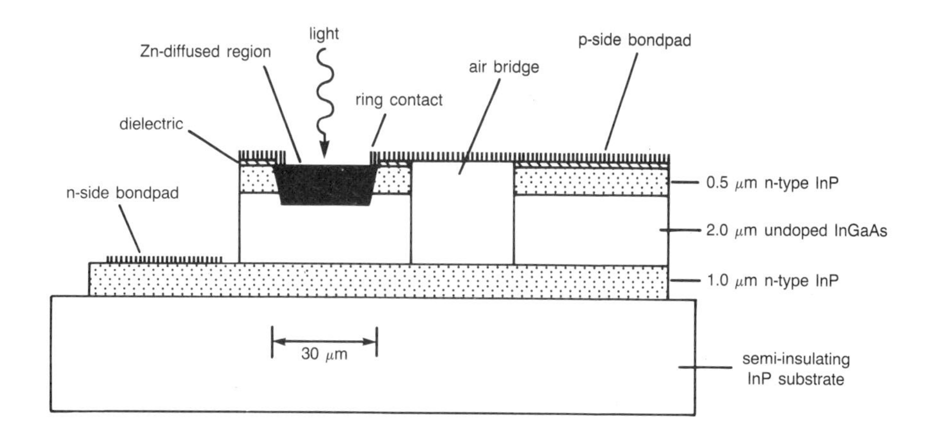

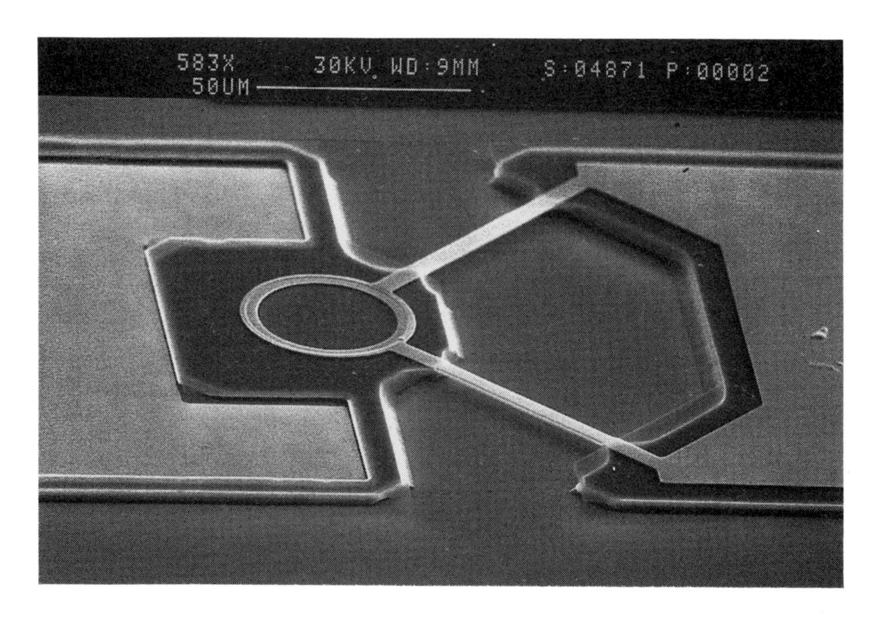

Fig. 8.7 Schematic cross-section and photograph of 30 μm diameter PIN detector showing air bridge to bonding pad.

satellite traffic between antennas and a central plant building. Currently, antennas are linked at an intermediate frequency using coaxial cable. The ability to co-locate most of the operating plant (up-converters, down-converters, etc) is expected to lead to significantly reduced maintenance costs and reduced capital costs as less plant needs to be duplicated. Call revenue from an international satellite antenna is very high per dish. Consequently, the improved speed at which antennas and antenna frequencies can be

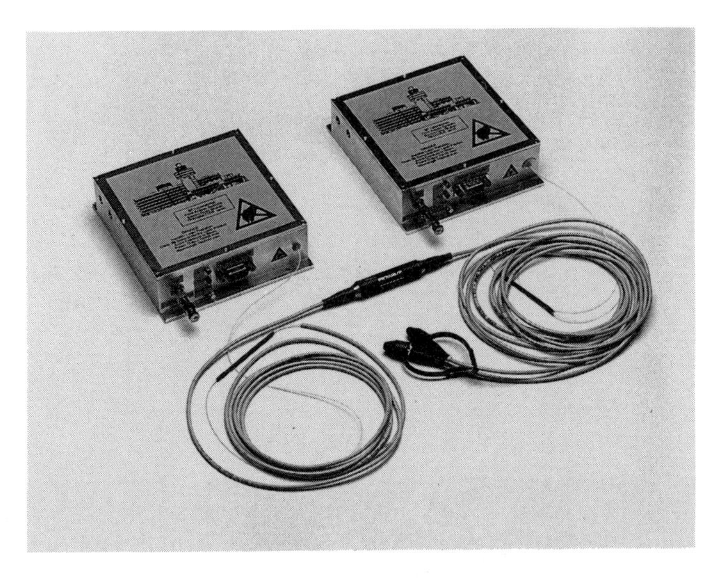

Fig. 8.8 K-band SCM modules.

configured using microwave links to the antenna can prevent the loss of substantial revenue in the event of an antenna failure.

The link specification equates to an intermodulation-free dynamic range of between 90 and 96 dB Hz depending on whether the link is up or down (to or from the satellite). 1550 nm Fabry-Perot (multi-longitudinal mode) 16 quantum well lasers are used with dispersion-shifted fibre since the link length is less than 1 km. Residual fibre dispersion and modal partition noise ultimately limit link length and a single mode Ku band distributed feedback (DFB) laser is currently under development for applications requiring transmission over 1 km. Ultimately, these may be used to remotely connect complete antenna sites to a nearby urban exchange building.

8.3 MILLIMETRE-WAVE RADIO FIBRE

8.3.1 Introduction

The previous section dealt with systems that are currently being implemented. Radio-fibre systems in the millimetre-wave band are not so well developed, partly due to the difficulty of obtaining suitable optoelectronic components. Such systems may find important application areas as a result of small antenna size, looser regulation and higher atmospheric absorption (for frequency reuse when contiguous coverage is needed). Work aimed at realizing millimetre-wave optoelectronic components for such systems has already yielded fruitful results. This work has not necessarily concentrated on new components with enhanced performance or functionality. Indeed, much progress has been made by taking existing components and using innovative techniques to give the required performance advantage.

This work is of a highly speculative nature, but initial results have been very encouraging. In section 8.3.2, the progress towards a practical millimetre-wave optical source is described, and in section 8.3.3, options for millimetre-wave photodetectors are outlined. This work had culminated in a demonstration of video transmission over a 40 GHz radio-fibre link, and this is described in section 8.3.4.

8.3.2 Sources

As mentioned previously, the highest frequency at which laser diodes can be modulated is in the region of 20 GHz. To realize a millimetre-wave radio-fibre system operating in the 40-60 GHz regime using present technology, it has been necessary to transmit the radio information over the optical fibre at a lower frequency, and then up-convert to the millimetre-wave radio frequency at the fibre-to-radio interface. If it were possible to transmit over the fibre directly at millimetre-wave frequencies then the fibre-to-radio interface would be simplified considerably, and the up-conversion equipment could be located with the optical transmitter, preferably in an exchange environment. An alternative approach would be to use an external optical modulator in combination with an unmodulated laser. However, modulators that operate at millimetre-wave frequencies are not yet available, although significant developments can be expected in the future.

This section describes techniques where the need for electrical up-conversion at either the transmitter or receiver is avoided. This is achieved by making use of the nonlinear properties of the optical link itself to realize the up-conversion process. A transmitter laser (or external modulator) is then driven at a relatively modest frequency (4 GHz in the example below), while high-frequency harmonics (60 GHz) are detected at the optical receiver.

This method of generating millimetre-wave radio signals makes use of a laser designed such that its optical frequency can be modulated by application of a drive signal to one of its terminals [17]. The optical spectrum of a frequency-modulated laser contains lines spaced by the drive frequency, and millimetre-wave signals are generated by the beating between widely spaced sidebands on a photodetector. A pure FM signal has constant intensity, and would not induce any photocurrent at harmonics of the drive frequency.

However, if the light is propagated over dispersive optical fibre, then the relative phasing of the optical sidebands is altered, and the light acquires intensity fluctuations at harmonics of the laser drive frequency.

The process of modifying an optical FM spectrum through dispersion in single-mode fibre can be analysed theoretically, and an expression derived for the intensity modulation depth of each harmonic at the output of the photodetector. The modulation depth M_p, of the pth harmonic is defined as the ratio of the amplitude of the alternating photocurrent at the pth harmonic to the d.c. photocurrent. The resulting expression for the modulation depth is:

$$M_p = |2J_p(2\beta \sin(p\phi)|$$

where J_p is the pth Bessel function of the first kind, β is the FM index of the laser, and ϕ is an angle parameterizing the fibre dispersion given by:

$$\phi = -\frac{\omega^2}{4\pi}\frac{D\lambda^2}{c}z$$

where λ is the free-space wavelength of the laser, c is the speed of light, z is the fibre length, D is the fibre group velocity dispersion parameter (normally quoted in ps/(km nm)), and ω is the angular frequency (rads/s) of the drive signal applied to the laser. The greatest modulation depth that can be obtained for the pth harmonic is therefore equal to twice the greatest value of the pth Bessel function ($p \geq 1$), and occurs when:

$$2\beta \sin(p\phi) = j'_{p,1}$$

where $j'_{p,1}$ is the first zero of the derivative of the pth Bessel function. Thus to obtain the greatest modulation depth of the pth harmonic at the receiver it is necessary to adjust the FM index β to an optimum value, which is itself minimized if the fibre length is chosen such that $p\phi = \pi/2$. The theoretical dependence of the modulation depth of the 10th to 15th harmonics on the FM index is shown in Fig. 8.9(a), where a value of $\phi = 0.027$ has been assumed. It is possible to achieve up to 60% modulation depth for the 10th harmonic, and the corresponding values fall off quite slowly for higher harmonics. Therefore, the technique of FM-IM conversion offers the possibility of providing an efficient up-conversion process.

The arrangement shown in Fig. 8.10 was set up experimentally, using a three-section, BH DFB laser with an FM efficiency of approximately 1 GHz/mA [18] as the source. A 4 GHz drive signal was applied to the centre section, and the output of the laser was launched into 12.5 km of conventional step index fibre. Various high-speed photodetectors (described below) were employed as the optical receiver. A spectrum analyser was used to measure the magnitudes of the photocurrent at harmonics of the drive frequency applied to the laser, and one of the photodetectors was calibrated so that the modulation depth of each harmonic could be calculated. The values of modulation depth measured by this technique are shown in Fig. 8.9(b) as a function of the drive current applied to the laser. While the qualitative agreement with the theoretical plot is good, the actual modulation depths

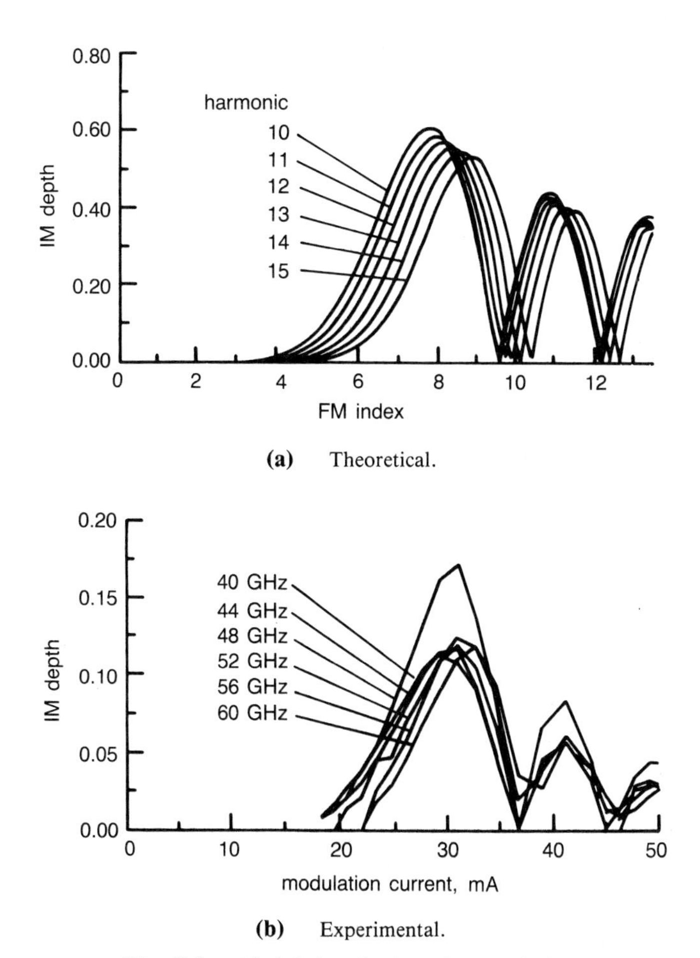

(a) Theoretical.

(b) Experimental.

Fig. 8.9 Modulation depth against FM index.

measured were in the region of 13%, which compares with their theoretical optimum of 60%. This discrepancy has been attributed to the large intensity modulation present at the output of the laser itself, which significantly alters the optical spectrum from that of a pure FM signal.

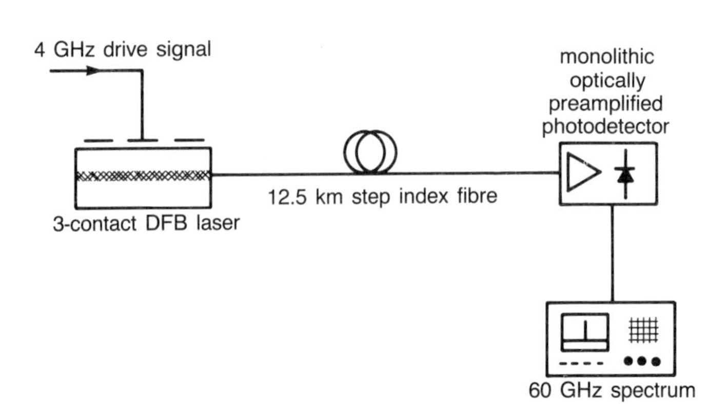

Fig. 8.10 Experimental set-up for modulation depth measurements.

The technique of FM-IM conversion in conjunction with a high-speed photodiode has been shown in these experiments to be an effective method for producing millimetre-wave signals for a fibre-to-radio interface. However, it relies on the availability of a laser capable of wide optical frequency deviation at microwave rates; the peak-to-peak frequency deviation must be at least as great as the millimetre-wave frequency it is required to generate. This has been achieved in our experiments through the use of a three-contact laser, although even with this device only 13% modulation depth was obtained. It would, therefore, be desirable to find an alternative source which could improve the modulation depth, and hence the overall efficiency of the fibre-to-radio conversion process.

A possible alternative to the FM source described above is derived from a sinusoidally phase-modulated signal produced using an external phase modulator. The theory presented above is also valid for phase modulation, since phase and frequency modulation can be considered as equivalent. Another potential method of millimetre-wave generation makes use of the nonlinear transfer characteristic of an electrooptic Mach-Zehnder amplitude modulator, which does not require dispersive fibre, and is likely to be attractive for short links.

Discussion has not yet covered how the radio signal information can be impressed on the millimetre-wave carrier generated at the fibre-to-radio interface. There are various ways in which this could be done. In section 8.3.4 below a video transmission experiment is described in which the frequency applied to the laser is itself modulated with the video baseband signal. Thus an electrical FM signal is used as the input to the FM laser. While this approach works well for single-channel transmission, it is not clear how it would be extended to enable multichannel transmission. The use of an external modulator to effect up-conversion through phase or amplitude modulation may lend itself more readily to multichannel transmission as the laser can then be independently intensity modulated with a composite multichannel signal. Experiments are under way to investigate such possibilities.

8.3.3 Photodetectors

The photodetector is the fundamental radio-to-fibre interface. Requirements will differ according to a particular application, but common to all is the need for a high absolute response at millimetre-wave frequencies. Various types and styles of photodetector can be envisaged for this work which fulfil this basic requirement, and some which go further. The most promising of these have been designed, fabricated and assessed at BT Laboratories, and are discussed below.

The surface-illuminated PIN photodiode that forms the basis of the microwave radio-to-fibre systems outlined in the previous section can be modified for use at millimetre-wave frequencies by reducing the size of the photosensitive area (for smaller capacitance) and also by thinning the absorber layer (for shorter carrier transit time). The latter modification has the unfortunate side-effect of reducing the sensitivity since less of the incident light is absorbed in a thinner layer. Consequently, for this type of geometry there is a trade-off between speed and sensitivity. Changing the geometry of the device to edge-entry removes this trade-off because the direction of light entry is decoupled from the thickness axis, i.e. thickness has little effect on sensitivity. For this geometry, the sensitivity is limited by the coupling efficiency of the optical input into the edge of the device. For relatively low frequencies, this limitation makes edge-entry photodetectors uncompetitive in terms of sensitivity. However, as the frequency of operation goes up, edge-entry sensitivity remains fixed whereas surface-entry sensitivity is reduced. Therefore at some break frequency, the edge-entry photodetector is the more sensitive option. From the results obtained with these two types of device experimentally, and also from literature surveys, a break frequency of around 40-50 GHz is suggested.

In the edge-entry photodetector fabricated at BT Laboratories, light is coupled into an optical waveguide that includes an absorbing layer [19]. Parameters such as the length and width of this waveguide and the absorber layer thickness have been carefully optimized to provide the right balance between the various trade-offs involved in the design. Frequency response was the primary design consideration, but efforts were made to maximize the external quantum efficiency for a given bandwidth target. Coupling efficiency was enhanced by the insertion of a thick waveguide layer into the structure, analogous to the separate confinement heterostructure (SCH) used successfully in semiconductor lasers for increasing the optical mode size. Bondpad capacitance was reduced by placing the bondpad on a thick dielectric layer, which ensures a total chip capacitance low enough to provide an adequate frequency response when combined with packaging parasitics.

Packaging is of critical importance for high-frequency operation. The design of the photodetector chip cannot be considered in isolation because of the close interaction of chip parameters with package parasitics. In particular, series inductance must be optimized for a given value of chip capacitance, which in practice involves the variation of length and type of bondwire. An important requirement of any component is user-friendliness, and in this respect, inclusion of a fibre pigtail is of great benefit because precision optical alignment is not necessary. This is especially important for edge-entry devices due to the more stringent alignment tolerances involved. Furthermore, an internal bias network built into the package avoids the need for an external bias tee. Packaged edge-entry photodetectors such as this have been constructed, and exhibit high-frequency operation with moderate sensitivity. Figure 8.11 shows the frequency response of one such photodetector, measured at a wavelength of 1556 nm using an optical heterodyne technique. The 3 dB bandwidth of this device is 40 GHz. The response beyond 50 GHz is uncalibrated, but the device is obviously useful at least to frequencies up to 60 GHz. At d.c. the responsivity (ratio of photocurrent to optical power) for this device was 0.28 A/W, resulting in a responsivity at 40 GHz of 0.20 A/W. Although higher values of bandwidth and responsivity have been obtained with free-space mounted chips (50 GHz bandwidth and 0.5 A/W responsivity at d.c.), the advantages of having a ready-to-use component with no optical alignment or bias tee requirements are readily apparent.

The major drawback of the photodetectors described so far is responsivity. Although the values given above for this parameter are state-of-the-art, higher values are certainly desirable. Optical preamplification is recognized as an efficient means of improving the signal-to-noise ratio in many systems. A monolithic photodiode and optical preamplifier can be thought of as a photodetector with a high effective responsivity, where the increased responsivity comes from the optical gain in the preamplifier. Such a component has been designed and fabricated at BT Laboratories [20]. With this device, high-frequency operation has been combined with high responsivity (best results obtained for free-space optical input are a bandwidth of 33 GHz and a responsivity of 89 A/W). The structure is based upon the ridge waveguide laser, with the photodiode absorbing layer being an unpumped extension of the preamplifier gain layer in an edge-entry configuration. This arrangement ensures good optical coupling and low

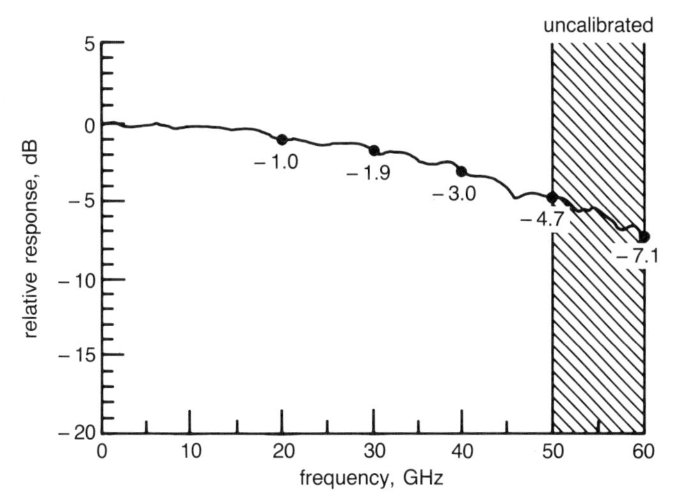

Fig. 8.11 Frequency response of edge-entry photodiode with fibre pigtail and internal bias tee.

optical feedback at the interface between preamplifier and photodiode. Figure 8.12(a) shows the longitudinal cross-section of this device in schematic form. The principal design feature is the use of selective epitaxy of p-type ridges to produce localized-type conversion in underlying planar n-type material using a patterned dielectric mask. Figure 8.12(b) shows the topology of the device immediately after this stage, also in schematic form. This design technique creates a dopant distribution that ensures low-junction capacitance (for high-frequency operation) and good electrical isolation. As in the edge-entry photodiode described above, the bondpad sits on a thick dielectric layer to give a low capacitance. Further design features have been incorporated that give this device useful characteristics such as travelling wave amplification, polarization insensitivity, wide optical bandwidth, and high saturated output power. The major drawback, however, is noise, which is due to the spontaneous emission from the optical amplifier. In some low-noise applications, the gain must be lowered by reducing the drive current to the preamplifier section, to bring the noise down to acceptable levels (since noise is proportional to gain).

These devices have also been packaged with fibre pigtails, but so far the performance has not matched that of free-space packaging. However, responsivity values of over 1 A/W at 40 GHz have been achieved in pigtailed devices, which is an order of magnitude higher than the values achieved for discrete photodetectors.

The heterojunction phototransistor (HPT) is another example of a photodetector with gain. In this case, the gain is electrical, and takes place through transistor action. This device is simply a heterojunction bipolar transistor (HBT) made with a window area in the emitter metallization for

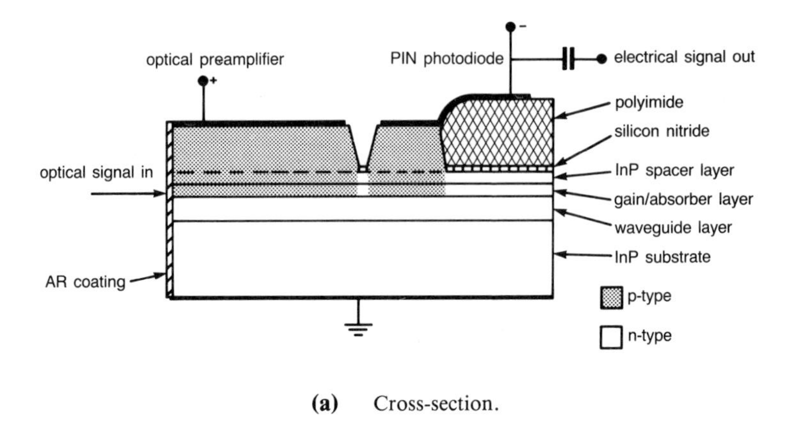

(a) Cross-section.

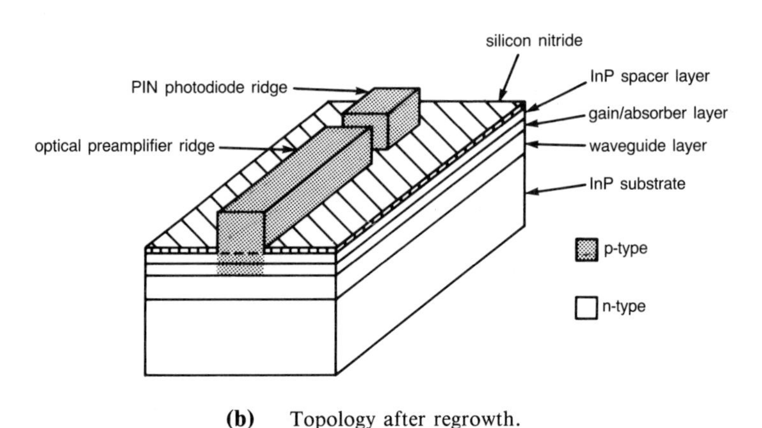

(b) Topology after regrowth.

Fig. 8.12 Monolithic optically preamplified photodiode.

optical input. The InGaAs layer used in the base-collector of InP-based devices ensures strong absorption at the desired wavelength of around 1550 nm. Although this device can be used simply as a photodetector, it is more efficient to use it for applications where advantage can be taken of its dual role as photodetector and active electronic circuit element. An example of this approach is a remotely optical injection-locked oscillator, which could become important in future radio-fibre systems because it allows high-power and low-cost oscillators to be designed without being constrained by purity requirements. In this configuration a locking signal is transmitted over an optical link to transfer the purity of a reference source located centrally in a benign environment to the remote oscillator. The heterojunction phototransistor is an ideal device for this application because it allows good optical coupling from single-mode fibre, and because it can be made with high gain and good frequency response. Prototype devices have been made at BT Laboratories which exhibit high optical responsivity (27 A/W at d.c.) and high maximum oscillation frequency (>20 GHz), even for sizes and structures not deliberately optimized for this application. In the near future, oscillators will be fabricated with optimized phototransistor design so that the possibilities and limits of optical injection-locking can be tested.

Enhancement to the absolute responsivity of a photodetector at frequencies of interest is an obvious direction for future work, but not at the expense of power consumption or complexity. The idea of a lightweight radio transmitter requiring reduced d.c. electrical power is an attractive proposition, and much of the future work on photodetectors is directed this way. Direct mounting of a photodiode in an electrical waveguide will be a step in this direction, because it allows the effective integration of this device with a horn antenna. In addition, packaging parasitics will be virtually removed, enabling very high-frequency performance. The aim will be to radiate the output of the photodiode directly into the waveguide, i.e. no electrical connection to the chip will be required. An on-chip d.c. short circuit must be provided to allow a recombination path for d.c. photocurrent. Taking this concept to its logical conclusion will see the monolithic integration of a photodetector and antenna.

8.3.4 Video transmission experiment

A video transmission experiment was assembled using many of the components and techniques described earlier as a demonstration of feasibility [21]. This experiment involved the transmission of a video signal superimposed on a millimetre-wave carrier over a length of optical fibre and a radio path. The layout of this system is shown in Fig. 8.13. The optical source used for this demonstration was the FM sideband laser described earlier, which was driven by a microwave oscillator at a frequency of 4 GHz. Superimposed on to this was a video signal, which was applied to the FM input of the oscillator. The isolated output of the laser, which consisted of optical FM sidebands generated by the composite signal, was transmitted along 12.5 km of standard single-mode fibre to the photodetector. The dispersion arising from this fibre was sufficient to perturb the phase relationships between individual sidebands so that the output of the photodetector consisted of harmonics of the drive signal with high power levels. This effective multiplication of the composite drive signal meant that

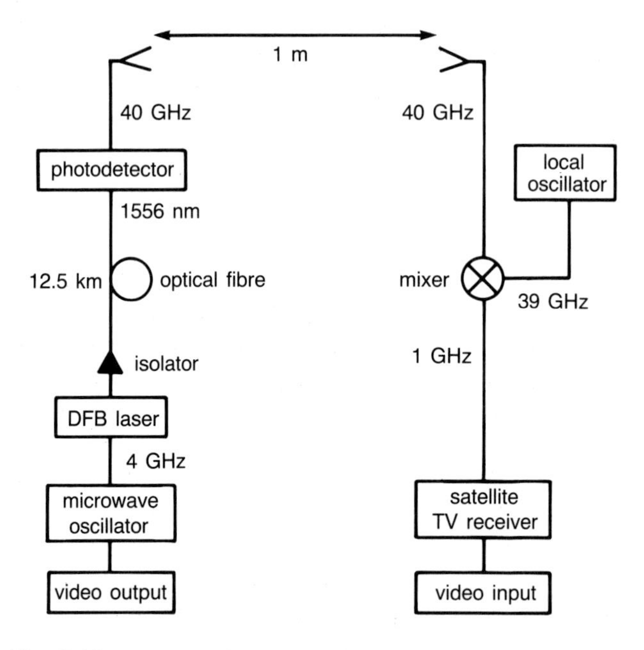

Fig. 8.13 Layout of 40 GHz radio-fibre transmission experiment.

the FM deviation of the input video signal had to be divided by the number of the selected harmonic to compensate for the deviation multiplication.

A variety of high-frequency photodetectors has been used in this demonstration. Package configurations included an edge-entry photodiode integrated with a monolithic millimetre-wave integrated circuit (MMWIC) amplifier. Versions having only a monolithic optically preamplified photodiode in the package have also been used. The tenth harmonic (40 GHz) was chosen for these systems since it was the highest frequency for which many of the radio components were available. The radio path was 1 m long, which was considered to be adequate for a proof-of-principle experiment. The radio receiver was tuned to the particular harmonic number of interest to produce an intermediate frequency suitable for a satellite TV receiver, the output of which was displayed on a video monitor. Good-quality video transmission was observed, with a comfortable power margin, even though the radio receiver was not optimized for low-noise operation. Although the range was only 1 m, future enhancements to the photodetector (higher responsivity) and optical source (greater modulation depth, higher power) are expected to provide a radio range of at least several hundred metres.

The harmonic generation technique used in this demonstration is not thought to be suited to multichannel operation because of intermixing between the channels. Other techniques within the scope of this work, such as up-conversion using optical modulators, are under investigation with the aim of demonstrating multichannel transmission.

8.4 CONCLUSIONS

The potential advantages of using SCM optical links to feed radio transmitters/receivers are becoming clear to both manufacturers and operators, and such systems are now being installed. The work described in this chapter addresses some of the limitations of SCM and outlines several techniques which can be used to overcome them in future systems. In summary:

- SCM is a very flexible technique but is limited in frequency by fundamental device physics — the techniques described here can help to overcome these limits;
- very simple, cheap and reliable radio transmitters can be realized via remote optical control of highly efficient oscillators;
- d.c. power requirements are often excessive for radio nodes using SCM — the novel techniques described earlier can be employed to reduce the power consumption of some of the key components.

The fresh approach, described in this chapter, to the implementation of future radio-to-fibre systems has already paid dividends in terms of the components and techniques that have been developed so far, and which have been covered in previous sections. Although the emphasis of this work has been at the higher frequencies, in many instances the components and techniques are equally suited to applications in the 1-2 GHz range. To a large extent, the techniques described here relate to one-way systems, i.e. radio transmit only. The return path can be considered as a duplicate of the send path in some cases, whereas in others a simple down-conversion and baseband/IF return path would be more appropriate. However, much of the future research effort in the radio-to-fibre area will be aimed at identifying and demonstrating new components and techniques specifically for the return path.

The potential advantages of the new techniques in terms of cost, reliability and flexibility of service provision are clear, but it is vital that today's system and network architectures do not act as constraints when evaluating the possibilities that this work may open up in the future.

REFERENCES

1. Lau K and Yariv A: 'Ultra-high speed lasers', J Quantum Electron, 21, pp 121-138 (1985).

2. Darcie T E, Dixon M E, Kasper B L and Burrus C A: 'Lightwave system using microwave subcarrier multiplexing', Electron Lett, 22, pp 774-775 (1986).

3. Way W, Wolff R S and Krain M: 'A 1.3 μm 35 km fibre optic microwave transmission system for satellite earth stations', J Lightwave Technol, LT5, pp 1325-1332 (1987).

4. Bowers J E, Chipaloski A C, Boodaghians S and Carlin J W: 'Long distance fibre-optic transmission of C band microwave signals to and from a satellite antenna', J Lightwave Technol, 5, pp 1733-1741 (1987).

5. Olshansky R, Lanzisera V, Su C B, Powazinik W and Lauer R B: 'Frequency response of an InGaAsP vapor phase regrown buried heterostructure laser with 18 GHz bandwidth', J Appl Phys Lett, 49, pp 128-130 (1986).

6. Bowers J E, Hemenway B R, Gnauck A H and Wilt D P: 'High-speed constricted mesa lasers', J Quantum Electron, 22, pp 833-844 (1986).

7. Olshansky R, Lanzisera V A and Hill P M: 'Subcarrier multiplexed lightwave systems for broadband distribution', J Lightwave Technol, 7, pp 1329-1341 (1989).

8. Way W I: 'Subcarrier multiplexed lightwave system design considerations for subscriber loop applications', J Lightwave Technol, 7, pp 1806-1817 (1989).

9. Darcie T E: 'Subcarrier multiplexing for lightwave networks and video distribution systems', J Selected Areas Commun, 8, pp 1240-1258 (1990).

10. Lipson J, Chainulu U, Huang S, Roxlo C B, Flynn E J, Nitzsche P M, McGrath C J, Fenerson G L and Scaefer M: 'High fidelity lightwave transmission of multiple am-vsb ntsc signals', Trans Microwave Theory Technol, 38, pp 483-492 (1990).

11. Cox C, Betts G E and Johnson L M: 'An analytic and experimental comparison of direct and external modulation in analogue fibre optic links', Trans Microwave Theory Tech, 38, pp 501-508 (1990).

12. Lealman I F, Bagley M, Cooper D C, Fletcher N, Harlow M, Perrin S D, Walling R H and Westbrook L D: 'Wide bandwidth multiple quantum well 1.55 μm lasers', Electron Lett, 27, pp 1191-1193 (1991).

13. Tatham M C, Lealman I F, Seltzer C P, Westbrook L D and Cooper D M: 'Resonance frequency, damping and differential gain in 1.5 μm multiple quantum well lasers', J Quantum Electron, 28, pp 408-414 (1992).

14. Bowers J E and Burrus C A: 'Ultra wideband long wavelength p-i-n photodetectors', J Lightwave Technol, 5, pp 1339-1350 (1987).

15. Wake D, Walling R H and Henning I D: 'Planar junction, top-illuminated GaAs/InP pin photodiode with bandwidth of 25 GHz', Electron Lett, 25 , pp 967-968 (1989).

16. Demonstration at 9th International Conference on Digital Satellite Communications, Copenhagen (1992).

17. Walker N G, Wake D and Smith I C: 'Efficient millimetre-wave signal generation through FM-IM conversion in dispersive optical fibre links', Electron Lett, 28 , pp 2027-2028 (1992).

18. Sherlock G, Wickes H J, Hunter C A and Walker N G: 'High speed, high efficiency, tunable DFB lasers for high density WDM applications', ECOC'92, paper Tu P1.1 (1992).

19. Wake D, Spooner T P, Perrin S D and Henning I D: '50 GHz InGaAs edge-coupled pin photodetector', Electron Lett, 27 , pp 1073-1074 (1991).

20. Wake D: 'A 1550 nm millimetre-wave photodetector with a bandwidth-efficiency product of 2.4 THz', J Lightwave Technol, 10 , pp 908-912 (1992).

21. Wake D, Smith I C, Walker N G, Henning I D and Carver R D: 'Video transmission over a 40 GHz radio-fibre link', Electron Lett, 28 , pp 2024-2025 (1992).

9

NOVEL COMPONENTS FOR OPTICAL SWITCHING

M J Adams, P E Barnsley, J D Burton, D A O Davies,
P J Fiddyment, M A Fisher, D A Mace, P S Mudhar,
M J Robertson, G Sherlock, J Singh and H J Wickes

9.1 INTRODUCTION

Efficient utilization of high-capacity optical networks using time-division or wavelength-division multiplexing (TDM or WDM) will be achieved by the flexible use of space, time and wavelength switching. This chapter will focus on semiconductor components to perform these tasks. The devices are all based on laser amplifier technology, but here the nonlinear behaviour is utilized for performing complex functions in addition to straightforward linear amplification. It should be noted that the term 'laser amplifier' is used throughout the present chapter with reference to **diode** laser amplifiers; erbium-doped fibre amplifiers, which have significant advantages for linear amplification, are not considered here, since this chapter is exclusively concerned with the exploitation of the nonlinear properties of semiconductor amplifiers. The specific functions to be addressed are wavelength conversion, space and time switching, and clock extraction. In each case the functional behaviour will be discussed with brief details of the device and its relevant performance, but with the emphasis firmly on the applications potential. The concluding section will look at the possible future developments in this area and survey some of the performance targets which could be set.

9.2 WAVELENGTH CONVERSION

The demand for wavelength conversion arises in wavelength-routed networks [1]; a simplified illustration of such a network is shown in Fig. 9.1. At a node, signals are either dropped to a terminal or routed through to another node. The addition of switching at the node gives extra flexibility for re-routeing to accommodate varying traffic or for network protection or maintenance.

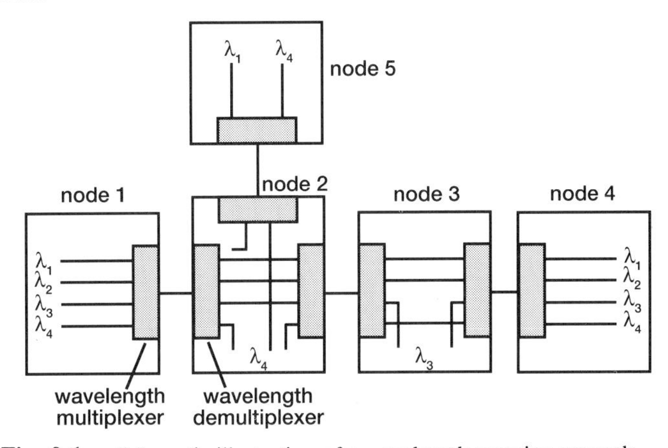

Fig. 9.1 Schematic illustration of a wavelength-routeing network.

Figure 9.2 shows the interconnection field for a node with some switching capacity. The switching might take the form of a space switch, in which case each wavelength on each input fibre could be switched to any output fibre. Alternatively, wavelength conversion might be required where, for example, networks operating at incompatible wavelengths might be interconnected. In addition, WDM systems which adopt wavelength reuse (in order to reduce the total number of wavelengths) can maintain flexibility by the use of wavelength conversion at gateways between network regions which are restricted to specific wavelength sets (see Chapter 3).

Many schemes for wavelength translation have been demonstrated, based on a variety of physical processes, including four-wave mixing and other nonlinear effects [2-4]. However, if a wide conversion range with good efficiency and high speed is required, then the most promising candidate appears to be the use of gain and/or absorption saturation in semiconductor laser amplifiers [5, 6]. A wavelength conversion device of the latter type can be implemented with a two-section laser amplifier, as schematically illustrated in Fig. 9.3. In this device the conventional stripe contact has been

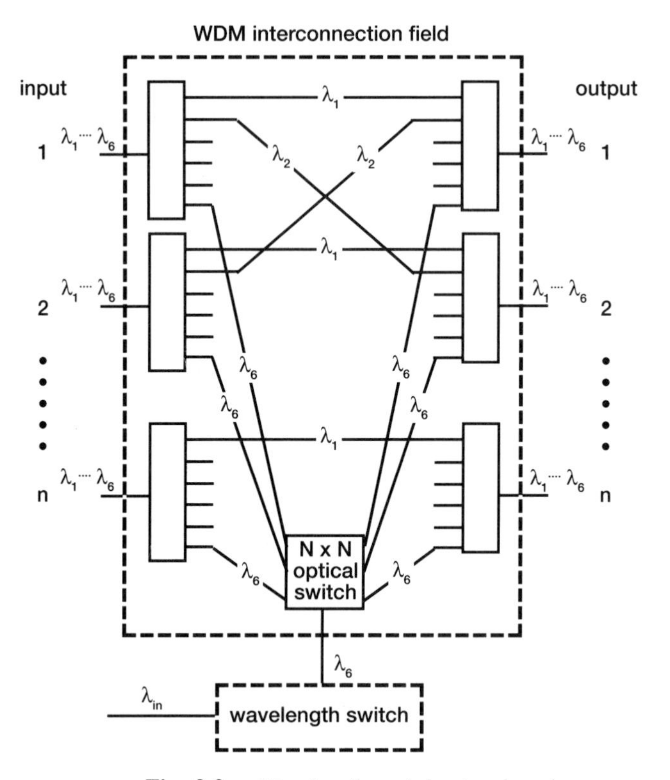

Fig. 9.2 Wavelength-routed network node.

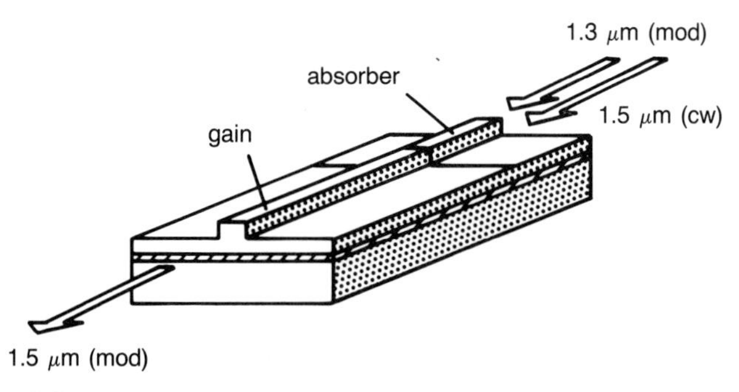

Fig. 9.3 The two-section laser amplifier used for wavelength conversion.

split into two sections, arranged such that one section is driven with sufficient current to provide amplification of an optical signal while the other section acts as an absorber. Two optical inputs are incident on the absorber, the first being the modulated signal at a wavelength of 1.3 μm, and the second a constant intensity at 1.5 μm. The device function is to convert the data on the incident signal at 1.3 μm to the signal at 1.5 μm using the modulation of the electron density in the absorbing section of the device.

The operating principle of the two-section wavelength conversion device is shown in more detail in Fig. 9.4, where the theoretical gain per unit length (in the amplifying section) and the loss per unit length (in the absorber section) are plotted versus wavelength. The absorption of the input signal at 1.3 μm is sufficient to saturate the absorber and thus alter the loss spectrum over the entire wavelength range. Thus the loss for '1's in the modulated signal is lower than that for '0's, and this difference appears as a modulation on the transmission of the signal at 1.5 μm. Hence the modulated data is transferred from 1.3 to 1.5 μm. The resultant contrast ratio between '1's and '0's will vary exponentially with the product of the device length and the loss difference (for the simple travelling-wave amplifier). Significant values of contrast ratio (of order 10:1) are achievable, as will be seen later. Since the device structure is identical to standard laser sources, there is no new technology development required and thus high yields and low fabrication costs are ensured.

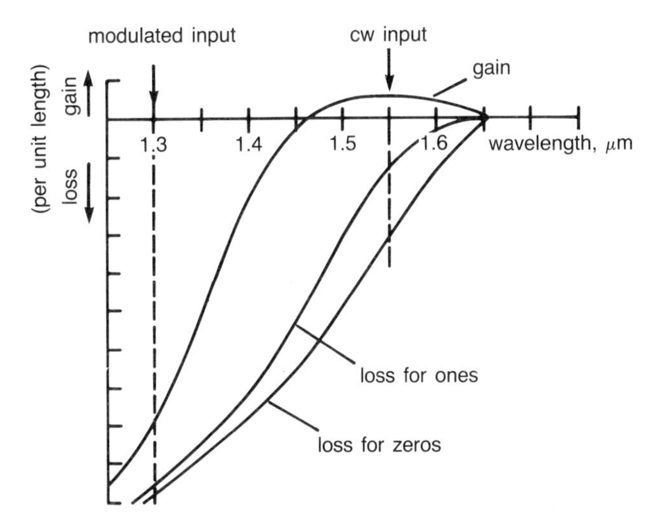

Fig. 9.4 Operating principle of the wavelength-conversion device, showing spectral variation of the gain per unit length (in the gain section) and loss per unit length (in the absorber).

The device structure is based on a buried heterostructure laser grown entirely by metal-organic vapour phase epitaxy (MOVPE) [7], as illustrated in cross-section in Fig. 9.5. All semiconductor lasers contain a region, called the active region, where electrical carriers (electrons and holes) recombine to create photons. There are two further requirements:

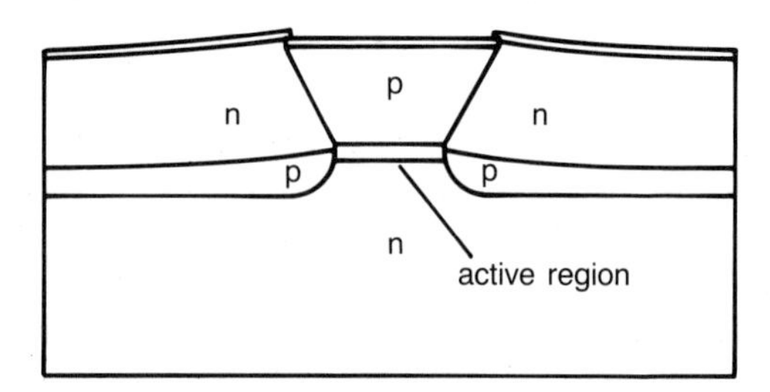

Fig. 9.5 Schematic cross-section of the buried heterostructure laser.

- an optical waveguide to confine the photons to the active region;
- a means to channel the current through a narrow region of active material to improve the electrical-to-optical conversion efficiency.

The buried heterostructure is a good example of an efficient laser structure where a narrow active stripe is formed from a continuous active layer by etching and subsequent overgrowth of material with electrical blocking characteristics. The overgrown material has a lower refractive index than the active layer, and in this way both the above requirements are satisfied. This type of laser structure is commonly used by manufacturers world-wide.

For the two-section wavelength-conversion device, the structure is a little more complex than that described above, with the difference being in the final stages of fabrication. Following p metallization with Ti/Au, the devices are patterned and the metal, SiO_2 and p-ternary layers are etched to provide isolation of $\sim 400\ \Omega$ between the two contacts for an etched width of 18 μm. Etching through the p-InP layer was found to improve isolation further (~ 19 kΩ), but results in spectral patterning from the extra reflection at the etched interface, an undesirable feature for the operation of the device. Threshold current for a 500 μm long device with a split contact was 15 mA with both contacts shorted together. A significant enhancement to the functionality of the device has been achieved with little deviation in design from a standard component. This makes it an attractive manufacturable solution.

This simple wavelength-conversion device has been demonstrated at bit rates up to 400 Mbit/s with a maximum unfiltered contrast ratio of 3:1 and a minimum power for switching at 1.3 μm of $\sim 60\ \mu$W [6]. The device has been used to switch 155 Mbit/s data from 1.3 μm into a 1.55 μm wavelength-division multiplexed (WDM) network [8]. A contrast ratio of 10:1 in the device output was achieved, using optical filtering of the spontaneous emission, with 30 μW power and a 1 dB power penalty at a bit error rate

(BER) of 10^{-9}. When transmitted over 63 km of step-index fibre, measurements on the full system showed a receiver sensitivity penalty of only 2 dB for 10^{-9} BER.

In summary, the two-section laser amplifier offers a convenient, reliable, low-cost means of wavelength conversion using established technology. It should be able to convert data to wavelengths spanning the 30 nm spectral range covered by the erbium-doped fibre amplifier, and requires only modest powers (1-10 μW) in this wavelength range. Future developments will utilize multiple quantum well (MQW) devices with the possibilities of enhanced performance and higher speed. In addition, similar devices have been demonstrated as tuneable wavelength filters [9, 10], thus offering additional functionality from the same technology.

9.3 SPATIAL SWITCHING

Spatial switching will be needed in a wide range of network applications as a consequence of the increasing requirement for optical systems which are transparent to bit rate. Optical switching in a transparent optical network (TON) is desirable for a reduction of electronic switching and improved maintenance, upgradability and management (see Chapter 3). One example is an $N \times N$ space switch for re-routeing and reconfiguration at a node in a WDM network as illustrated earlier in Fig. 9.2. A second example is that of a network protection switch to enable signal data to be re-routed in the event of transmission path failure or during periods of maintenance. Further ahead lies the possibility of large matrix switches using a variety of possible architectures [11].

The basic building block for guided-wave matrix switching is a 2×2 or 1×2 space switch. The requirements on such a switch for use in a TON are that it should possess a wide spectral bandwidth (≥ 40 nm), low polarization sensitivity (< 1 dB), low crosstalk (< -20 dB), fast switching speed (for route restoration), and very low-loss (ideally zero) including coupling to fibres. This last requirement can really only be satisfied by the use of laser amplifiers within the crosspoint switch, either by hybrid or monolithic integration. The hybrid approach involves the use either of passive splitters with amplifiers as gates [12], or of lossy switches with amplifiers to compensate for the losses, and quite large switch arrays (up to 128×128 [13]) have been constructed by these means.

A monolithically integrated switch is one where the splitter/combiner and the switch are integrated into one small, stable unit [14]; two strategies for constructing such a switch will be discussed here. In the first the input optical signal is split and routed in passive waveguides, and gated under electrical control in amplifiers which are integrated with the passive guides [15]. Such a switch is termed an amplifier gate switch matrix. In the second approach the switching is accomplished by varying the currents in a twin-guide amplifier (a directional coupler with gain) [16], with the additional option of gating in the flared output guides [17]. Both of these approaches are capable of lossless switching with crosstalk below −30 dB over a bandwidth of 40 nm.

Figure 9.6 shows a plan view of a 2×2 amplifier gate switch matrix [18] which uses novel passive waveguide splitters and combiners together with total internal reflection (TIR) mirrors. These components, which are shown in more detail in Fig. 9.7, have the effect of avoiding large bending radii and thus ensuring a compact optical circuit. The splitter incorporates a taper

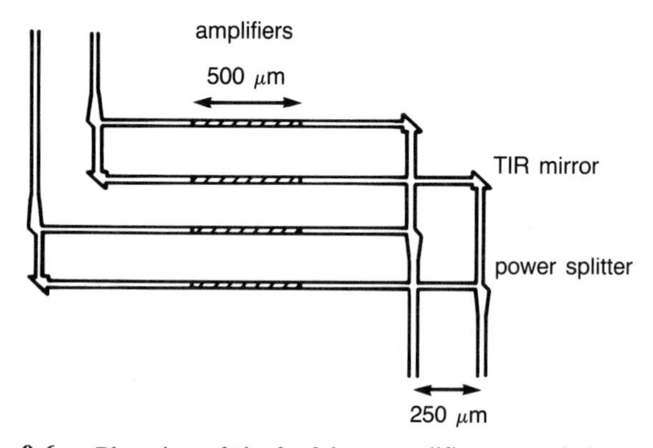

Fig. 9.6 Plan view of the 2×2 laser amplifier gate switch matrix.

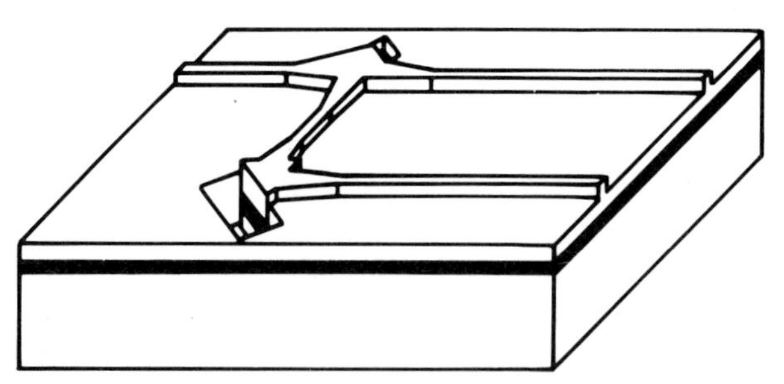

Fig. 9.7 Schematic diagram of the power splitter and total internal reflection mirror used in the gate switch matrix.

to expand the beam, followed by a 45° TIR mirror which reflects half the light into a perpendicular arm, with the other half transmitted, undeflected. The laser amplifiers are monolithically integrated with the passive waveguides using the structure shown in Fig. 9.8. The structure incorporates a 0.4 μm-

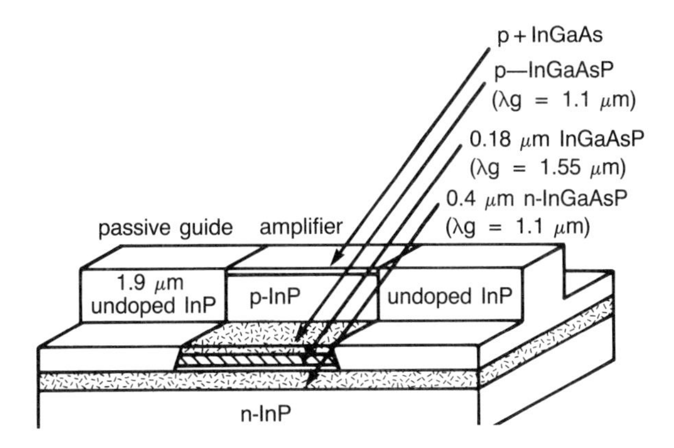

Fig. 9.8 Schematic diagram of the layer structure for the gate switch matrix, showing the active-passive interface.

thick InGaAsP guide layer, the band gap of which is substantially larger than that of the active layer in order to allow low-loss transmission (a 'passive' waveguide). The passive regions are formed by etching away the layers above the guide layer, and then selectively overgrowing with undoped InP.

The 2×2 switch shown in Fig. 9.6 has been characterized using antireflection coatings and lensed fibres for input and output coupling. To date a fibre-to-fibre gain of 2 dB at an injection current of 200 mA has been achieved for all four paths for TM input polarization. The fibre coupling losses amount to 6-9 dB, giving a chip gain of around 10 dB. The measured gain for TE polarization was 3 dB lower and this polarization sensitivity could be reduced with minor changes to the layer structure. The optical bandwidth is typically 50-60 nm centred around 1.55 μm, with less than 0.5 dB gain ripple at 200 mA current. The crosstalk between channels is less than -45 dB, with an on-off extinction ratio in excess of 45 dB. Currently, work is concentrated on reducing the level of drive current required.

The alternative approach, using twin-guide amplifiers, has been used to construct the 1×4 switch [19] whose layout is illustrated in Fig. 9.9. The device uses 3 μm-wide active ridge waveguides throughout, and the shaded areas in the figure show the current injection contact pads. There is a total of three twin-guide amplifiers, connected by active curved waveguides, and these two types of components are shown in more detail in Fig. 9.10. In this case the twin-guide amplifier is simply used as a power splitter with a single current contact, in contrast to earlier versions [16, 17] where separate current contacts were used for switching. For the 1×4 switch of Fig. 9.9, the routeing is achieved by switching the connecting waveguide amplifier sections between

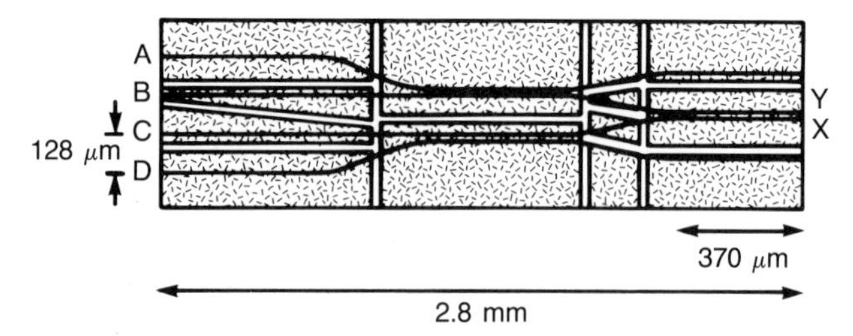

Fig. 9.9 Plan view of the 1 × 4 switch based on twin-guide laser amplifiers.

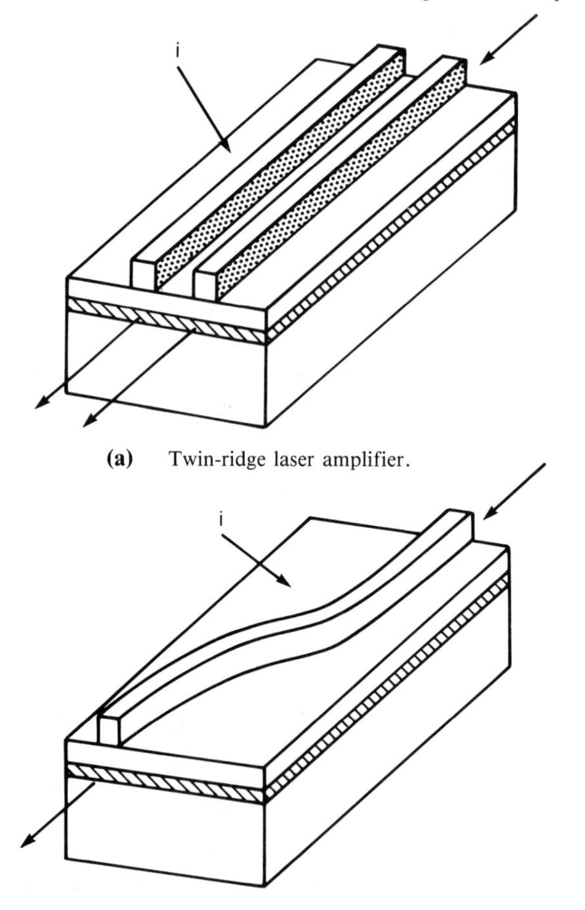

(a) Twin-ridge laser amplifier.

(b) Curved active ridge waveguide.

Fig. 9.10 Components used in the 1 × 4 switch.

loss and gain, as for the amplifier gate switch matrix. Two different etch depths were employed, a relatively deep one for the interconnecting guides (in order to ensure sufficient guiding for the curves), and a somewhat shallower one between the guides in the twin-guide sections (in order to allow good coupling).

The 1×4 switch shown in Fig. 9.9 has been characterized using anti-reflection coatings and lensed fibres for input and output coupling. Switching in the device is reciprocal and takes place between either port X or port Y, and ports A, B, C and D. For a particular path to transmit, all the amplifying sections along its length must be pumped, and in practice it is envisaged that only the four pads next to ports A, B, C and D would be used as control gates. Lossless fibre-to-fibre switching has been demonstrated for all four paths, and each path can be controlled by a single current of less than 90 mA for TE input polarization. The broadband spontaneous emission noise outputs from ports A-D were less than -10 dBm, and the optical 3 dB bandwidth was measured as 20 nm.

While the results discussed above for the 1×4 switch are extremely encouraging, it must be emphasized that they were obtained only with TE polarization. The performance for TM polarized light is considerably worse, with a detectable signal only found in the path C-X, where the peak fibre-to-fibre gain was 6 dB, compared with 18 dB for TE. This implies that paths which cross the couplers incur a higher loss for TM than for TE, which is expected to be the case since the coupling length for TM in the twin-guide amplifier is considerably longer than that for TE. It follows that it would be difficult to optimize the device for low polarization sensitivity, since a compromise would then have to be made for the lengths of coupling region, which would imply reduced performance for the device as a whole. This polarization problem, together with the difficulty of scaling the device to large array sizes, militates against the use of the twin-guide amplifier technology for switches larger than 1×4. By contrast, it is likely that the amplifier gate switch matrix approach discussed previously (Figs. 9.6-9.8) could be scaled to much larger array sizes, without any significant polarization sensitivity. Indeed, variations of this generic approach have already been reported in versions of 1×16 [20] and 4×4 [21]. Simple 1×2 and 2×2 switches of this type were used in a RACE optical switching systems demonstrator at an exhibition associated with the European Conference on Optical Communications held in Berlin in September 1992. Some simple comparisons between the amplifier gate switch matrix and twin-guide amplifier approaches for a 2×2 switch are given in Table 9.1.

Table 9.1 Comparison between gate switch matrix and twin-guide amplifier approaches for a 2×2 spatial switch.

Parameter	*Gate switch matrix*	*Twin-guide amplifier*
Fibre-to-fibre gain	2 dB	−1 dB
Crosstalk	< -45 dB	< -28 dB
Polarization sensitivity	3 dB	2 dB (but crosstalk degraded to −8 dB)
3 dB bandwidth	50 nm	30 nm

9.4 TIME SWITCHING

The potential requirement for Tbit/s capacities in future telecommunications networks has led to the concept of transparent optical networks which are insensitive to the bit rate. Such networks will use all-optical processing at the nodes in order to avoid electronic bottlenecks. In order to synchronize the optical processing elements to the incoming line traffic, there is a need for non-invasive all-optical clock recovery. A two-section laser amplifier can be used (with single wavelength input) both as an all-optical switch, i.e. a power-dependent gate, and (under different bias conditions) as a self-pulsing laser. In the latter case the pulsation frequency can be locked to an incoming data stream and used for clock extraction. Thus the two-section amplifier discussed earlier (Fig. 9.3) in the context of wavelength translation has two other applications in the time-switching domain.

The operating principle of the time switch is illustrated in Fig. 9.11, where the gain per unit length (in the amplifying section) and the loss per unit length (in the absorber section) are plotted versus wavelength. In this case, when the input signal is low the device is in a low transmittance state, but, when the input signal reaches a critical power, then the absorber saturates and the transmittance switches to a high value. This all-optical switching can sometimes exhibit hysteresis in the plot of output versus input signal. A similar bistable behaviour can be produced in the plot of light output versus current, and this was the original absorptive bistability which initiated interest in the two-section laser nearly three decades ago [22].

In order to characterize the time switching, a sinusoidal modulation is applied to the input signal, and the output waveform monitored, as illustrated in Fig. 9.12. The characteristic overshoot spike and rectified pulse shape are typical of the switching response, and their variation with input power,

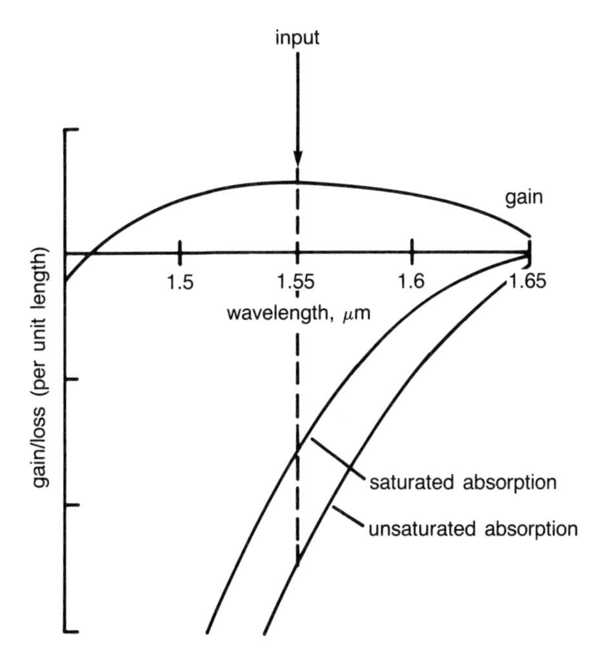

Fig. 9.11 Operating principle of the time-switching device, showing spectral variation of the gain per unit length (in the gain section) and loss per unit length (in the absorber).

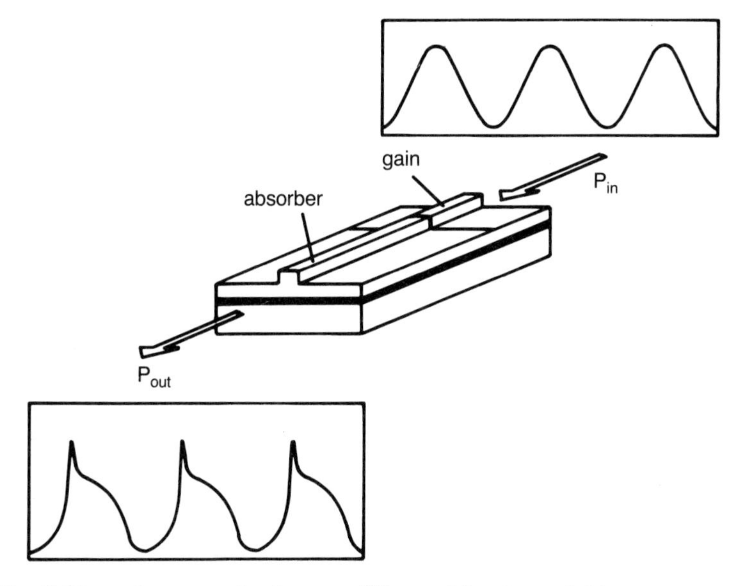

Fig. 9.12 The two-section laser amplifier used for time switching.

repetition frequency, and wavelength tuning give valuable information about the internal device dynamics [23, 24]. Such a response for a power-dependent gate has potential applications in pulse-shaping for optical regeneration. Nonlinear switching has been demonstrated for input powers as low as −51 dBm with minimum repetition time of 700 ps and a maximum gain of 26 dB [25]. Similar devices have been used as optical memories in time-division switches [26], for optical drop and insert functions in access nodes [27], and as 'AND' gates in optical regenerators [28, 29]. They are expected to find further applications in all-optical header recognition and as optical gates for packet switching in asynchronous transfer mode (ATM).

The two-section device can also be biased to function as a self-pulsing laser [30], where the absorber is repeatedly saturated by high-power pulses emitted from the gain section. The principle is illustrated in Fig. 9.13, which gives schematic plots of output power, gain per unit length (in the gain section) and loss per unit length (in the absorber) versus time. Between output pulses, the loss recovers from its saturated value while, at the same time, the gain builds up. At the point where the gain exceeds the loss by a sufficient amount, the rapid emission of an intense light pulse occurs. As a consequence, the gain reduces rapidly (since the electron population is consumed), the loss also saturates, and the light emission is quenched. Thereafter the cycle commences again. In order for the pulsation to be periodic, the lifetime of electrons in the absorber must be less than that in the gain section [30]. This can be achieved by selectively doping the absorber section with zinc [31], which

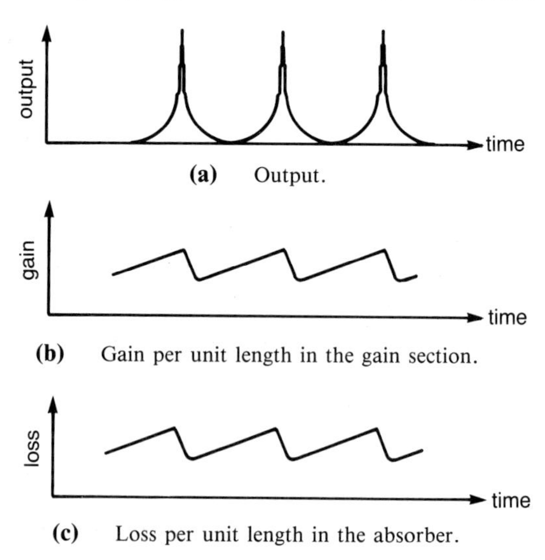

Fig. 9.13 Operating principle of the self-pulsing laser, showing temporal variations.

acts as a centre for nonradiative recombination and thus reduces the electron lifetime. Devices made in this way have been demonstrated to have pulsation frequencies which can be tuned by varying the current in the gain section over ranges from 3 to 5 GHz [31] and from 1 to 3 GHz [32].

Self-pulsating lasers have been used to perform all-optical clock recovery [33] by injecting the optical data into the gain region of the device in order to synchronize the self-pulsations to the data clock rate. Synchronization from 5 Gbit/s return-to-zero (RZ) data [31] occurs for about 10 μW of coupled input power, and is not dependent on the input polarization state. This technique has been used successfully as part of a 20 Gbit/s optical time-division multiplexing (OTDM) transmission system [34] for pseudorandom bit sequences between $2^7 - 1$ and $2^{31} - 1$. BER measurements using the extracted clock signal showed no receiver penalty when compared to measurements using the transmitter clock. Clock recovery from non-return-to-zero (NRZ) data has also been accomplished using self-pulsating lasers for a data rate of 2.5 Gbit/s [32]. Combining the optical-switching function of one two-section device (to generate the missing clock components) with the self-pulsations of a second device has also enabled clock extraction to be performed from NRZ data at 3.2 Gbit/s [35]. This type of clock recovery technique could have a major impact on future gigabit per second telecommunications systems. The applications include all-optical regeneration in nonlinear transmission systems (see Chapter 3).

9.5 FUTURE PROSPECTS

One area of common concern to all the switching components discussed is that of bit rate. The ultimate limit on the speed of switching in laser amplifiers is set by the electron recombination time, and this is usually in the nanosecond range for 1.5 μm devices, thus imposing bit rate limits of the order of gigabits per second. Clearly there is some scope for reducing the recombination time and thus increasing the maximum bit rate by using heavy Zn doping, as discussed above, but techniques such as these are unlikely to achieve improvements by more than a factor of about 10. In order to consider semiconductor switching components at higher bit rates a radical new approach is required, and one possibility is the use of nonlinearities based on intraband relaxation mechanisms whose time constants may be of the order of picoseconds or less. Recently, the size of such mechanisms has been measured for laser amplifiers biased at transparency [36, 37], and it has been shown that the optical power levels required may not be excessive. If the appropriate device developments can be made then this technique might offer a route towards all-optical switching at bit rates approaching 1 Tbit/s at optical power levels available from semiconductor laser sources.

Another way forward for the future development of optical switching technology lies in parallel processing using two-dimensional arrays of laser amplifiers [38]. Recent progress in vertical-cavity surface-emitting laser (VCSEL) technology has been spectacular [39], and further development of the same structures towards long-wavelength operation [40] and monolithic integration with other components [41] are in hand. For the amplifier-based switches discussed here, the surface-emitting geometry offers the potential for two-dimensional arrays, and large matrix switches and multifunction switches, using the basic building blocks described, can be expected to appear. These components will offer solutions to future network problems associated with traffic bottlenecks at the nodes, as well as stimulating novel architectures for optical processing and interconnects.

REFERENCES

1. Hill G R: 'A wavelength routeing approach to optical communication networks', BT Technol J, 6, No 3, pp 24-31 (1988).

2. Großkopf G, Ludwig R and Weber H G: '140 Mbit/s DPSK transmission using an all-optical frequency convertor with a 4000 GHz conversion range', Electron Lett, 24, No 17, pp 1106-1107 (1988).

3. Kawaguchi H, Oe K, Yasaka H, Magari K, Fukuda M and Itaya Y: 'Tunable optical wavelength conversion using a multielectrode distributed-feedback laser diode with a saturable absorber', Electron Lett, 23, No 20, pp 1088-1090 (1987).

4. Inoue K and Takato N: 'Wavelength conversion for FM light using light injection induced frequency shift in DFB-LD', Electron Lett, 25, No 20, pp 1360-1362 (1989).

5. Glance B, Wiesenfeld J M, Koren U, Gnauk A H, Presby H M and Jourdan A: 'High performance optical wavelength shifter', Electron Lett, 28, No 18 pp 1714-1715 (1992).

6. Barnsley P E and Fiddyment P J: 'Wavelength conversion from 1.3-1.55 μm using split contact optical amplifiers', IEEE Photon Technol Lett, 3, No 3, pp 256-258 (1991).

7. Nelson A W, Devlin W J, Hobbs R E, Lenton C G D and Wong S: 'High-power low-threshold BH lasers operating at 1.52 μm grown entirely by MOVPE', Electron Lett, 21, No 20, pp 888-889 (1985).

8. Barnsley P E and Chidgey P J: 'All-optical wavelength switching from 1.3—1.55 μm WDM wavelength routed network: system results', IEEE Photon Technol Lett, 4, No 1, pp 91-94 (1992).

9. Magari K, Hawaguchi H, Oe K and Fukuda M: 'Optical narrow-band filters using optical amplification with distributed feedback', IEEE J Quantum Electron, 24, No 11, pp 2178-2190 (1988).

10. Numai T: '1.5 μm optical filter using a two-section Fabry-Perot laser diode with wide tuning range and high constant gain', IEEE Photon Technol Lett, 2, No 6, pp 401-403 (1990).

11. Reason R J: 'Optical space switch architectures based upon lithium niobate crosspoints', BT Technol J, 7, No 1, pp 83-91 (1989).

12. Terui H, Kominato T and Kobayashi M: 'Lossless 1×4 laser diode optical gate switch', J Lightwave Technol, 9, No 11, pp 1518-1522 (1991).

13. Burke C, Fujiwara M, Yamaguchi M, Nishimoto H and Honmou H: 'Studies of a 128 line space division switch using lithium niobate switch matrices and optical amplifiers', Topical Meeting on Photonic Switching, Salt Lake City, Utah (March 1991).

14. White I H, Watts J J S, Carroll J E, Armistead C J, Moule D J and Champelovier J A: 'InGaAsP 400 μm $\times$ 200 μm active crosspoint switch operating at 1.5 μm using novel reflective Y-coupler components', Electron Lett, 26, No 10, pp 617-618 (1990).

15. Lindgren S, Oberg M G, Andre J, Nulsson S, Brobert B, Holmberg B and Backbom L: 'Loss compensated optical Y-branch switch in InGaAsP-InP', J Lightwave Technol, 8, No 10, pp 1591-1594 (1990).

16. Mace D A H, Adams M J, Singh J, Fisher M A, Henning I D and Duncan W J: 'Twin-ridge laser amplifier crosspoint switch', Electron Lett, 25, No 15, pp 987-988 (1989).

17. Mace D A H, Adams M J, Singh J, Fisher M A and Henning I D: '1×2 lossless semiconductor optical switch', Electron Lett, 27, No 3, pp 198-199 (1991).

18. Burton J D, Fiddyment P J, Robertson M J and Sully P: 'Low loss monolithic 2×2 laser amplifier gate switch matrix', 13th IEEE International Semiconductor Laser Conference, Takamatsu, Japan (September 1992).

19. Davies D A O, Mudhar P S, Fisher M A, Mace D A H and Adams M J: 'Integrated lossless InP/InGaAsP 1 to 4 optical switch', Electron Lett, 28, No 16, pp 1521-1522 (1992).

20. Young M G, Koren U, Miller B I, Newkirk M A, Chien M, Zirngibl M, Dragone C, Glance B, Koch T L, Tell B, Brown-Goebeler K and Raybon G: 'A 1×16 photonic switch operating at 1.55 μm wavelength based on optical amplifiers and a passive optical splitter', Postdeadline paper PD8 presented at the Topical Meeting on 'Optical Amplifiers and their Applications', Santa Fe, New Mexico (June 1992).

21. Gustavsson M, Lagerstrom B, Thylen L, Janson M, Lundgren L, Morner A C, Rask M and Stoltz B: 'Monolithically integrated 4×4 InGaAsP/InP laser amplifier gate switch arrays', Postdeadline paper PD9 presented at the Topical Meeting on 'Optical Amplifiers and their Applications', Santa Fe, New Mexico (June 1992).

22. Lasher G J: 'Analysis of a proposed bistable injection laser', Solid State Electron, 7, pp 707-716 (1964).

23. Barnsley P E, Marshall I W, Wickes H J, Fiddyment P J, Regnault J C and Devlin W J: 'Absorptive and dispersive switching in a three region InGaAsP semiconductor laser amplifier at 1.57 μm', J Modern Optics, 37, No 4, pp 575-583 (1990).

24. Adams M J and Barnsley P E: 'Theory of optical switching in two-section semiconductor laser amplifiers', Topical Meeting on 'Optical Amplifiers and their Applications', Santa Fe, New Mexico (June 1992).

25. Marshall I W, O'Mahony M J, Cooper D M, Fiddyment P J, Regnault J C and Devlin W J: 'Gain characteristics of a 1.5 μm split contact laser amplifier', Appl Phys Lett, 53, No 17, pp 1577-1579 (1988).

26. Suzuki S, Terakado T, Komatsu K, Nagashima K, Suzuki A and Kondo M: 'An experiment on high speed optical time-division switching', J Lightwave Technol, LT-4, No 7, pp 894-899 (1986).

27. Masuda S, Fujimoto N, Rokugawa H, Yamaguchi K and Yamakoshi S: 'Experiments on optical drop/insert function using bistable laser diodes for optical access nodes', Topical Meeting on Photonic Switching, Salt Lake City, Utah (March 1989).

28. Webb R P: 'Error-rate measurements on an all-optically regenerated signal', Optical and Quantum Electron, 19, pp S57-S60 (1987).

29. Jinno M and Matsumoto T: 'Optical timing regenerator using 1.5 μm wavelength multielectrode DFB LDs', Electron Lett, 25, No 20, pp 1332-1333 (1989).

30. Ueno M and Lang R: 'Conditions for self-sustained pulsation and bistability in semiconductor laser', J Appl Phys, 58, No 4, pp 1689-1692 (1985).

31. Barnsley P E, Wickes H J, Wickens G E and Spirit D M: 'All-optical clock recovery from 5 Gbit/s RZ data using a self-pulsating 1.56 μm laser diode', IEEE Photon Technol Lett, 3, No 10, pp 942-945 (1991).

32. Barnsley P E and Wickes H J: 'All-optical clock recovery from 2.5 Gbit/s NRZ data using self-pulsating 1.58 μm laser diode', Electron Lett, 28, No 1, pp 4-6 (1992).

33. Jinno M and Matsumoto T: 'All-optical timing extraction using a 1.5 μm self-pulsating multielectrode DFB LD', Electron Lett, 24, No 23, pp 1426-1427 (1988).

34. Barnsley P E, Wickens G E, Wickes H J and Spirit D M: 'A 4×5 Gbit/s transmission system with all-optical clock recovery', IEEE Photon Technol Lett, 4, No 1, pp 83-86 (1992).

35. Barnsley P E: 'NRZ format all-optical clock extraction at 3.2 Gbit/s using two-contact semiconductor devices', Electron Lett, 28, No 13, pp 1253-1255 (1992).

36. Grant R S and Sibbett W: 'Observations of ultrafast nonlinear refraction in an InGaAsP optical amplifier', Appl Phys Lett, 58, No 11, pp 1119-1121 (1991).

37. Hultgren C T and Ippen E P: 'Ultrafast refractive index dynamics in AlGaAs diode laser amplifiers', Appl Phys Lett, 59 , No 6, pp 635-637 (1991).

38. Sharfin W F and Dagenais M: 'The role of nonlinear diode laser amplifiers in optical processors and interconnects', Optical and Quantum Electron, 19 , S47-S56 (1987).

39. Jewell J L, Huang K F, Tai K, Lee Y H, Fischer R J, McCall S L and Cho A Y: 'Vertical cavity single quantum well laser', Appl Phys Lett, 55 , No 5, pp 424-426 (1989).

40. Tadokoro T, Okamoto H, Kohama Y, Kawakami T and Kurokawa T: 'Room temperature pulsed operation of 1.5 μm GaInAsP/InP vertical-cavity surface-emitting laser', IEEE Photon Technol Lett, 4 , No 5, pp 409-411 (1992).

41. Zhou P, Cheng J, Schaus C F, Sun S Z, Hains C, Armour E, Myers D R and Vawter G A: 'Inverting and latching optical logic gates based on the integration of vertical-cavity surface-emitting lasers and photothyristors', IEEE Photon Technol Lett, 4 , No 2, pp 157-159 (1992).

10

NONLINEAR LOOP MIRROR DEVICES AND APPLICATIONS

K J Blow and K Smith

10.1 OPTICAL PROCESSING

In the last ten years optics has achieved a great penetration into the UK network. So far this penetration has been limited to transmission, mostly in the core network but with some application in the access network. Optical fibres are the most efficient transmission medium yet discovered and together with the advent of the doped fibre amplifier are now capable of sending data at rates of many gigabits per second over thousands of kilometres. The major challenge facing optics today is to make some inroad into the switching domain, and this chapter discusses one approach to the problem.

Over the last six years investigations have been made into the potential of nonlinear optics for achieving all optical processing functions in optical fibres. This approach, if successful, would eliminate optoelectronic devices from a communications network apart from getting the data in and getting it out again. Admittedly such a network is still a long way off but already many of the basic functions which would be required have been demonstrated in the laboratory.

The main requirement can be expressed simply as 'controlling light with light'. In other words, instead of an electronic signal determining which path a light beam would take through the network the beam is steered in response to another light beam present in the same optical fibre. One of the fundamental problems is that light beams are extremely good at passing through each other with absolutely nothing happening. In order to be able to control light with light the beams must interact with each other and this can be achieved by using the nonlinear response of a medium. Nonlinearity, in this case, simply means that the properties of the medium in which the light propagates are altered by the presence of the light itself. This simple fact leads to a rich variety of optical phenomena, some of which limit the performance of light propagation and some of which can be used to achieve advanced functionality. In this chapter the simplest form of nonlinearity will be described and it will be shown how it can be harnessed to achieve all-optical processing functions.

10.2 NONLINEAR PHYSICS

The two fundamental properties which underpin the whole mathematical analysis of linear systems can be expressed simply through the following symbolic relations:

superposition: $f(A+B)=f(A)+f(B)$;
scale invariance: $f(\lambda A)=\lambda f(A)$.

The first property, superposition, states that if the solution of a problem for the two cases A and B is known, then the solution for the case $A+B$ is also known, and this leads naturally to the concept of normal modes of a system. The second property, scale invariance, says that there is no need to worry about the absolute magnitude of the initial state A, and if, for example, the size of the input state is doubled then the output is simply doubled as well. Neither of these properties is available when nonlinearity is introduced into a system. In general, there is very little one can say about nonlinear systems. The properties of each system have to be studied separately and for each possible initial condition one must study how the output changes when its magnitude is increased.

Interest lies in a particular nonlinearity known as the Kerr effect where the refractive index of the medium depends on the intensity of light propagating in it. Since the speed of light in a medium depends on the refractive index, a situation now arises where the phase velocity of a light beam depends both on its own intensity and the intensity of any other beams present. This phenomenon can be expressed through the simple relationship:

$$n = n_0 + n_2 I \qquad \text{... (10.1)}$$

where n represents the total refractive index of the medium, n_0 is the linear part (i.e. the refractive index in the absence of light), n_2 is the Kerr coefficient and I is the intensity of the light. Although this looks like a very simple modification to the linear theory it leads, in optical fibres, to a number of new effects including bright solitons, dark solitons, enhanced pulse broadening, spectral broadening and modulation instability. In the presence of other beams, equation (10.1) is slightly modified but the same basic effect takes place — the refractive index seen by one beam depends on the intensity of a second beam. These two cases are often referred to as self-phase modulation (SPM) and cross-phase modulation (XPM) respectively.

For our purposes, only the simplest effect of the intensity dependence of the refractive index, which is the change in phase of an optical beam, need be considered. The phase shift experienced by a light beam in propagating a distance, L, in a medium is:

$$\phi = \frac{2\pi}{\lambda} nL \qquad \text{... (10.2)}$$

where λ is the wavelength of the light. If the expression for the Kerr effect is now inserted into this formula, the following is obtained (neglecting absorption in the fibre):

$$\phi = \frac{2\pi}{\lambda} n_0 L + \frac{2\pi}{\lambda} n_2 L I \qquad \text{... (10.3)}$$

Thus the phase shift experienced by the beam is modified by the intensity of the light. If this phenomenon is to be of use, a device must be used whose output depends in some way on the phase of the optical field — one such device is an interferometer.

Figure 10.1 shows a Mach-Zehnder interferometer in which the beam is split into two paths by a beam splitter and then recombined by a second beam splitter. The output of the device is given by:

$$P_{\text{out}} = P_{\text{in}} \frac{1}{2} (1 + \cos(\Delta\Theta))$$

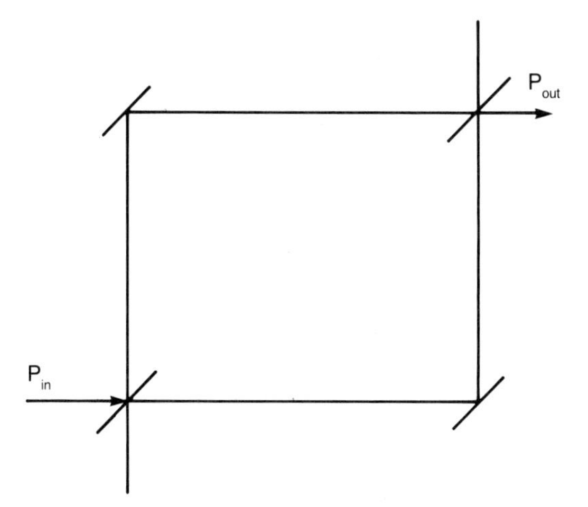

Fig. 10.1 A Mach-Zehnder interferometer constructed from two beam splitters and two mirrors. The light travels over two physically separate paths.

where $\Delta\Theta$ is the difference between the phase changes in the two paths. In order to make the device show some nonlinear response, the Kerr coefficients could be made to be different in the two paths or the input beam could be split asymmetrically between the two paths so that the two paths carried different intensities.

The major problem in dealing with interferometers is that they are somewhat unstable as a result of interactions with the environment. The two beams need to propagate on two different paths whose lengths need to be stable to a small fraction of a micrometre. However, there is one form of interferometer where these effects are effectively suppressed and this is shown in Fig. 10.2. The Sagnac interferometer sends the beams in opposite directions along the same path. Thus, provided the environment does not change in the time taken for the light to traverse the interferometer, it affects both beams equally and does not affect the response. Since the majority of environmental influences are acoustic or thermal this enables optical fibre interferometers to be constructed which can be several kilometres long.

Finally, here are a few comments about nonlinear optics in silica fibres. The Kerr coefficient in silica is $3.2 \times 10^{-20} \mathrm{m}^2/\mathrm{W}$ which is just about the smallest known value for any solid-state material (incidentally, this is a direct consequence of the low loss of silica fibres). It would therefore appear that silica is the worst candidate for a nonlinear optics material but the low absorption coefficient enables the use of very long path lengths and the achievement of results so far only speculated about in 'high' n_2 materials.

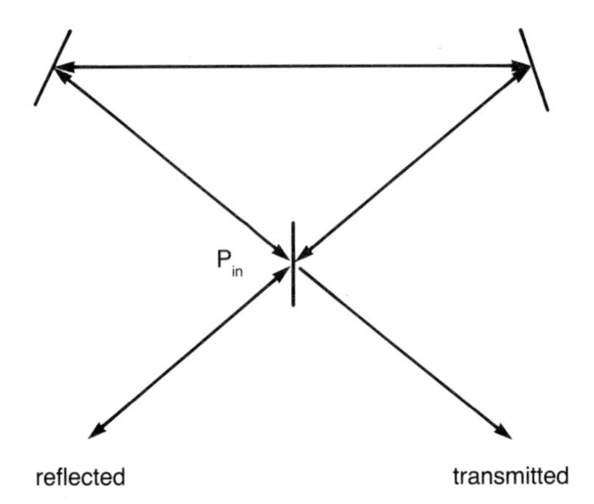

Fig. 10.2 A Sagnac interferometer constructed from one beam splitter and two mirrors. The light travels in opposite directions over the same path giving some immunity to environmental instability.

As a rule of thumb, about 1 W of peak power in 1 km of optical fibre is required to make an optical switch.

10.3 OPTICAL SWITCHING

The Sagnac interferometer is the device used in almost all of our demonstrations of optical switching. In its original form the coupler (see Fig. 10.2) was designed to give equal powers in the two arms. If the light travels on identical paths then the same phase shift is guaranteed (technically, this is only true in the absence of non-reciprocal effects) and there can be no nonlinear response. To put it another way, the device is completely symmetric in both its linear and nonlinear properties and changing the input intensity will not alter the output state. If a nonlinear response is required, then the interferometer must be unbalanced in some way. In our early experiments a coupler was used that gave an unequal power split between the two counter-propagating beams. The experimental results are shown in Fig. 10.3 for the transmitted power as a function of the input power. These results demonstrated that light could switch itself but we still need to be able

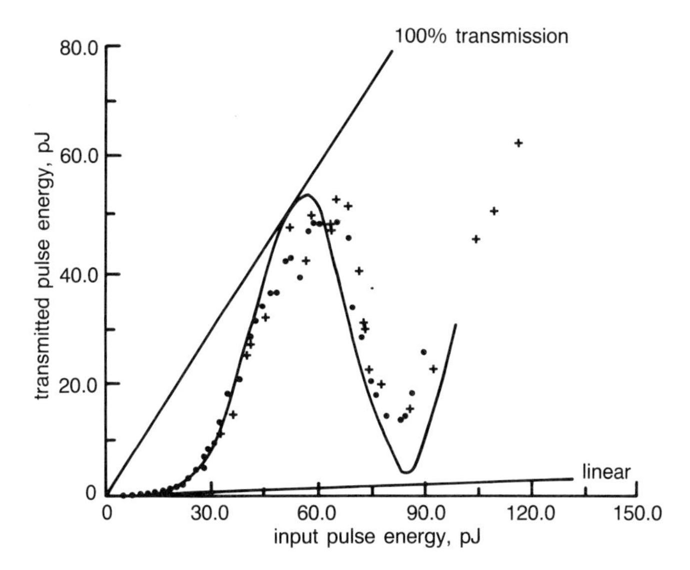

Fig. 10.3 Self-switching of light in an unbalanced Sagnac interferometer.

to control one beam by another. To do this the cross-phase modulation effect is used whereby phase shift can be imposed on a signal beam by a stronger pump beam.

The Sagnac interferometer is modified in such a way that a second beam of light, at a different wavelength to the signal, propagates around the loop in one direction only. When the signal and pump both consist of pulses of light, as in a digital data stream, it can be arranged that the pump only overlaps with the signal travelling in the same direction. When the pump is present it imposes a π phase shift on the signal, which is sufficient to switch the interferometer from reflecting to transmitting. Figure 10.4 shows the experimental results demonstrating this effect, in which light controls light. The switching window created by the pump pulse depends both on the shape of the pump and the difference in group velocity between the pump and signal pulses. The best window can be obtained by using a very short pump pulse. In this case the window consists of a flat region, whose width is determined by the differential velocity, with sharp edges determined by the pump pulse width.

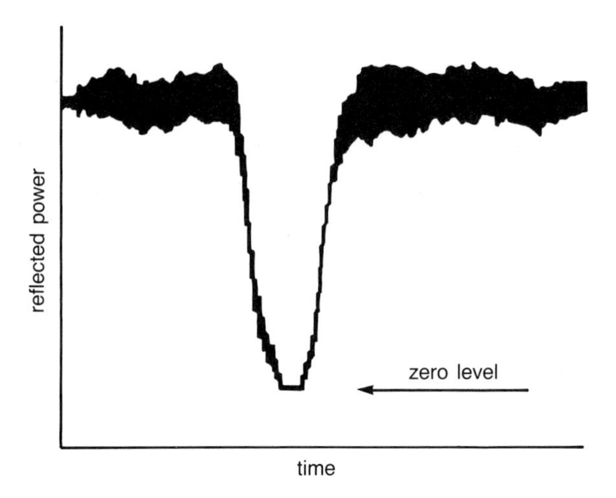

Fig. 10.4 Switching one light beam with another at a different wavelength.

10.4 APPLICATIONS OF NONLINEAR FIBRE MODULATORS

Some of the potential applications of nonlinear fibre devices in telecommunications are outlined in this section.

10.4.1 Gigabits per second switching using a nonlinear loop mirror — drop and insert

The two-wavelength nonlinear loop mirror has proved to be extremely successful in demonstrations of all-optical switching whereby a stream of optical pulses at one wavelength can switch those at a second wavelength. One of the most recent demonstrations [1] involved the switching of a 20 Gbit/s pulse train at 2.5 Gbit/s. The schematic diagram of the experiment is shown in Fig. 10.5. The loop employed was a 6.4 km length of dispersion-shifted fibre. The loop fibre was chosen such that the 1.53 μm switching source (a gain-switched DFB laser operating at 2.5 GHz) and the 1.56 μm signal source (a 10 GHz mode-locked semiconductor laser interleaved to 20 GHz) straddle the dispersion minimum (at 1.545 μm) to minimize the group delay difference between the two wavelengths. With this loop the peak power for complete switching was ~160 mW which, for a 10 ps pulse, translates to a switching energy of only 1.6 pJ. The gain-switched DFB pulses were compressed to ~16 ps (via the 700 m length of negative group-delay dispersion fibre) and amplified (using a diode-pumped erbium amplifier) to

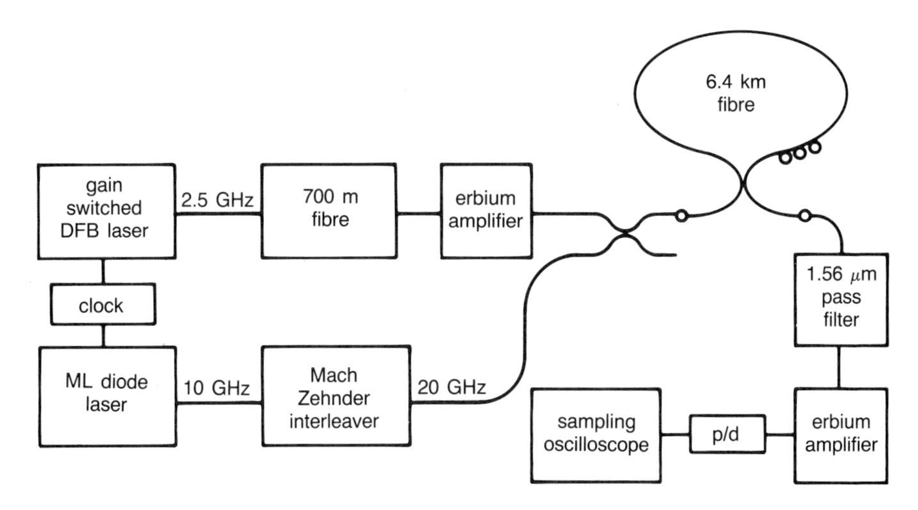

Fig. 10.5 Schematic diagram of an all-optical Gbit/s switching experiment.

a maximum mean power of ~20 mW. The two wavelengths were then combined using a WDM coupler and launched into the loop. The loop was constructed from a coupler which had a 50:50 coupling ratio for the 1.56 μm signal and 100:0 for the 1.53 μm switching signal. An important feature of the loop is the set of polarization controllers, which provided an adjustable phase bias between the counter-propagating signal beams and thereby allowed the operation of the loop in 'reflecting' or 'transmitting' mode [2]. In the reflecting mode, the output of the loop was zero in the absence of switching pulses, i.e. the whole signal was reflected. The injection of switching pulses served to switch the signal pulses to the output. In the transmitting mode, the situation was reversed and the switched pulses were reflected.

Figure 10.6(a) shows the 20 GHz interleaved mode-locked 1.56 μm train of pulses (displayed on a high-speed photodiode and sampling oscilloscope) incident on the loop mirror. Figure 10.6(b) shows the switched output with the loop in 'reflecting' mode, i.e. only signal pulses coincident with the switching pulses are transmitted (every eighth pulse). Figure 10.6(c) shows the loop biased in transmitting mode and clearly shows every eighth pulse switched out. In this configuration the loop mirror is demonstrating the 'drop' function where one channel in a stream is removed. The inverse of this, 'insert', can also be achieved. If the loop is set in 'reflecting' mode it can be arranged for a signal pulse to be inserted to enter the mirror through the 'transmitted' arm in coincidence with the empty channel. In the absence of any pumping the signal is reflected and the empty channel remains empty. However, if a pump pulse is applied coincident with the empty channel then the signal pulse will be transmitted to the 'reflected' arm and thus will have been inserted.

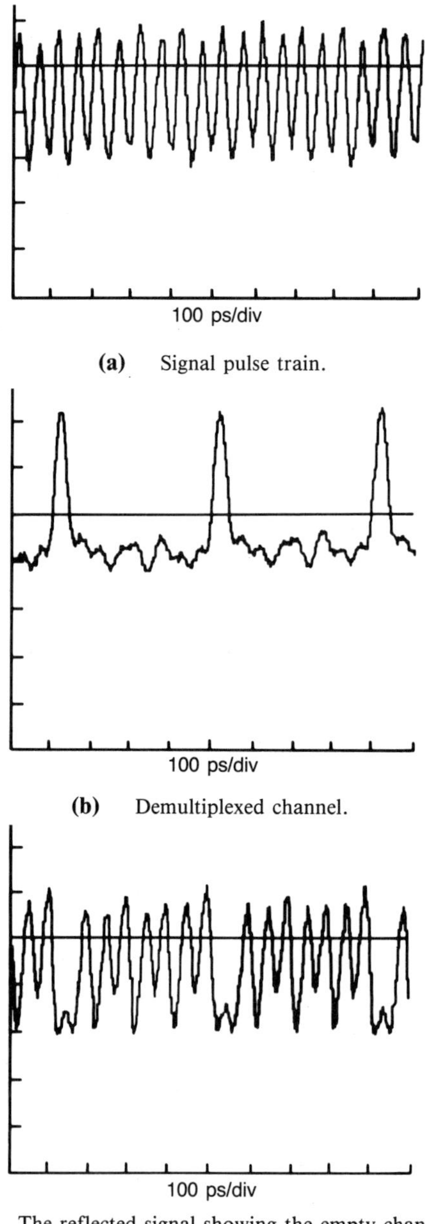

(a) Signal pulse train.

(b) Demultiplexed channel.

(c) The reflected signal showing the empty channel.

Fig. 10.6 Experimental results obtained using the arrangement of Fig. 10.5.

10.4.2 Optical soliton memory

A major goal of all-optical processing is speed, which requires rapid access to a high-capacity, relatively short-term memory (or buffer). By employing long lengths of single-mode fibre as optical delay lines, relatively long storage times can be achieved. For example, a kilometre of fibre gives ~5 μs of optical storage time. Therefore, in order to achieve milliseconds of storage, a delay line of some thousand kilometres would be necessary. It is therefore sensible to think in terms of a recirculating loop or a multiple-pass fibre delay line where data can be stored until required in further processing. An essential feature of the memory is amplification (e.g. Raman or doped-fibre amplifiers), since this compensates for the loss which otherwise restricts the long-term storage time. A further important feature is the use of soliton pulses as the data storage bits in the optical delay line. In this way, the essentially distortionless propagation properties of solitons are exploited. As an example, for a data rate of 10 GHz a fibre length of 50 km has a storage capacity of approximately 2.5 Mbits and an access time of ~250 μs. By recirculating the signal pulses many times along the delay line considerable storage times are available, e.g. 400 round-trips give rise to a storage time of 0.1 s (equivalent to 20 000 km of fibre path length).

For a variety of all-optical processing functions, the memory must be effectively linked with ultrafast switching elements. Figure 10.7 shows a basic architecture which permits the memory to be addressed and updated directly via the ultrafast loop-mirror switches. A 25 km length of single-mode fibre and a single erbium-doped fibre amplifier constitute the linear fibre delay line, which is sandwiched between the two loop-mirror switches (LM1 and LM2). The loop mirrors described here are both operated in two-wavelength

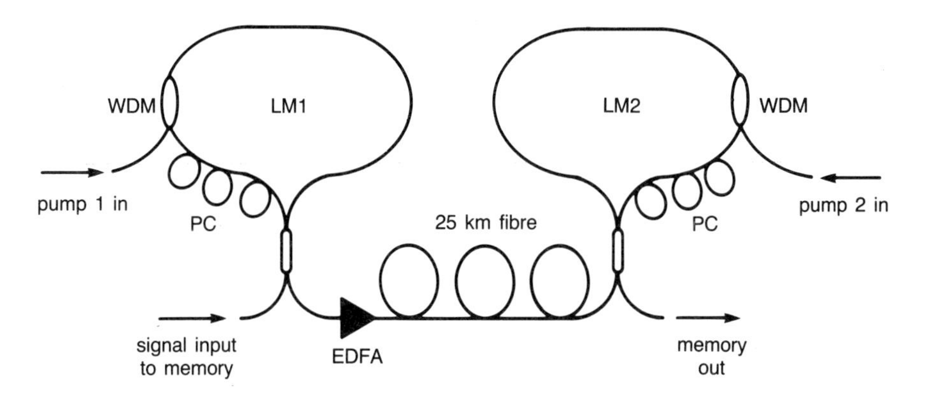

Fig. 10.7 Soliton memory configuration.

mode. In this case, the loop mirror fibre coupler is chosen to have a coupling ratio of 50:50 at the signal wavelength. The pump pulses are introduced into the loop mirror via the separate wavelength dependent fibre couplers (WDM). Both loop mirrors are set up in the reflecting mode through adjustment of the polarization controllers (PC). Therefore, the incoming data (signal pulse) to LM1 is only transmitted through the loop mirror (and into the delay line) when a pump pulse is coincident. This first loop mirror can also be used to modulate (encode) the incoming data for storage in the delay line. The signal is gated into the delay line (for a time up to the round-trip time of the delay fibre) by gating the pump pulses. When the data has been admitted to the delay line, multiple pass circulation is achieved between the LM1 and LM2 reflectors. Ultimately, stored data is degraded through pulse jitter effects (the so called Gordon-Haus limit [3]), which accumulate over the thousands of kilometres of effective path-length propagation. This may limit the total storage time to a duration of the order of hundreds of milliseconds. In order to extract data from the delay line after the required storage time has elapsed, either the second loop-mirror switch LM2 (an identical configuration of LM1) can be used or the data can be returned to the input port via switching at LM1.

As noted previously, the accumulated effects of amplified spontaneous emission and the associated timing jitter tend to limit the storage time of the memories. However, this memory degradation may be suppressed by the use of signal regeneration techniques within the optical delay line (see section 10.4.5). In this case, one can imagine storing data rates well into the tens of gigabits range (giving a memory capacity of >10 Mbits) for times in excess of ~ 10 s.

10.4.3 Ultra-fast mode-locked laser sources

In addition to the processing elements described, it is essential to have the capability of generating ultra-fast pulse streams at repetition rates well into the gigahertz region. One method is that of active laser mode-locking whereby a periodic (amplitude or phase) perturbation of a laser cavity acts to couple the cavity modes and thereby force the laser to produce a continuous train of ultra-short optical pulses. Generally, mode-locked operation of fibre lasers at gigahertz repetition rates has been achieved using integrated-optic $LiNbO_3$ modulators. In fact, our work at BT Laboratories has led to the development of mode-locked erbium-fibre lasers as sources for soliton systems with the generation of high-quality (transform-limited) picosecond pulse streams at repetition rates up to 15 GHz [4]. The mode-locking potential of all-optical fibre modulators has recently been investigated experimentally. For mode-locking applications, such devices have advantages of speed, low loss, large optical bandwidths and low polarization sensitivity over $LiNbO_3$ modulators.

The first experimental results [5] were obtained with the configuration shown in Fig. 10.8(a). A loop modulator (shown in the dashed box) formed one of the end reflectors of an erbium-fibre (Er) laser which was pumped by a high-power MQW laser diode (LD) at $\sim 1.48\ \mu m$. The other end reflector was a 7% reflecting (at 1.556 μm) optically written fibre Bragg grating (FG) [6]. The introduction of a switching pulse to the loop therefore served to modulate the reflectivity of the end reflector. Providing that the modulation frequency coincided with a harmonic of the fundamental cavity frequency, mode-locking followed. A particularly simple actively mode-locked arrangement [7] is shown in Fig. 10.8(b). Here mode-locking was achieved by using a stream of optical pulses propagated along a simple length of optical fibre to provide a periodic phase perturbation of the laser cavity. A tuneable filter (F) permitted wavelength control of the laser output, and a Faraday

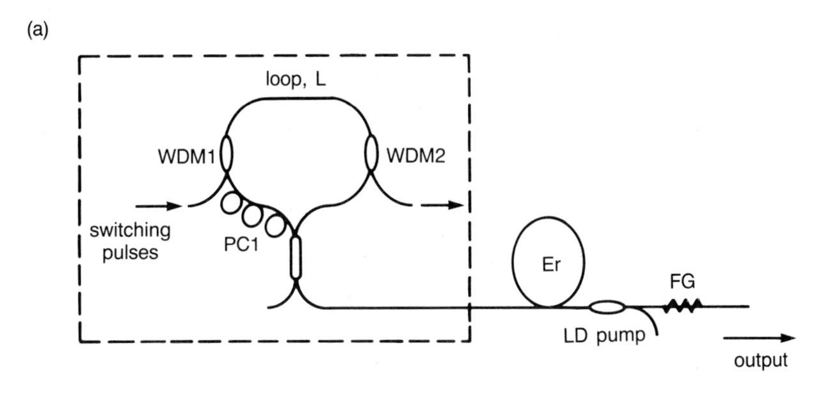

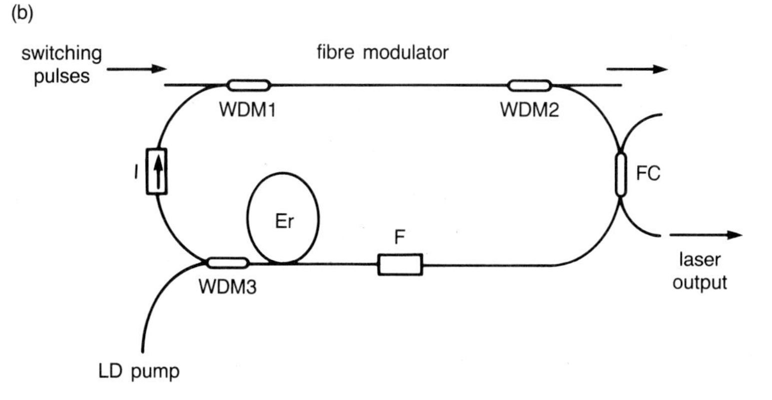

Fig. 10.8 All-optical actively mode-locked laser configurations.

isolator (I) ensured that the full laser power could be accessed in a single pulse train via the ~20% output fibre coupler (FC). For the laser configurations in Fig. 10.8, pulse durations of ~10 ps were typically generated (Fig. 10.9) at repetition rates up to ~40 GHz.

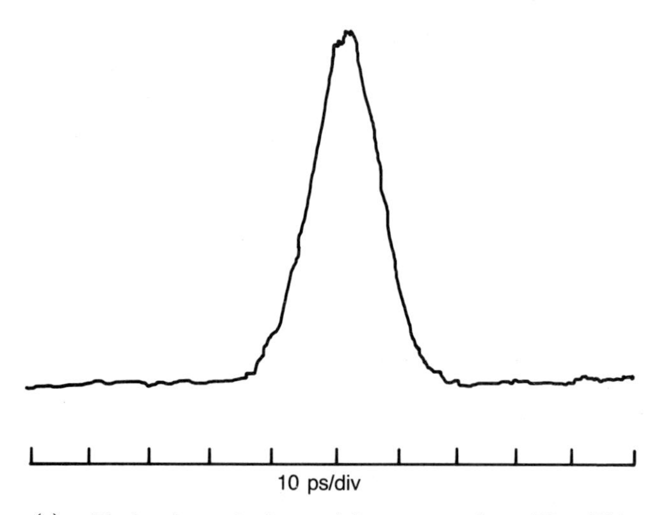

(a) Displayed on a background-free autocorrelator (10 ps/div).

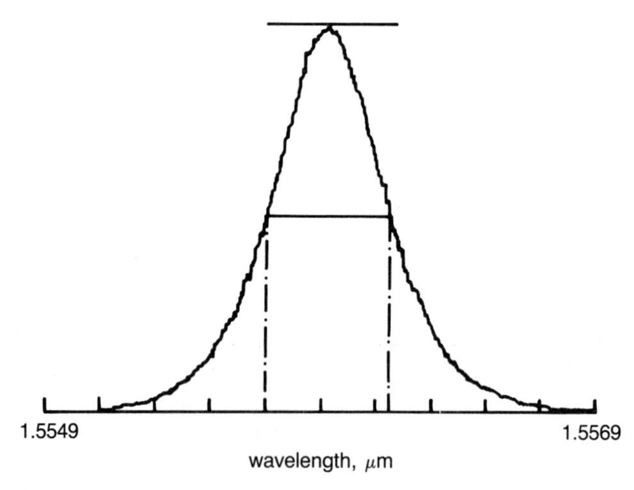

(b) Displayed on corresponding optical spectrum (0.2 wm/div).

Fig. 10.9 Output of all-optical mode-locked laser. These pulses have a time-bandwidth product within 50% of the transform limit.

10.4.4 Clock recovery

The above actively mode-locked lasers use a stream of optical pulses to excite a modulator, which in turn forces a laser to mode-lock. Following on from this, a novel method of all-optical clock recovery has been demonstrated. The basic idea is illustrated in Fig. 10.10. Since a laser cavity and transmission fibre share a nonlinear optical modulator (NOM), the data itself can serve to modulate either the amplitude or the phase of the light in the laser cavity. Providing this modulation takes place with a period (T) equal to (or an integer multiple of) the laser round-trip time, mode-locking of the laser follows. The ensuing continuous stream of high-quality pulses (at a repetition rate $1/T$) can be accessed via the output coupler, and then used directly in further all-optical processing functions.

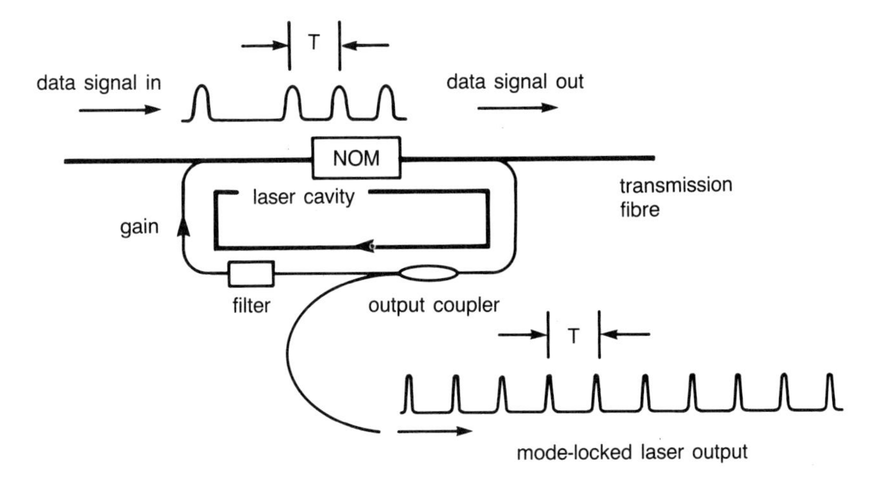

Fig. 10.10 All-optical clock recovery circuit.

In our recent experimental demonstration [8], the clock recovery circuit was based on the mode-locked fibre laser in Fig. 10.8(b). In this way the NOM was a simple length of fibre in which cross-phase modulation of the laser cavity by the data stream drives the mode-locking. Again, the signal source was a 1 GHz gain-switched DFB laser producing 30 ps pulses and subsequently modulated by a 1 Gbit/s data pattern using a $LiNbO_3$ modulator and data generator.

Figure 10.11(a) shows the 1 GHz output of the mode-locked laser (clock) at a wavelength of ~ 1.56 μm displayed on a photodiode/sampling oscilloscope (28 ps response) when driven by a $2^{15}-1$ pseudorandom data sequence. Figure 10.11(b) shows the output displayed on a scale of 40 ps per

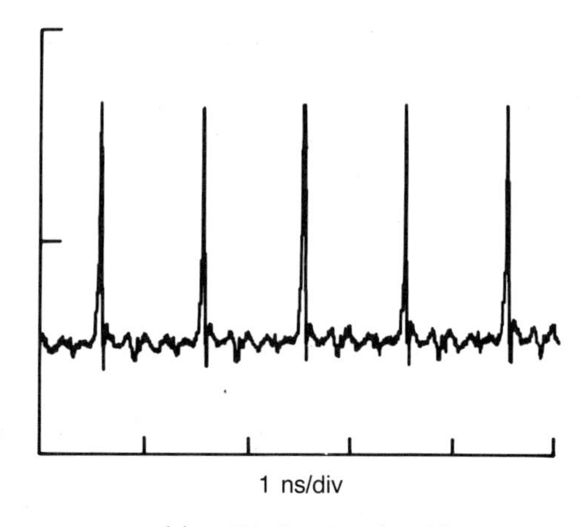

(a) Displayed at 1 ns/div.

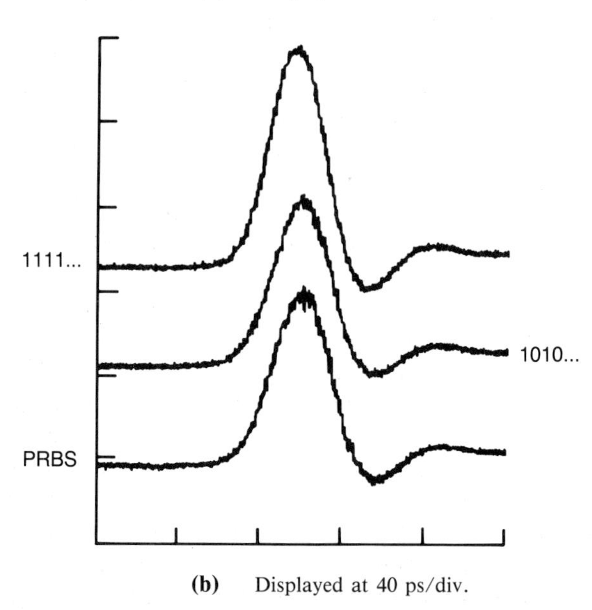

(b) Displayed at 40 ps/div.

Fig. 10.11 The recovered clock from a $2^{15}-1$ PRBS displayed on a high-speed photodiode and sampling oscilloscope.

horizontal division. For reference, the laser output when driven by a straightforward 1 GHz pulse stream (i.e. 1111...) is shown. In Fig. 10.11(b), the laser is mode-locked on an 'odd' order ring cavity mode, such that when the driving pattern is switched to 1010..., the laser output remains a

1 GHz stream (i.e. the first harmonic of the 1010... pattern, at 500 MHz, falls exactly halfway between cavity ring modes). In this way, a clock pulse only 'sees' the modulator every two transits of the cavity filter and hence is slightly broader ($\sim 1.2\tau$) than the 1111... result. When the laser is driven by a pseudo-random bit stream (PRBS) with equal probabilities of 1s and 0s, a clock pulse sees the modulator on average every two filter transits and therefore looks very similar to the 1010... result.

10.4.5 Optical signal regeneration and wavelength translation

All-optical networks will include an abundance of fibre amplifiers in order to maintain the signal strength. To a large extent this bypasses the need for high-speed electronics and potentially offers data rates of many tens of gigahertz. However, such systems are still limited in performance by the effects of timing jitter through the build-up of amplifier spontaneous emission. In order to overcome such limits, we proposed that all-optical regeneration schemes are used which perform both amplitude and timing restoration. In general, simple regeneration schemes can be designed which consist of a clock recovery circuit and fibre amplifiers together with a nonlinear optical processing element.

One example of such a configuration [9] is shown in Fig. 10.12. This particular device not only regenerates the data but also performs a wavelength translation (to that defined by the clock). The incoming data stream (at 1.53 μm) first drives a mode-locked laser clock recovery circuit. The subsequent continuous stream of optical pulses (at ~ 1.56 μm) then serves as the source for the regenerated stream of data. By amplifying (in an erbium-doped fibre amplifier) the 1.53 μm data exiting the clock circuit, it can be ensured that the data is sufficiently intense to fully modulate the clock output via the nonlinear loop configuration. It is worth noting that both the clock recovery circuit and the loop-mirror modulator are tolerant to timing and amplitude fluctuations on the signal stream. Figure 10.13 shows the experimental results demonstrating the operation of the scheme in Fig. 10.12. A 1 GHz signal stream was derived from a gain-switched DFB laser producing ~ 30 ps pulses and then modulated by a 1 Gbit/s data generator. In this particular experiment, the input data consisted of a repetitive 8-bit word, i.e. 10100110 (Fig. 10.13(a)). The recovered clock at 1 GHz is shown in Fig. 10.13(b) and the subsequent regenerated word at 1.56 μm is shown in Fig. 10.13(c). It is worth noting that by suitable adjustment of polarization controllers within the loop (as described earlier) it was straightforward to generate the complement of the input word (i.e. 01011001).

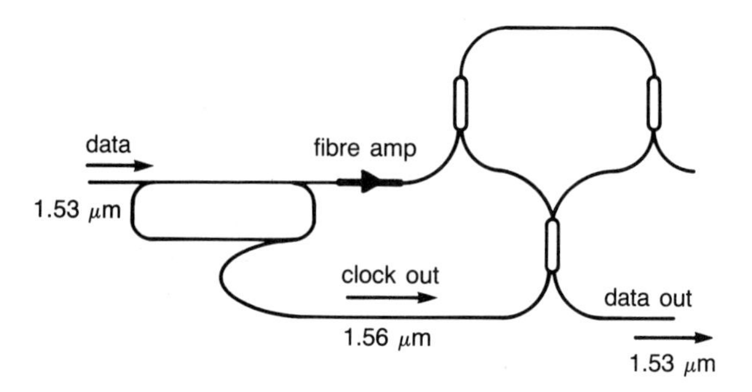

Fig. 10.12 The all-optical signal regenerator.

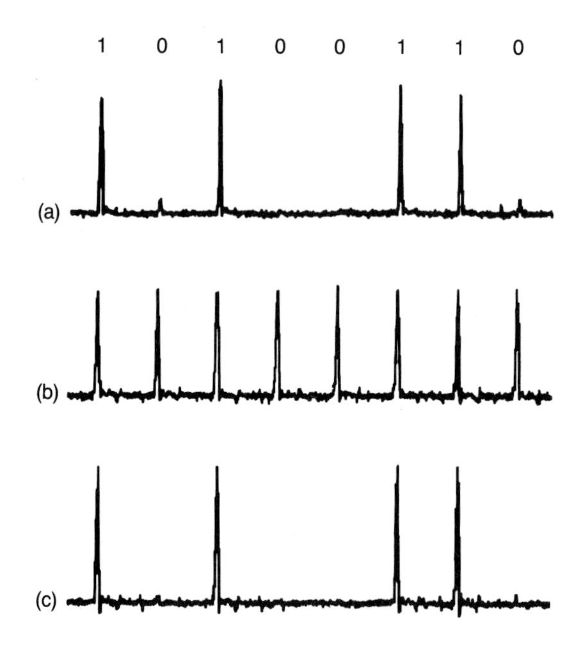

Fig. 10.13 Experimental results of all-optical signal regeneration: (a) the 8-bit input pattern (10100110) to the regenerator (at 1.53 μm), (b) the recovered clock (at 1.56 μm) and (c) the regenerated data (at 1.56 μm).

10.5 CONCLUSIONS

The application of nonlinear optical techniques to switching functions in an all-fibre form offers the potential for processing information while it remains within the fibre. Many of the functions required to make an all-optical system have been demonstrated including switching, memory, clock recovery and regeneration. In silica fibres these processes can work at bit rates currently in use but can also work at much higher rates. Indeed, in early experiments pulses were used with durations below 1 ps. This work is still some way off achieving the all-optical network but there are nearer-term specialist applications, such as transatlantic communications, where all-optical processing could soon find employment.

REFERENCES

1. Nelson B P, Blow K J, Constantine P D, Doran N J, Lucek J K, Marshall I W and Smith K: 'All-optical Gbit/s switching using nonlinear optical loop mirror', Electron Lett, 27, p 704 (1991).

2. Mortimore D B: 'Optical fibre loop mirrors', IEEE J Lightwave Technol, 6 (1988).

3. Gordon J P and Haus H A: 'Random walk of coherently amplified solitons in optical fibre transmission', Opt Lett, 11, pp 665-667 (1986).

4. Davey R P, Smith K and McGuire A: 'High speed, mode-locked, tunable, integrated erbium fibre laser', Electron Lett, 28, p 482 (1992).

5. Nelson B P, Smith K and Blow K J: 'Mode-locked erbium fibre laser using all-optical nonlinear loop mirror', Electron Lett, 28, p 656 (1992).

6. Kashyap R, Armitage J R, Wyatt R, Davey S T and Williams D L: 'All-optical narrowband reflection gratings at 1500 nm', Electron Lett, 26, p 730 (1990).

7. Greer E J and Smith K: 'All-optical FM mode-locking of a fibre laser', Electron Lett, 28, p 1741 (1992).

8. Smith K and Lucek J K: 'All-optical clock recovery using a mode-locked laser', Electron Lett, 28, p 1814 (1992).

9. Lucek J K and Smith K: 'All-optical signal regenerator', Opt Lett, 18, p 1226 (1993).

11

ANALYSIS OF AN OPTICALLY AMPLIFIED TWO-SECTION LINK SPECIFICATION INCORPORATING AN ERBIUM-DOPED FIBRE AMPLIFIER

A W O'Neill and T G Hodgkinson

11.1 INTRODUCTION

Erbium-doped fibre amplifiers [1] are a technological alternative to regenerators for increasing the distance over which 1.55 μm optical fibre transmission systems can be operated. The regeneration technique relies on the use of optoelectronic interfaces (photodetector and laser) and integrated electronics to detect, regenerate and retransmit optical signals. A regenerated link is composed of a number of independent optical sections, isolated by electronics, with each section comprising a laser, some transmission loss and a photodetector. A network-wide system specification can therefore be used to define the operational requirements (maximum loss, minimum launch power, etc) of the various elements of a section so that arbitrary-length links can be installed using components from a range of suppliers. The major limitations of using regenerators are that the embedded optoelectronics and electronics are complex, costly, reduce reliability and are functionally dependent on the format of the optical signals (modulation, coding, etc). This electronic bottleneck reduces transmission flexibility and it also requires that the regenerators are replaced whenever the link capacity is upgraded.

In contrast, the fibre amplifier is an optically transparent device as it contains no in-line optoelectronic interfaces. This offers a number of operational advantages compared with regenerators. In particular, it can handle a range of transmission formats and data rates, and can also accommodate a number of channels, separated in wavelength, simultaneously. However, the fibre amplifier is a complex analogue device — its gain is dependent on the average power of the input signal (gain saturation) [2], as is shown by the plots in Fig. 11.1, and it generates spontaneous noise [3], which degrades the gain process in subsequent amplifiers and the photodetection process at the receiver. Therefore the deployment of amplified links could involve complex planning and design procedures because there are a wide range of amplifier positions and associated input/output transmission loss combinations, even within simple two-section links. In addition, satisfying this wide range of operating conditions could require the availability of a variety of different amplifier designs (power, line, preamplifier) and/or amplifier control techniques. Consequently, optically amplified links might not be considered practical on a network-wide basis unless it could be shown that a simple amplified link specification exists which is compatible, in a performance sense, with its regenerated equivalent. In addition, the amplified link would then need to offer some unique operational advantage(s) to justify the costs associated with the introduction of a new technology into the field.

The aims of this study are to investigate the existence of a network-wide specification for an amplified, two-section link, and to then assess its potential performance advantage(s) over the equivalent regenerated link. Initially, a qualitative discussion of regenerated and optically amplified links is presented, and a network-wide amplified link specification derived from that of the equivalent regenerated link. This network-wide specification is then investigated by comparing the calculated performance of a regenerated and an amplified, two-section, 622 Mbit/s link for the full range of possible input and output section loss combinations. The potential for increasing the capacity of the amplified link is then investigated for three types of capacity upgrade, namely 2.5 Gbit/s time division multiplexing (TDM), four-channel 622 Mbit/s wavelength division multiplexing (WDM) [4] and four-channel 2.5 Gbit/s WDM. This study shows shows that it is possible to use a network-wide specification for an amplified, two-section link, and that this amplified link can always outperform the equivalent regenerated link. It also shows that the amplified link provides increased positional flexibility and an inherent upgrade capability.

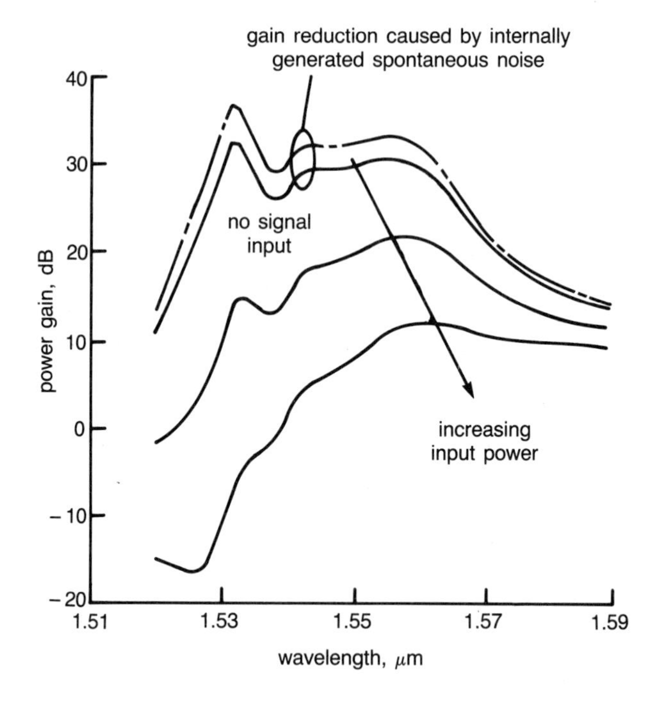

Fig. 11.1 Impact of input power and spontaneous noise on the gain profile of an erbium-doped fibre amplifier (these plots were derived using the improved average power analysis [5]).

11.2 TWO-SECTION OPTICAL LINKS

11.2.1 Regenerated links

A regenerated two-section link is shown in Fig. 11.2, and a typical 622 Mbit/s design specification for this link is given in Table 11.1. The system margin in this specification is intended to take account of reasonable degradations in section loss and/or launch power during the lifetime of the link. For the purposes of this investigation, this degradation has been arbitrarily associated wholly with the transmitter laser so as to reflect the worst case situation.

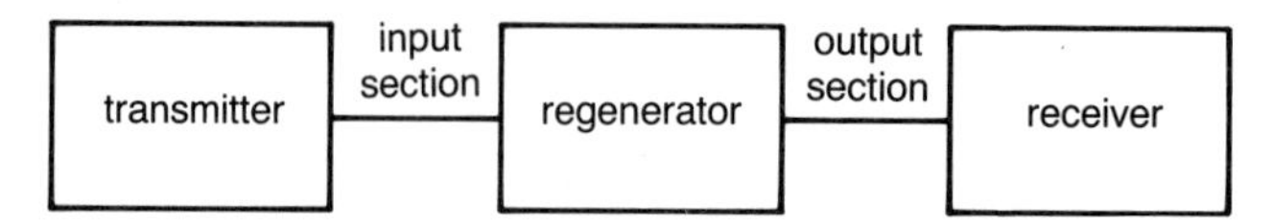

Fig. 11.2 A regenerated, two-section, optical link.

Table 11.1 622 Mbit/s link specification.

Maximum launch power	-2 dBm
Maximum section loss	20 dB
Minimum receiver power	-28 dBm @ 10^{-10} BER
System margin	6 dB

A graph can now be produced which indicates all possible combinations of input and output section losses that can be accommodated by the regenerator at day one. This operating window, plotted in Fig. 11.3, guarantees that the link would still be operational if the transmitter and regenerator laser launch powers were both reduced by an amount equal to the system margin (6 dB). The shape of the window reflects the fact that the regeneration process is not optically transparent, and that the input and output section losses are therefore independent of each other. Consequently, reduced input loss does not facilitate an equivalent increase in output loss, and vice versa (i.e. the maximum tolerable total link loss, or maximum reach, varies).

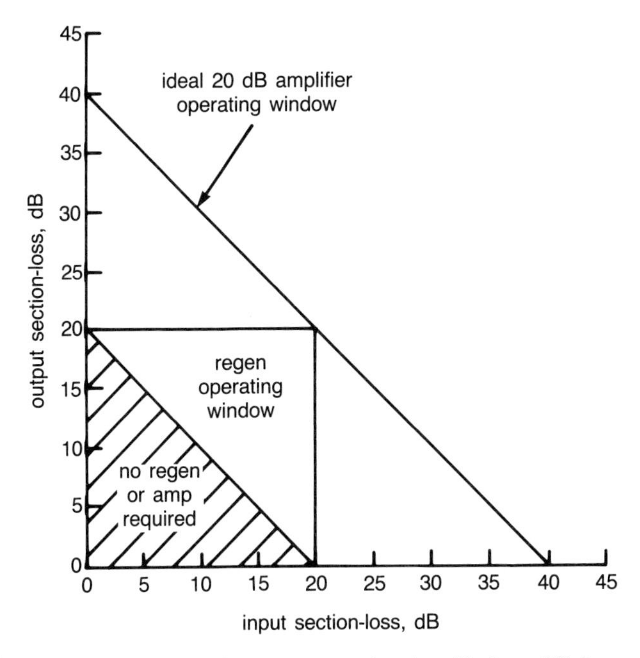

Fig. 11.3 Operating windows for a regenerated and an ideal amplified, two-section 622 Mbit/s link.

11.2.2 Optically amplified links

If an ideal amplifier (no gain saturation and no spontaneous noise generation), with 20 dB gain, was used to replace the regenerator in Fig. 11.2, the operating window for the 622 Mbit/s link would be as shown in Fig. 11.3. It is immediately obvious from this figure that this operating window completely encloses that of the regenerated link. In addition, the shape of this window shows that the input and output section losses are now dependent; hence reduced input loss now facilitates an equivalent increase in output loss (constant maximum reach), and vice versa. Consequently, the possibility exists for an amplifier to offer range and positional flexibilities that are far greater than those offered by the regenerator. However, the extent to which these will be available in practice will depend on the performance limitations imposed by the non-ideal amplifier characteristics, namely gain saturation and the generation of spontaneous noise within the amplifier. The limitation imposed by these non-ideal characteristics is that they both prevent the maximum reach of the link from having a constant value. For example, at low input section losses the reach is reduced because the increased input power to the amplifier causes a significant reduction in amplifier gain (gain saturation). Conversely, at low output section losses, the reach is now reduced because the receiver performance is dominated by amplifier-generated noise which is not reduced by increasing input section loss (see Appendix C). In addition, the amplifier noise reduces the optimum decision threshold within the receiver.

For the complete range of section loss combinations in the network, the consequences of gain saturation, amplifier noise and suboptimal decision thresholds can only be assessed with the aid of comprehensive computer models of the fibre amplifier [5] and the receiver. Therefore, a precursor to this study has been the development of such models (see the Appendix).

11.2.3 Network-wide, amplified link specification

The procedure adopted for deriving the network-wide specifications for the amplified link scenarios considered in this study began by taking the specifications for the equivalent regenerated links. Then it was assumed that the regenerator had been replaced by an amplifier, and that the values of transmitter launch power and section losses were equal to the maximum specified values. Finally, using the laser, fibre amplifier and receiver models discussed in the Appendix, the pump power was iteratively adjusted until the amplifier power gain was just sufficient to produce the specified system margin. This value of power gain and the associated value of input power were then incorporated into the regenerated link specification, effectively giving an amplified link specification which is based on using a constant value of pump power. The resulting amplified link specifications for the scenarios considered in this study are given in Table 11.2. Although not considered in this study, it should in principle be possible to extend this specification procedure to cover amplified, multisection optical transmission links.

Table 11.2 Amplified, two-section link specifications.

	622 Mbit/s TDM	*2.5 Gbit/s TDM*	*4×622 Mbit/s WDM*	*4×2.5 Gbit/s WDM*
Maximum launch power	−2 dBm	−2 dBm	−2 dBm/channel*	−2 dBm/channel*
Maximum section loss	20 dB	20 dB	20 dB	20 dB
Minimum receiver power @ 10^{-10} BER	−28 dBm	−25 dBm†	−28 dBm	−25 dBm†
System margin	6 dB	6 dB	6 dB	6 dB
Demux loss	0 dB	0 dB	6 dB	6 dB
Amplifier power gain	20 dB (−22 dBm)●	23 dB (−22 dBm)●	26 dB (−13 dBm)●	29 dB (−13 dBm)●

* the power per channel after having assembled the multiplex
† 1 dB dispersion penalty assumed
● total average power input to the amplifier at the specified power gain

11.3 AMPLIFIED, TWO-SECTION LINK ANALYSIS

11.3.1 Assumptions and procedure

It has been assumed that the amplifier pump power would be maintained at a constant level by a local stabilization loop, that the transmitter performance is such that fibre dispersion produces only a small performance penalty, and that isolators are included within the amplifier. To satisfy worst-case analysis conditions, the system margin was allocated totally to transmitter output power degradation. Because it is impractical to optimize the decision threshold within the receiver on a link-by-link basis, for the purposes of this study the normalized threshold was assumed fixed at 0.5 (see Appendix C), this being the standard setting for present network receivers (coincidentally, this setting gives best immunity from the range of amplifier noise levels that could be encountered).

Prior to analysing a link, the amplifier pump power was first set so that the power gain satisfied the two-section link specification (Table 11.2), and then the transmitter launch power was reduced to -8 dBm, i.e. the specified maximum value reduced by an amount equal to the system margin. Having set these parameter values, the worst-case operating window was then derived by repeatedly analysing the link for various combinations of input and output section losses. For the purposes of this study, it was decided that the operating window should enclose those loss combinations for which the link was just operational, i.e. tolerant of the 6 dB drop in launch power. Throughout the analysis, the total power at the receiver was monitored to ensure that receiver saturation was avoided.

Although the amplified link analysis uses typical erbium-doped fibre cross-section data [5] and typical regenerated link specifications, it is not anticipated that this would significantly affect the generality of the analysis results. Therefore, the analysis trends and the overall conclusions should be representative of most, if not all, amplified, two-section link designs.

11.3.2 622 Mbit/s link analysis

The operating window for the amplified link is plotted in Fig. 11.4, and it can be seen that it completely encloses the operating window for the equivalent regenerated link. The line defining the operating window does not pass through the 20 dB/20 dB point because, prior to analysis, the launch power was reduced by 6 dB. The associated increase in amplifier gain then facilitates an increase in output loss. It can also be seen that there has been a closure at the two extremes of the window when compared with that of the ideal amplified link, as predicted in section 11.2.2. These closures highlight the inherent nonlinear performance of the amplified link, which, as previously mentioned, is due to gain saturation and/or spontaneous noise. The operating regions lost due to this nonlinearity could be recovered by either over-pumping the fibre amplifier (power amplifier) or by optically filtering prior to the receiver (preamplifier). However, the former would require a significant increase in pump power, affecting the reliability, stability and cost of the amplifier, and the latter would require the provision of optical noise filters, which would increase link cost and impose more stringent wavelength stability requirements on the source lasers. It is apparent from Fig. 11.4 that the amount of the window which might be recovered with these techniques would probably not justify the complexities of moving from a network-wide specification.

To illustrate the extended operating range offered by the amplifier, Fig. 11.5 presents the results from Fig.11. 4 in a form which compares the reach

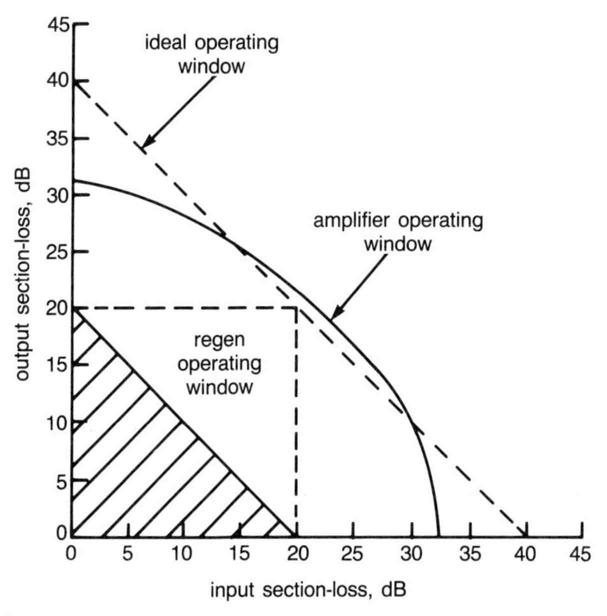

Fig. 11.4 Operating window for an amplified, two-section, 622 Mbit/s link.

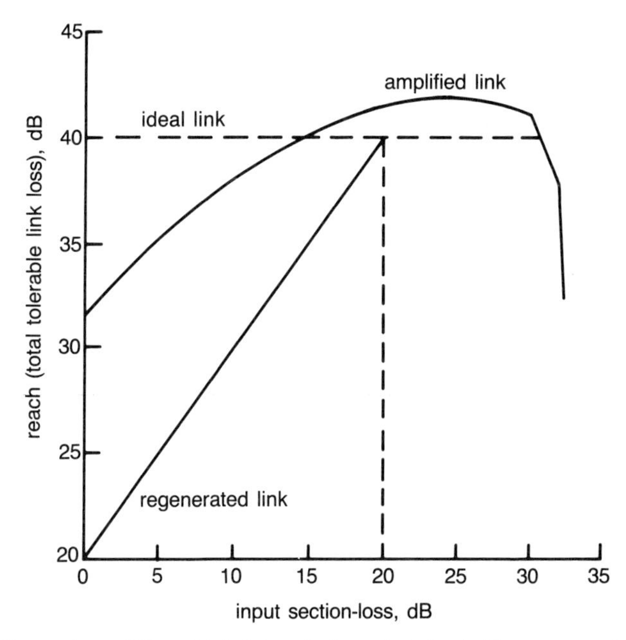

Fig. 11.5 Comparison of the total link loss that can be tolerated by a regenerated and an amplified, two-section, 622 Mbit/s link as a function of the input section loss.

of the regenerated link with that of the amplified one as a function of the input section loss.

11.3.3 Upgrade options

The 622 Mbit/s amplified link specification requires only that the amplifier provides a specific power gain. Therefore, if the fibre amplifier design is able to provide sufficient additional gain, it would then be possible to upgrade an amplified link simply by increasing the pump power and updating the terminal equipment. Using this approach, upgrades to 2.5 Gbit/s, 4×622 Mbit/s and 4×2.5 Gbit/s WDM (the four channels being located at 1.547, 1.550, 1.553 and 1.556 μm) have been analysed for the link specifications given in Table 11.2. The specified amplifier gains for the WDM upgrades were derived for the worst-case wavelength (lowest gain), with all launch powers at their maximum specified values.

With the launch power of the channel being analysed reduced to -8 dBm, i.e. the maximum specified value reduced by an amount equal to the system margin, the operating windows were derived for each channel, and Fig. 11.6

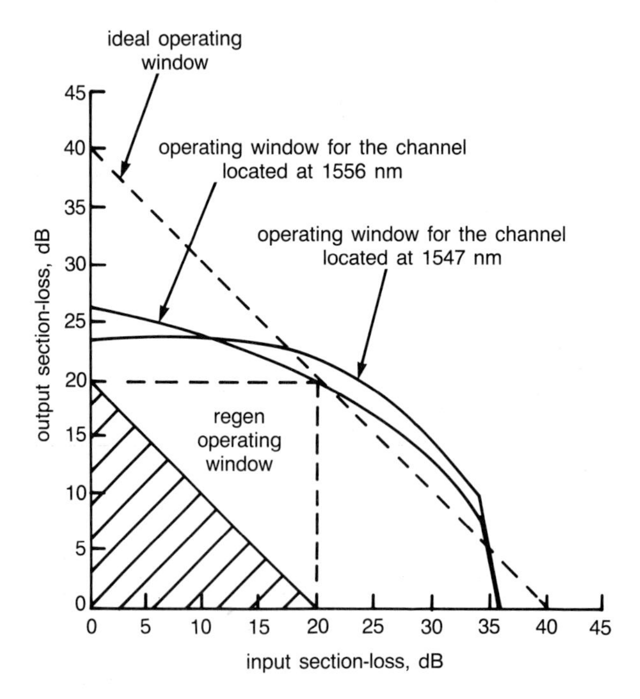

Fig. 11.6 Operating windows for the two extreme wavelengths of an amplified, two-section, 4×622 Mbit/s WDM link.

shows the results for the two extreme multiplex wavelengths in the 4× 622 Mbit/s upgrade. The slight shape difference between these two windows is due to the gain saturation characteristic of the amplifier being wavelength-dependent. Compared with the operating window for the amplified 622 Mbit/s link, the WDM results show that the higher total input power into the amplifier significantly increases gain saturation. In addition, the optical filters used to extract the desired WDM channel suppress the spontaneous noise at the receiver. Upgrading to 2.5 Gbit/s per channel would require an additional increase in amplifier gain to overcome both the reduced receiver sensitivity and the small dispersion penalty. This increase in gain results in there being more spontaneous noise at the receiver, which reduces the maximum input section loss that can be used with a given output section loss.

By superimposing the individual operating windows for all of the various amplified link scenarios (Table 11.2), and then ignoring all loss combinations not enclosed by all such windows, the overall operating window produced is as shown in Fig. 11.7. This window shows that a simple, amplified, two-section link can upgrade any regenerated 622 Mbit/s link, and that it can also accommodate a larger range of input/output section loss combinations.

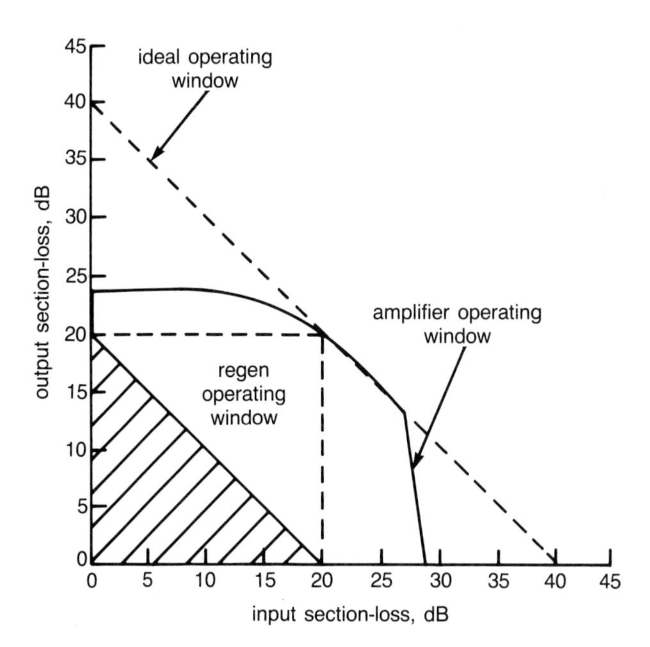

Fig. 11.7 Overall operating window for the various amplified, two-section link scenarios given in Table 11.2.

11.4 CONCLUSIONS

A detailed theoretical analysis of a network-wide specification for an amplified, two-section, 622 Mbit/s link has shown that its operating window completely encloses that of the equivalent regenerated link. Investigation of its upgradability has shown that with the appropriate terminal equipment upgrades, the information-carrying capacity of the link can be increased by up to 16-fold simply by using step increases in the amplifier pump power. These results show that the network-wide specification for the amplified link inherently accommodates both the undesirable performance characteristics of the fibre amplifier and the wide range of loss combinations which could be encountered within two-section links. The existence of a network-wide specification significantly simplifies the design of amplified links because it removes the need for complex planning and design procedures. It also removes the need for sophisticated input power/wavelength-dependent amplifier control techniques and offers simple, inherent capacity upgrades.

APPENDIX

A. Fibre amplifier model

To analyse the amplified, two-section link it was essential that the amplifier model should be able to accurately predict the performance of the fibre amplifier under all operating conditions, and that it should also give a reasonably fast computation run time. The accurate performance requirement inhibited the use of existing empirical and simplified-analytic models [6], because both of these sacrifice accuracy for simplicity. On the other hand, the operating speed of most computers is not sufficiently fast to be able to use the accurate, full-numerical analysis model [6]. The average power analysis technique [5], which predicts the performance of the amplifier by using the total average-core-power along the amplifier, provides an excellent compromise. It is an iterative analysis technique which has a speed advantage over the full numerical analysis (it takes a few seconds on a Macintosh SE/30), and it overcomes the limitations of the empirical and simplified-analytic models by taking account of gain profile changes caused by both the level and the wavelength distribution of the input power, and gain saturation caused by internally generated spontaneous noise (see Fig. 11.1).

B. Laser spectral profile

For analysis purposes, the lasers are modelled as single-mode, narrow linewidth devices, with the transmitter wavelengths being located in the 1.55 μm low loss transmission window, and the pump wavelength located at 1.49 μm. When applied to the pump, this narrow linewidth approximation gives results which are slightly different from those given when using the actual power spectral density. However, it has been verified that for cases where the amplifier power gain is specified, this approximation only results in there being a small difference between the predicted and the actual value of pump power.

C. Receiver model

The receiver model includes all noise terms, namely receiver thermal noise, signal and spontaneous shot noise, signal times spontaneous noise ($s \times n$) and spontaneous noise times spontaneous noise ($n \times n$). All of the noise terms are assumed to have Gaussian noise statistics which, for slightly reduced accuracy, avoids unwieldy computations. The receiver model also assumes that the received pulses are rectangular, and that these are equalized such that at the input to the decision element they have a 100% raised cosine spectrum. The normalized decision threshold (normalized with respect to the peak pulse height at the input to the decision element) is fixed at 0.5 for all link scenarios. This is necessary because, although the total noise power can be larger for binary 'ones' than binary 'noughts', the value of normalized threshold (≤ 0.5) which optimizes performance would be dependent on the link configuration, thus introducing a further level of design complexity (influenced by output section loss, amplifier noise, etc).

In the receiver model, the double-sided noise spectral density expression used for a binary 'one' is:

$$S_1(f) = S_{\text{th}} + S_{\text{shot}}^{\text{sig}} + S_{\text{shot}}^{\text{spont}} + S_{\text{s}\times\text{n}}(f) + S_{\text{n}\times\text{n}}(f) \qquad \text{... (11.1)}$$

and that used for a binary 'nought' is:

$$S_0(f) = S_{\text{th}} + S_{\text{shot}}^{\text{spont}} + S_{\text{n}\times\text{n}}(f) \qquad \text{... (11.2)}$$

where (f) indicates a function of frequency, and the individual noise spectral densities represent receiver thermal noise (S_{th}), signal shot noise ($S_{\text{shot}}^{\text{sig}}$), spontaneous shot noise ($S_{\text{shot}}^{\text{spont}}$), signal times spontaneous noise ($S_{\text{s}\times\text{n}}(f)$) and spontaneous times spontaneous noise ($S_{\text{n}\times\text{n}}(f)$).

For analysis purposes, the thermal noise spectral density was derived from the specified value of minimum receiver power, and the other spectral densities were derived using the expressions below:

$$S_{shot}^{sig} = \frac{\eta e^2 \lambda_{sig} L}{hc} P_{sig}^{peak} \qquad \text{... (11.3)}$$

$$S_{shot}^{spont} = \frac{\eta e^2 \lambda_{sig} L}{hc} P_{spont}^{total} \qquad \text{... (11.4)}$$

(P_{spont}^{total} is the sum of the spontaneous noise power from both polarization states.)

$$S_{s \times n}(f) = 2\left(\frac{\eta e \lambda_{sig} L}{hc}\right)^2 P_{sig}^{peak} \left\{ S_{spont}(f + f_{sig}) + S_{spont}(f - f_{sig}) \right\} \qquad \text{... (11.5)}$$

(When deriving this expression the signal was assumed to be polarized, so only one spontaneous noise polarization state was used.)

$$S_{n \times n}(f) = 4\left(\frac{\eta e \lambda_{sig} L}{hc}\right)^2 \left\{ S_{spont}(f) * S_{spont}(f) \right\} \qquad \text{... (11.6)}$$

(When deriving this expression both polarization states were used because spontaneous noise is randomly polarized.)

In the above expressions λ is optical wavelength, e is electronic change, η is photodiode quantum efficiency, h is Plank's constant, c is the velocity of light, L is output section loss, P is optical power at the output of the amplifier, $S_{spont}(f)$ is the spontaneous noise power spectral density profile at the output of the amplifier, $S_{spont}(f \pm f_{sig})$ represents a frequency translated version of $S_{spont}(f)$, * indicates convolution, 'peak' and 'total' superscripts distinguish between peak and total values, respectively, and 'sig' and 'spont' subscripts distinguish between signal and spontaneous terms, respectively.

The above noise expressions show that, because the improved average power analysis technique for the fibre amplifier inherently gives the spectral density of the spontaneous noise output power, the $s \times n$ and $n \times n$ terms can be derived without having to either approximate the profile of the spontaneous noise power or use the amplifier noise figure [7], both of which have their limitations.

REFERENCES

1. Millar C A: 'Fibre amplifiers in optical telecommunications — a perspective', BT Technol J, 9, No 4, pp 5-11 (1991).

2. Desurvire E, Giles C R and Simpson J R: 'Gain saturation effects in high-speed multichannel erbium-doped fibre amplifiers at $\lambda = 1.53\ \mu m$', IEEE J Lightwave Technol, 7, pp 2095-2104 (1989).

3. Laming R I et al: 'Noise in erbium doped fibre amplifiers', Conf Proc ECOC '88, 1 (1988).

4. Goodman M S et al: 'Application of wavelength division multiplexing to communication networks', Conf Proc ICC '86, Toronto (1986).

5. Hodgkinson T G: 'Improved average power analysis technique for erbium-doped fibre amplifiers', IEEE Photonics Technol Lett, 4, pp 1273-1275 (1992).

6. Giles C R and Desurvire E: 'Modelling erbium-doped fibre amplifiers', IEEE J Lightwave Technol, 9, pp 271-283 (1991).

7. Olshansky R: 'Noise figure for erbium-doped optical fibre amplifiers', Electron Lett, 24, pp 1363-1365 (1988).

12

TOWARDS A PRACTICAL 1.3 μm OPTICAL FIBRE AMPLIFIER

T J Whitley, R Wyatt, D Szebesta and S T Davey

12.1 INTRODUCTION

The past few years have seen an unprecedented change in the capabilities of optical fibre communications systems [1-3]. Surprisingly, this change can be ascribed almost uniquely to the emergence of a single optical component — the erbium-doped fibre amplifier [4, 5]. The great impact of erbium-doped fibre amplifiers is usefully illustrated by considering a state-of-the-art, point-to-point transmission system circa 1990.

In Fig. 12.1, information, in the form of optical pulses, is transmitted over a length of low-loss optical fibre, and then converted back into an electrical signal at the destination. Unfortunately, because even the purest glass scatters light to some degree, the optical signal is attenuated as it travels along the optical fibre. This, combined with the fact that there is a minimum level to which the signal can fall after which it becomes unintelligible, means that there is a limit to the distance over which information can be transmitted. A 1990 state-of-the-art system would thus be limited in span to around 70 km when sending data at a rate of 2.5 Gbit/s (2.5×10^9 bit/s). In order to transmit data over longer distances, electronic regeneration of the signal is thus required every 70 km. The need for electronic regeneration, apart from being expensive and increasingly difficult as data rates increase, imposes many restrictions on the network design. What was clearly needed was a means

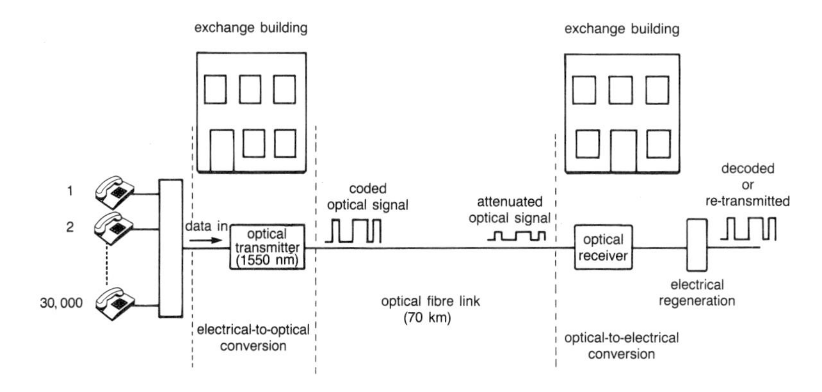

Fig. 12.1 High data rate transmission circa 1990 (state of the art).

of providing distortion-free amplification without the need to convert back into the electrical domain. The erbium-doped fibre amplifier provided this function with remarkable success giving the potential for transmission systems operating at tens of Gbit/s over transmission spans of many thousands of kilometres [6].

The erbium-doped fibre amplifier (EDFA) has had a similarly dramatic impact on the potential performance of multi-way distribution networks, increasing potential network size by two or three orders of magnitude and thus enabling literally millions of customers to be served from a single transmitter [7]. In short, the erbium-doped fibre amplifier has enabled network designers to throw away many of the power budget and line-rate constraints which previously took a dominant role in network design strategies. Interestingly, fibre amplifiers should also be beneficial from a network management viewpoint, as the simplified network topologies which EDFAs allow will reduce the management overhead of future communications systems (see Chapter 3).

However, erbium-doped fibre amplifiers only provide amplification at wavelengths around 1.5 μm, which means that all of the advances described above are limited to systems operating within the third telecommunications window[1]. Unfortunately, most installed terrestrial communications systems world-wide, including BT's own trunk network, operate at wavelengths around 1.3 μm in the so-called 'second telecommunications window'.

[1] The term 'telecommunications window' relates to a spectral region in which optical telecommunications systems have traditionally operated. These windows arise from the fact that the spectral attenuation characteristic of silica optical fibre exhibits three regions of low loss. The first window is centred at wavelengths around 0.85 μm, the second around 1.3 μm and the third around 1.55 μm.

Therefore, a practical optical amplifier operating in this region of the spectrum is of key importance as it would have a major role to play in 'near-term' network upgrade strategy. More importantly, the diversity of low- and high-bandwidth services which future networks will have to support make it likely that they will operate at wavelengths spanning all three of the current telecommunications windows [8]. As such the availability of efficient optical amplification over a wide range of wavelengths will be of critical importance as international telecommunications operators move towards the next century.

Though various options have been explored, in the search for a practical second window amplifier, the primary breakthrough came only two years ago with the almost simultaneous observation by three international research groups of optical gain in a new type of doped fibre amplifier [9-11]. This scheme, first proposed by Davey and France [12], was based upon doping the core region of fluoride optical fibre with the rare-earth element praseodymium (Pr^{3+}). This chapter aims to review the progress that has been made over the intervening two years towards a practical praseodymium-doped fluoride fibre amplifier (PDFFA), with particular emphasis on recent experimental results obtained at BT Laboratories [13, 14].

12.2 FIBRE AMPLIFIERS

In order to illustrate the operation of the PDFFA it is instructive to review the principles behind a general fibre laser amplifier. In its most rudimentary form this consists of:

- several metres of optical fibre, the core of which has been 'doped' with an appropriate rare-earth ion;
- an optical source to provide the energy required to achieve amplification — referred to as the 'pump' laser, this is usually in the form of a semiconductor laser diode capable of 'exciting' the rare-earth ion;
- a means of combining the light at the signal wavelength and the 'pump' wavelength.

Figure 12.2 illustrates the way in which these basic elements combine to provide optical amplification. The signal to be amplified is combined with the 'pump' light and then coupled into the doped optical fibre. As these two optical fields travel down the fibre the pump light is absorbed by the dopant ion and its energy stored (for a short time) in the fibre core. A proportion of this energy is then transferred to the signal. In this way the signal grows as it propagates through the fibre while the pump light decays. The

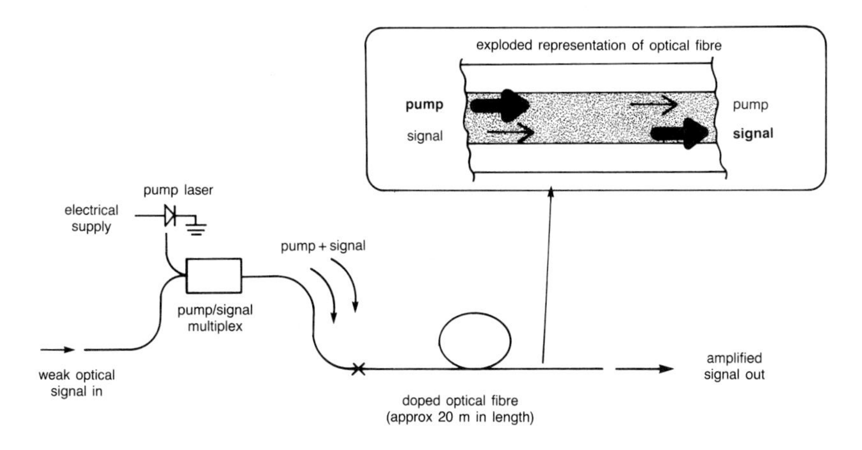

Fig.12.2 The basic elements of a doped fibre amplifier.

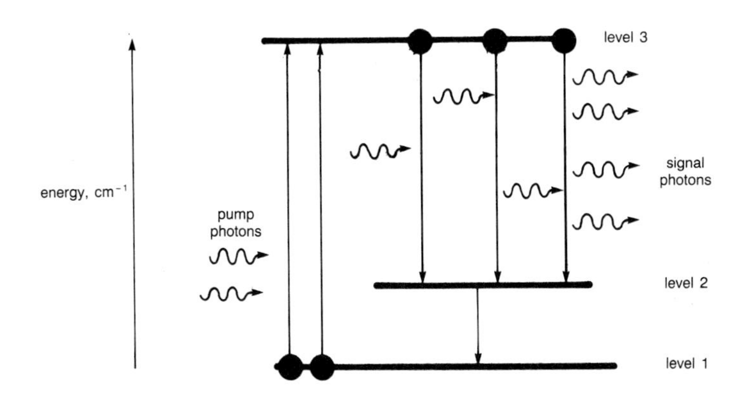

Fig. 12.3 Ensemble energy level structure of ideal dopant ions.

mechanism for energy transfer from pump to signal can be understood by reference to Fig. 12.3, which depicts the energy level diagram of the dopant ion, each level representing a particular 'energy state' in which the dopant ion can exist.

If the photon energy of the pump light is such that it matches the energy difference between levels 1 and 3, then as the pump light travels down the optical fibre it will be absorbed by the dopant ion. This absorption process excites a proportion of the dopant ions out of level 1, their ground state, and into level 3. If the energy of the signal photons matches the energy difference between levels 2 and 3, then signal photons traveling through the optical fibre can cause ions in level 3 to drop down into level 2, creating a signal photon as they do so. This coherent interaction, known as 'stimulated emission', results in the creation of more photons at the signal wavelength. As illustrated in Fig. 12.3, it is these interactions between the dopant ion and the optical fields that result in the absorption of the pump and the growth of the signal as they travel through the fibre. The probability for stimulated emission to occur is proportional to the likelihood that a photon will 'see' an excited ion. This implies that the gain coefficient in an amplifier is proportional to the number of ions in the excited state.

It can thus be seen how the process of stimulated emission can be used to amplify optical signals without the need to return to the electrical domain. Having examined the basic principles behind doped fibre amplifiers, attention is now turned to the specific question of a 'second window' amplifier.

12.3 THE SEARCH FOR A 1.3 μm FIBRE AMPLIFIER

12.3.1 Choice of dopant

In searching for an amplifier that will operate at a wavelength of 1.3 μm there are two primary criteria which must be satisfied. Firstly, a dopant ion must be found in which the energy difference between levels 2 and 3 corresponds to signal wavelengths around 1.3 μm. Secondly, the energy level structure of the ion must enable excitation of ground-state ions into level 3 through the use of a practically viable pump laser. A search through the energy levels of various rare earth ions quickly leads to the conclusion that praseodymium (Pr^{3+}) has the potential to satisfy both these requirements [15]. As is illustrated in Fig. 12.4, the $^1G_4$ state of praseodymium lies 7690 cm^{-1} above the $^3H_5$ state, an energy difference which corresponds almost exactly to a wavelength of 1.3 μm. It can also be seen from this figure that optical pumping at a wavelength around 1 μm will promote ions out of the ground state and into the desired $^1G_4$ level.

Having shown that praseodymium satisfies the two primary criteria, it is now necessary to consider in more detail the factors which make an amplification scheme viable. A question of primary importance in designing any amplifier is: 'How much power does it consume in order to provide a given level of amplification?'

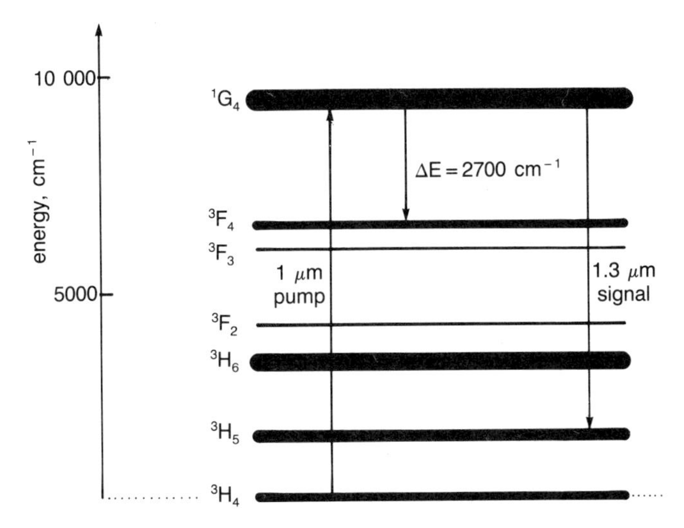

Fig. 12.4 Partial energy level diagram of the rare earth praseodymium (Pr^{3+}).

12.3.2 Pump power requirements

As discussed in section 12.2, the optical gain in a fibre amplifier is proportional to the number of dopant ions within the fibre core which are in the excited energy state (in the case of Pr^{3+}, the $^1G_4$ state). Thus, the power efficiency of a fibre amplifier is related to the amount of optical pump power that is required to maintain a significant number of ions in their excited state. Unfortunately, once an ion has been 'pumped' into an excited energy state there is a finite probability that it will 'spontaneously' decay back down into its ground state. Given a collection of excited ions, such as those in the core of a fibre amplifier, the rate at which these ions decay will be governed by the 'lifetime' of the excited state. Therefore, the pumping process within a fibre amplifier can be regarded as a competition between the rate at which ions are pumped into the upper level and the rate at which they spontaneously decay back down to the ground state. Thus the dynamics of the system result in an inverse dependence of pump power on lifetime. For instance, the excited state lifetime of erbium in silica glass is around 10 ms, which results in pump powers of a few milliwatts, producing gains in excess of 20 dB ($\times$ 100).

Therefore, the primary subject to be addressed, in the context of this potential 1.3 μm fibre amplifier, is the lifetime of the energy state labelled as $^1G_4$ in Fig. 12.4. Unfortunately, when praseodymium is incorporated into the core of an optical fibre fabricated from **silica** glass it turns out that the excited state lifetime is less than 1 μs. Remembering the inverse relationship between pump power and lifetime, this means that pump powers well in excess of 10 W would be needed to produce optical amplification! This clearly indicates that praseodymium-doped **silica** optical fibre will not provide a practical 1.3 μm amplifier.

12.3.3 Can the lifetime be changed?

Given that **erbium** in a silica glass has an upper state lifetime of 10 ms, why then is the lifetime of **praseodymium** in silica glass so short? To answer this question it is necessary to consider in more detail the factors governing the lifetime of an ionic energy level. The lifetime of a particular energy state is governed by the various ways in which the ion can give up its energy and thus return to its ground state. In erbium the primary mechanism by which the excited ions give up their energy is through a 'radiative' spontaneous decay process in which the energy contained in the excited ion is converted into a photon at the signal wavelength, thus returning the ion to its ground state. However, in some circumstances an excited ion can lose its energy without the production of a photon, in a process described as a non-radiative decay. Effectively, the energy of the excited ion is converted into vibrations within the glass fibre, known as phonons. The probability that an excited ion will lose energy in this way is critically dependent upon two parameters. The first of these is the energy gap between the excited state and the energy state immediately beneath it, indicated as ΔE in Fig. 12.4. The second parameter is a quantity known as the 'phonon energy', $\hbar\omega$, of the glass in which the ions are embedded. This quantity relates to the maximum amount of energy which can be dissipated as vibrations within the glass through the production of a single phonon.

The likelihood that the ion will undergo a non-radiative decay is related to the number of phonons that are required to bridge this energy gap. Generally speaking, the more phonons that are required to bridge an energy gap, then the less likely it is that there will be non-radiative decay [16]. Taking the specific example of praseodymium in silica glass, the energy gap between the $^1G_4$ state and the $^3F_4$ state is around 2700 cm^{-1}, while the highest phonon energy for silica glass is approximately 1100 cm^{-1}. Therefore, in a silica glass only three phonons are necessary to bridge this gap, leading to a high probability of non-radiative decay. This means that as ions are pumped into the $^1G_4$ level they will very rapidly decay back down to the ground state through a ladder of non-radiative decays. It is this non-radiative decay process that so drastically reduces the upper-state lifetime of praseodymium in silica glass fibre, with the consequent dramatic increase in pump power requirements.

The question which now arises is: 'Can an alternative glass be found in which the non-radiative decay process does not dominate the decay out of the excited state?'

12.3.4 Fluoride fibre as an alternative host glass

Optical fibres made out of the heavy metal fluoride glasses were originally developed for long distance optical telecommunications as they offered the potential for ultra-low-loss transmission at wavelengths around 2.55 μm [17]. The imperative behind this work was overtaken by the emergence of the EDFA which, as has been shown, effectively provides limitless transmission distance over existing silica fibre. Ironically, it is the fluoride glass technology developed for long-distance transmission that has proved to be critical to the viability of the 1.3 μm fibre amplifier.

The non-radiative decay process, so detrimental to Pr^{3+} in a silica glass, is strongly dependent on the way in which energy is dissipated as phonons within the glass structure of the fibre. This in turn is strongly dependent upon parameters such as the mass of the glass constituents and the strength of the bonds between these constituents. Because fluoride glass fibre is composed of heavy molecules, the amount of energy which can be supported by a single phonon within the glass is lower than in silica. Although many different fluorozirconate glass compositions have been investigated, the most stable discovered to date is known as ZBLAN and this glass or its derivatives are therefore preferred for fibre fabrication. The glass is based on the fluorides of zirconium, barium, lanthanum, aluminium and sodium, hence the acronym ZBLAN. The phonon energy of ZBLAN is around half that of silica glass. This reduced phonon energy means that approximately six phonons are now required to bridge the energy gap between the $^1G_4$ state and the $^3F_4$ state in praseodymium, thus drastically reducing the probability of non-radiative decay[1]. When Pr^{3+} is incorporated into a ZBLAN glass the lifetime of the $^1G_4$ energy state is around 100 μs which, though much shorter than the 10 ms lifetime of erbium, is sufficient to allow substantial amplification to be obtained.

The following section describes recent experimental results obtained at BT Laboratories to illustrate the performance now possible from state-of-the-art 1.3 μm fibre amplifiers in a laboratory environment.

---

[1] The reduced non-radiative decay probability in doped fluoride glasses leads to many long-lived states and therefore many radiative transitions. It also leads to a higher probability of pump excited state absorption. In this process ions in an excited state absorb a pump photon and are thus excited into an even higher energy state. This leads to many interesting frequency up-conversion schemes in which light is obtained at wavelengths significantly shorter than the pump wavelength. Many of the exciting new possibilities for fibre lasers operating in the visible region of the spectrum are reviewed in Chapter 13.

12.4 EXPERIMENTAL PROGRESS

12.4.1 Introduction

The first experimental results on praseodymium-doped fluoride fibre amplifiers confirmed that by optically exciting short lengths of Pr^{3+}-doped AN fibre, modest gain could indeed be achieved [9]. These early results also confirmed that the gain spectrum was ideally suited to telecommunications systems operating within the second telecommunications window, being centred almost exactly at a wavelength of 1.3 μm. However, the amount of pump power required to achieve even modest gains was relatively high, typically 180 mW of pump power producing around 5 dB gain. As discussed previously this comparatively poor efficiency is a direct result of the relatively short lifetime of the $^1G_4$ upper state. To increase the efficiency with which pump power produces gain requires either that the excited-state lifetime be increased or that the rate at which ions are pumped into the excited state be increased. Given that the lifetime is fixed by the glass composition, the approach that has been taken is to alter the parameters of the optical fibre in such a way that the pump rate is increased. The rate at which a given level of pump power excites ions into the excited state is related to the intensity (power/area) of the optical field within the fibre. If the optical power can be confined to a very small region of the core, then the intensity of the light will increase and thus the pump rate will increase. This can be achieved in practice by increasing a parameter known as the numerical aperture (NA) of the fibre [18]. The following section presents the results of a recent experimental investigation, carried out at BT Laboratories, into the performance of a PDFFA based on a high-NA fibre design.

12.4.2 Efficient amplification using high NA praseodymium-doped fibre

An amplifier's characteristics will depend upon the particular way in which it is deployed within a telecommunications system. For instance, if the amplifier is deployed immediately prior to the optical receiver, then the signal level entering the amplifier will be small and the amplifier is said to be operating in the 'small signal' regime. The primary metric of amplifier performance in this case is the amount of pump power required to achieve a given level of gain (i.e. dB gain per mW of pump power). If on the other hand an amplifier is deployed as a power-booster at the beginning of a transmission system then the level of input signal will be large and the amplifier is said to be operating in the saturated regime. In this case it is more useful to consider the efficiency with which the optical pump power is converted into signal rather than the actual gain of the amplifier. These important regimes of operation are now considered.

12.4.2.1 Experimental details

The fibre used during these measurements was fabricated within the network research division at BT Laboratories (BTL) and is based on a derivative of ZBLAN known as ZHBLAYLiNP. The optical properties of this high NA fibre are summarized in Table 12.1.

Table 12.1 Optical characteristics of high NA Pr^{3+}-doped fibre.

Doping density	500 ppmW
Cut-off wavelength	0.65 μm
Δn (= $[NA^2]/2n$)	0.05
Parasitic background loss at 1.01 μm	0.14 dB/m
Parasitic background loss at 1.3 μm	0.08 dB/m
Ionic loss due to Pr^{3+} at 1.01 μm	1 dB/m
Ionic loss due to Pr^{3+} at 1.3 μm	0.14 dB/m

A schematic diagram of the set-up used for this series of measurements is shown in Fig. 12.5. This has been designed to allow measurement of small signal and saturated performance, for both co-directional and counter-directional pumping. The 1.01 μm pump light, derived from a Ti:sapphire laser, and the 1.3 μm signal light, either from a DFB laser, or a grating tuned external cavity laser (LEC), were combined, in the case of a co-directionally pumped amplifier, or separated, in the case of a counter-directionally pumped amplifier, using a fused fibre multiplexer. The multiplexer was then fusion-spliced on to a short length of 0.04 Δn silica fibre which in turn was butt-jointed to the fluoride fibre using a precision alignment jig. In this way the large potential mode mismatch loss between the 0.05 Δn fluoride fibre and the low Δn silica multiplexer fibre (3-4 dB) was reduced, as the splice loss between the differing silica fibres can be much less than the direct butt loss when fused, due to the diffusion occurring during the fusing process. The total pump launch efficiency with this arrangement, from power entering the microscope objective to power in the doped fibre, was 58%. The signal loss from the input and output connectors to the doped fibre was around 1.2 dB. Isolators were used at both input and output in order to minimize the possibility of lasing from stray reflections, and to

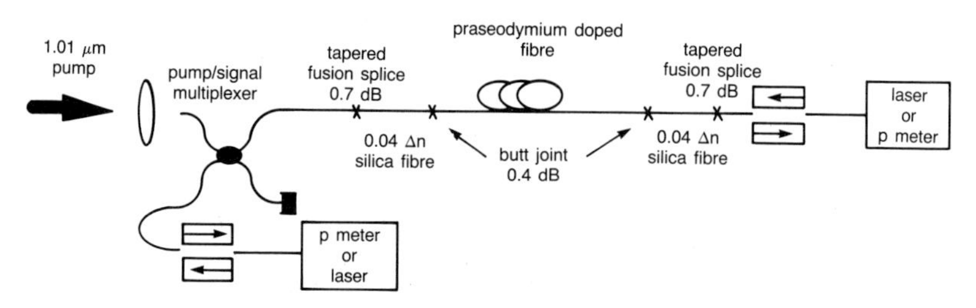

Fig. 12.5 Experimental amplifier configuration.

ensure that the signal laser remained stable under all conditions. For the measurements of small-signal gain, an optical spectrum analyser was used to monitor the signal amplification, and discriminate against the spontaneous background from the amplifier, while, for the power amplifier measurements, a calibrated thermopile measured the total output power, as the isolator blocks any remnant pump power, and the spontaneous background is negligible under these conditions. All measurements are referred to the input and output of the doped fibre, rather than the input and output connectors, as this gives an indication of the intrinsic capabilities of the fibre itself.

12.4.2.2 Small-signal gain measurements

The small-signal gain characteristics of a 23 m length of praseodymium-doped fibre were investigated for a range of launched pump powers, using the output from the tuneable LEC laser, which was tuned to 1.31 μm.

Figure 12.6 depicts the evolution of net gain with absorbed pump power for a pump wavelength of 1.01 μm. The input signal power during this measurement was 1 μW, to ensure operation was well into the small-signal regime. A maximum gain of 30 dB (× 1000) is obtained, for a launched pump power of 500 mW, corresponding to a connectorized and isolated gain of 26 dB, when coupling losses are accounted for.

An indication of intrinsic amplifier small-signal performance may be obtained by taking the initial slope of the gain curve; in this case, a gross fibre gain of 10 dB was achieved for an absorbed pump power of only 80 mW, representing a small-signal gain efficiency of 0.13 dB/mW. This result, although not as high as some reports [19], represents a considerable improvement over earlier work using lower-NA fibre [10], and is in reasonable agreement with a simple analytic-model [20], using our measured fibre parameters.

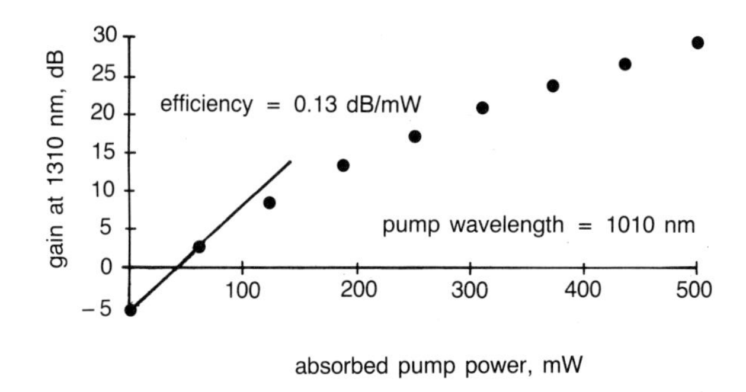

Fig. 12.6 Evolution of net gain at 1.3 μm with pump power.

Critical to the viability of any amplifier is the spectral range over which optical gain is available. The spectral gain characteristics were measured by replacing the single-wavelength laser input signal by a broad-spectrum ELED source. These results are depicted in Fig. 12.7, where the net gain spectrum is shown for a pump power of 490 mW. The peak net gain is seen to be 30 dB, with >20 dB obtainable over a 48 nm range from 1283 nm to 1331 nm. As can be seen from Fig. 12.7, this wavelength range covers a large portion of the second telecommunications window, which is also depicted. The 3 dB gain bandwidth is approximately 30 nm under these operating conditions.

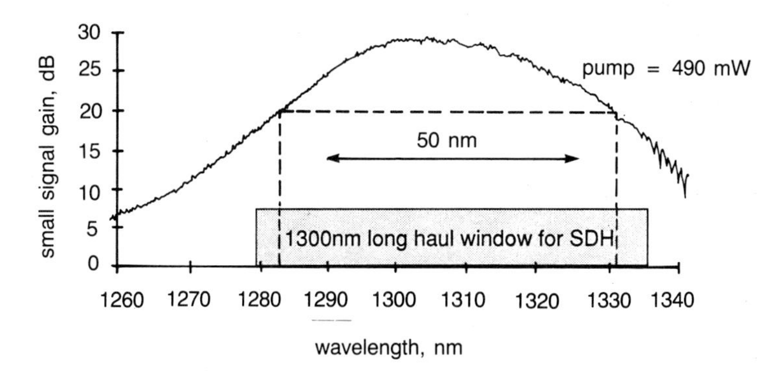

Fig. 12.7 Spectral variation of small signal gain.

12.4.2.3 Power amplifier measurements

If the input signal power to an optical amplifier is increased beyond a certain level the gain will start to reduce. This 'gain saturation' occurs when the number of ions stimulated into emission by the signal light becomes sizeable compared with the total number of excited ions within the fibre. Unlike an electrical amplifier, operation deep into saturation need not be a problem, as the amplification remains essentially linear for modulation frequencies higher than a few kilohertz. To investigate the saturation regime, the gain was measured for a range of input signal powers.

Figure 12.8 shows, for the case of counter-directional pumping, the variation of net gain as a function of output signal power, referenced to the output of the doped fibre. The launched pump power was fixed at 580 mW for these measurements. This graph indicates a maximum small signal net gain of 28 dB and a 3 dB saturation output power of +17 dBm. The maximum amplified output power for this level of pump power can be seen to be +20.7 dBm (just over 100 mW).

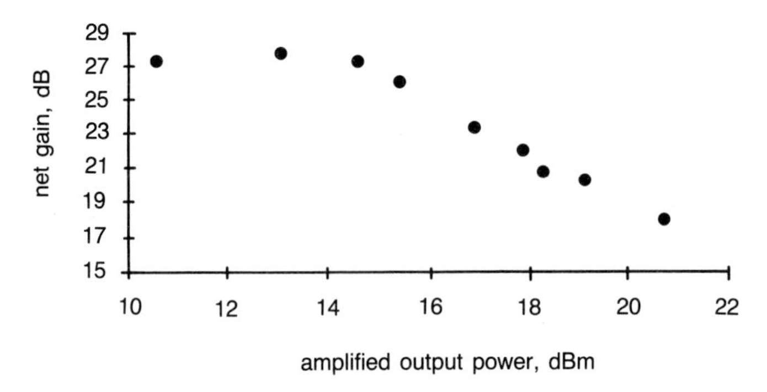

Fig. 12.8 Variation of gain with output signal power.

To assess likely performance as a power amplifier, it is usual to fix the input signal power level, and then vary the launched pump power up and down. In this way, a laser-like characteristic of amplified signal power against pump power is obtained, the slope of which is a measure of the intrinsic ability of the power amplifier to transfer energy from the pump source to the output signal.

Figure 12.9 depicts the evolution of output signal power with launched pump power in both the co- and counter-directionally pumped formats, with the input signal power fixed at approximately +2 dBm. In the counter-pumped case a maximum saturated output power of +23.2 dBm, or 212 mW, is obtained for a launched pump power of 900 mW representing a 24% conversion from pump light to signal light. In this format the slope efficiency with which pump is converted to signal is 30%, a value which is believed to be the highest yet obtained for a PDFFA. Allowing for the wavelength difference between pump and signal, this slope represents a differential 40% conversion from pump photons to signal photons. Given the 3 dB background loss in this length of fibre, this result emphatically confirms that efficient power amplification is possible even in an amplifier with comparatively low small signal efficiency. With a co-propagating pump the highest output signal power obtained was 116 mW for a pump power of 760 mW. The slope efficiency in this case was 21%. The higher efficiency obtained in the counter-propagating configuration is probably related to the presence of parasitic fibre-loss processes. These results are a considerable improvement on previously reported work [21, 22], and confirm that a low small signal gain efficiency is in no way a bar to the production of a high-efficiency power amplifier.

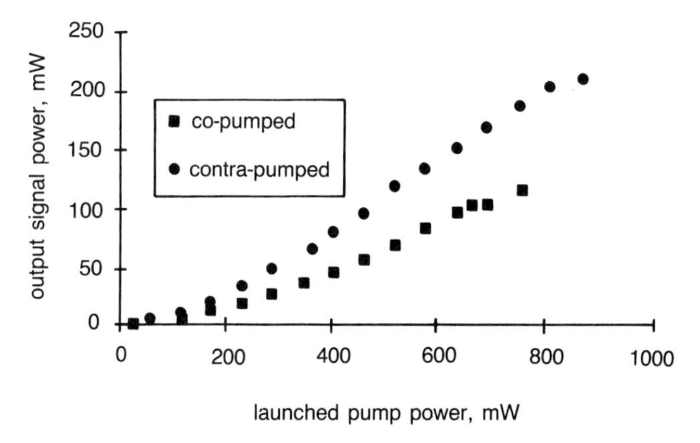

Fig. 12.9 Amplified output signal power versus launched pump power.

Having established that high output powers are possible, it is obvious that the next enquiry must be to establish the range of wavelengths over which these powers are available. To answer this question the tuneable LEC laser was used to measure the saturated gain spectrum. In this case, the signal input was set at + 1 dBm, with a pump power of 615 mW. These results are shown in Fig. 12.10, indicating that, for this pump power level, in excess of 100 mW amplified output can be obtained over a wavelength range of 33 nm, from 1284 nm to 1317 nm. This suggests application as a power amplifier for a wide range of signal wavelengths. For instance, such an amplifier could be used as a power booster at the start of a system — increasing the span in a point-to-point link, or dramatically increasing the possible number of customers in a multiway distribution network.

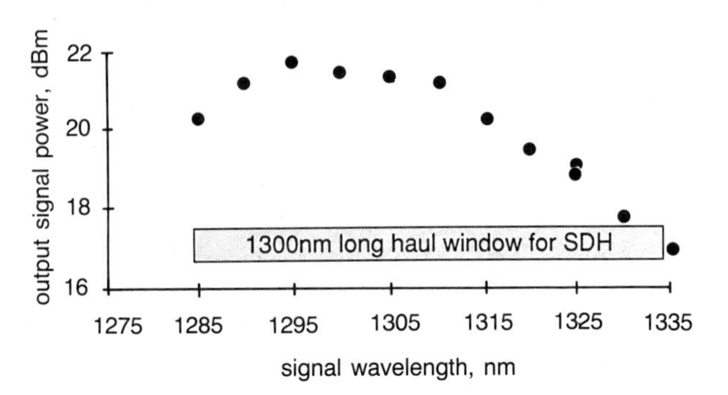

Fig. 12.10 Saturated gain spectrum.

12.4.3 Real-world pump sources for PDFFAs

12.4.3.1 General considerations

The results of the previous section, and those of other workers [21, 22], clearly indicate that the PDFFA has the potential to provide amplification over a key range of telecommunications wavelengths. The primary issue at present, as to the viability of this amplifier, relates to the pump source. In the previous section the optical pump light was obtained from a Ti:sapphire laser pumped by a large frame ion laser. This device consumes several tens of kilowatts of electrical power and as such could hardly be considered a practical real-world pump source. However, several options do exist which could provide sufficient optical power to enable useful amounts of gain. A recent report from NTT has demonstrated in excess of 15 dB small-signal gain from a PDFFA pumped using four semiconductor laser diodes [23]. This result already demonstrates the feasibility of a small-signal amplifier using practical pump technology.

In the power amplifier regime it is less clear that direct diode laser pumping could provide sufficient optical power, as a power amplifier providing 200 mW of amplified output is likely to require in excess of 500 mW of pump power. However, the ready commercial availability of high-power diode-pumped solid-state lasers suggests their use as a pump source for the PDFFA. In particular, the recent introduction of the Nd:YLF laser system [24], which operates at a wavelength of 1.047 μm, is of particular interest.

Figure 12.11 depicts the measured praseodymium pump absorption band, which is centred at 1.01 μm. This shows that, at 1.047 μm, the pump absorption coefficient is reduced to about 30% of its peak value, but nevertheless substantial absorption still remains, due to the extremely broad nature of the pump transition in praseodymium. In order to investigate the possibility of pumping PDFFAs at a wavelength of 1.047 μm, a series of measurements has been carried out in which the Ti:sapphire pump source was replaced by a commercial Nd:YLF laser. The reduction in pump absorption associated with pumping at a wavelength away from the peak can be counteracted by increasing the length of the amplifier. However, increasing the fibre length means that the background loss of the fibre will assume greater importance. For this reason, during these measurements, a fibre with an extremely low background loss of 0.03 dB/m was used, although its Δn, 0.014, was lower than that of the fibre described in the previous section.

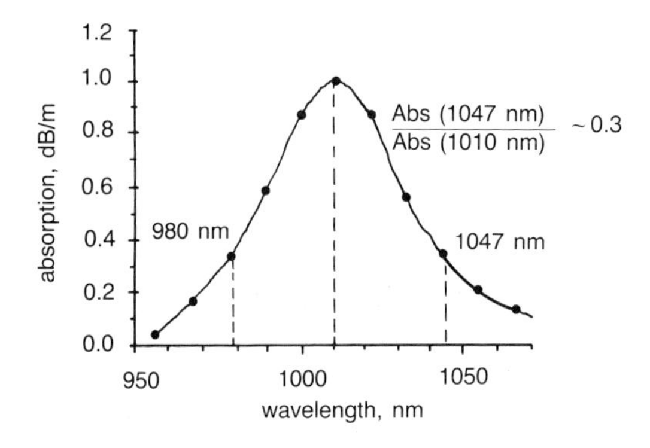

Fig. 12.11 Pump absorption band in praseodymium.

12.4.3.2 Experimental arrangement

Figure 12.12 shows the experimental configuration employed throughout this series of measurements. The amplifying fibre, doped to around 500 ppmw with praseodymium, was 30 m in length and had a cut-off wavelength around 800 nm. The pump source was a commercial diode-pumped Nd:YLF laser (Spectra-Physics TFR104C), providing >3 W TEM_{00} output at 1.047 μm. The output from the Nd:YLF laser was launched into a 50/50 fibre coupler to split the output into two, and then coupled into both ends of the doped fibre using two fibre multiplexers. This bidirectional pumping scheme was chosen to minimize thermal loading in the doped fibre. The interface between the fluoride fibre and silica multiplexers was a simple butt-joint, with a loss of 0.4 dB due to mode mismatch. Isolators were used at both input and output as before. The total unpumped loss at the pump wavelength over the 30 m length was around 11.7 dB of which ~0.9 dB was due to fibre background loss, implying an ionic loss of 0.35 dB/m. The ionic loss at the pump wavelength is therefore more than ten times larger than the background loss, and so little reduction in efficiency is expected from this cause. At a wavelength of 1.3 μm the background loss was again ~0.9 dB while the residual ionic loss was 1.6 dB.

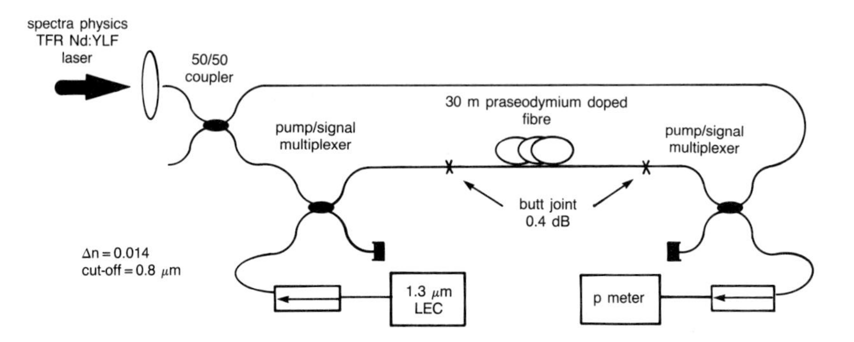

Fig. 12.12 Experimental arrangement for Nd:YLF pumped amplifier.

12.4.3.3 Small-signal performance

Figure 12.13 depicts the evolution of net small-signal gain at a signal wavelength of 1.3 μm as a function of total launched pump power. A maximum net gain of 30 dB, referred to the input and output of the doped fibre, is obtained for a launched pump power of 1.8 W, while 15 dB net gain is achieved for a pump power of 500 mW. The initial slope is 0.042 dB/mW, which is consistent with the value expected for this fibre, thus indicating that there is no significant penalty in small-signal gain efficiency due to pumping well into the long-wavelength wing of the pump absorption band.

The roll-over in gain seen at high pump powers is, it is believed, partially due to incomplete pump absorption, although this is difficult to quantify with a bidirectional pumping scheme.

Figure 12.14 shows the small-signal net gain spectrum for a fixed launched pump power of 1.1 W, measured by replacing the tuneable LEC laser by an ELED, as before. Net gains in excess of 20 dB are obtained over a range of 30 nm from 1290 nm to 1320 nm. This value is less than obtained in the previous section due to the peak gain being lower; the 3 dB gain bandwidth is unchanged.

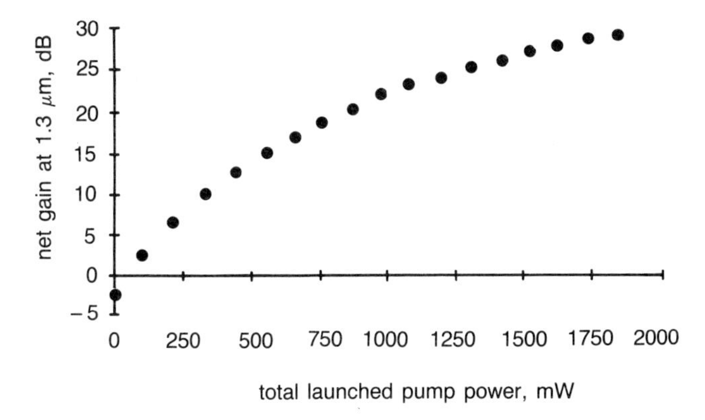

Fig. 12.13 Evolution of small-signal gain with launched pump power.

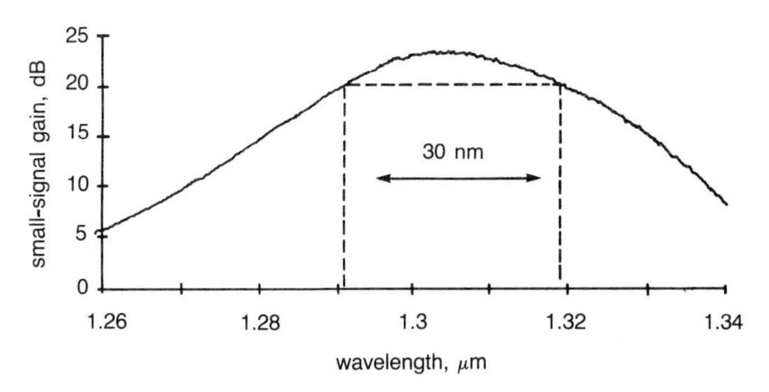

Fig. 12.14 Small-signal gain spectrum.

12.4.3.4 Power amplifier performance

Figure 12.15 shows the variation of output signal power with launched pump power. The signal input power in this case was +1 dBm at a wavelength of 1.3 μm.

A maximum output power of 250 mW was obtained for a launched pump power of 1.85 W. This represents the highest output power obtained from any PDFFA at 1.3 μm, irrespective of pump source. The power slope efficiency is around 18% while the extrapolated threshold is approximately 500 mW. This value of slope efficiency is less than achieved with the high-Δn fibre and Ti:sapphire pumping, as described above. Because of the lower background loss, it may be expected that the transfer efficiency would actually be higher in this fibre, rather than worse, as slope efficiency is not expected to depend on Δn. While part of the efficiency reduction is undoubtedly due to the fibre length being shorter than optimum (the graphs are plotted against launched pump power, rather than absorbed pump, as previously), the possibility remains that this pump wavelength is inherently less efficient than wavelengths nearer the peak absorption. This is currently the subject of a numerical study, together with more detailed measurements.

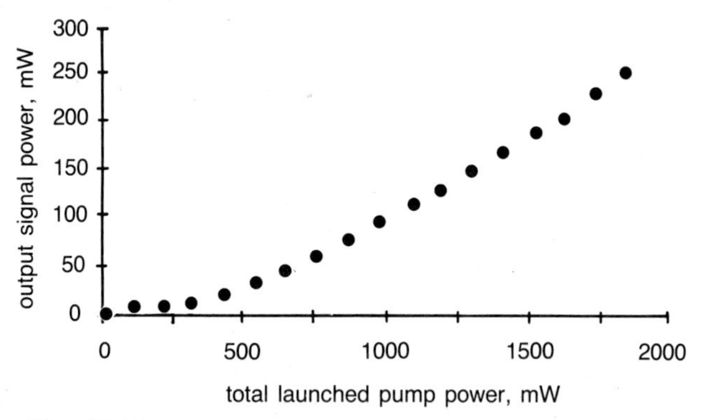

Fig. 12.15 Output signal power versus launched pump power.

With a slightly lower input signal power of 0 dBm and a fixed pump power of 1.6 W the saturated gain spectrum was measured. Figure 12.16 shows that output powers in excess of 100 mW can be achieved over a spectral region from 1285 nm to 1317 nm, a bandwidth >30 nm, again indicating applicability over a major part of the second telecommunications window. This result is very similar to that obtained with Ti:sapphire pumping, demonstrating once again high output power and broad spectral coverage.

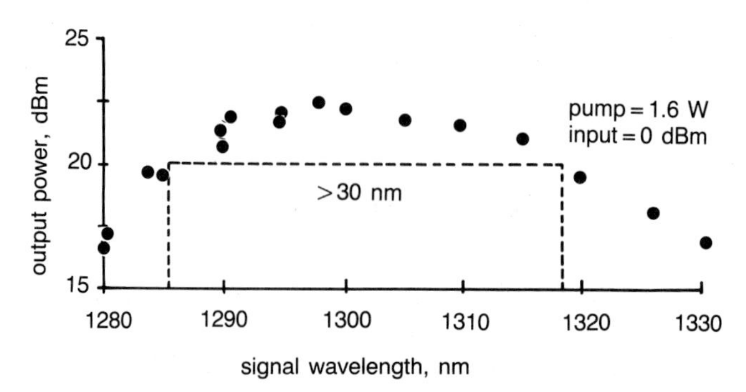

Fig. 12.16 Saturated gain spectrum.

The above results indicate that, with current fibre designs, diode-pumped solid-state lasers can provide sufficient pump power for practical amplifiers. Fibre-pigtailed Nd:YLF lasers are commercially available now, and have fibre coupled powers in excess of 1 W. On the basis of our measurements (optimized to make best use of the higher powers from the pump laser, and hence not necessarily optimum for lower pump powers), this is sufficient to provide >20 dB small-signal gain, and 100 mW output power, by using a compact pump source of this type. Optimization of the fibre design will improve performance still further, as discussed in section 12.5 below.

12.4.4 Preliminary systems experiments

In order to obtain an early indication of the way in which this amplifier will perform in a real telecommunications system, a preliminary digital system experiment was carried out, at a data rate of 2.5 Gbit/s. In this experiment the use of a PDFFA in three critical positions within a point-to-point transmission link was simulated. The use of the amplifier as a power amplifier, just after the transmitter, as a repeater, mid-way along the system, and as a preamplifier, just prior to the receiver, was simulated by using variable amounts of attenuation between the transmitter, amplifier and receiver. The amplifier, utilizing the 0.014 Δn fibre described in section 12.4.3, exhibited a connectorized small signal gain of 19 dB and was capable of delivering a maximum output of $+17.4$ dBm when used as a power amplifier (full details of amplifier and system measurements are given in Lobbett et al [25]).

When used as an in-line repeater the full 19 dB gain was available as increased system range. Configured as a power amplifier, bit error rate measurements confirmed that transient changes in the gain due to signal saturation happen too slowly to cause any significant penalty for a 2.5 Gbit/s digital system. Utilizing the amplifier as a preamplifier, deployed immediately prior to the standard commercial optical receiver, resulted in a sensitivity enhancement of 12 dB. This is consistent with the expected amplifier noise performance.

12.5 DISCUSSION

The previous section described the performance currently available from PDFFAs in a laboratory environment. It is useful here to consider what improvements in performance are likely in the near future, through fibre design optimization and pump source development, and to give some indication of future device expectations.

In the case where the host glass is kept essentially unchanged, but fibre losses are reduced to insignificant levels, then a small-signal gain coefficient of 0.2 dB/mW will be possible, resulting in a pump power requirement of 100 mW to achieve 20 dB gain. The saturated performance of such an amplifier would be characterized by a threshold power around 100 mW, and a slope efficiency of 60%.

The next stage of development would involve changing the actual glass composition of the optical fibre host. Such changes have already been shown to enhance the quantum efficiency of the 1.3 μm fluorescence by an order of magnitude in a praseodymium-doped bulk sample [26]. Assuming a modest doubling in quantum efficiency could be achieved via this route, then it should be possible to reduce the pump power necessary to achieve 20 dB gain to near 50 mW, a value comparable to the pump powers used in present-day commercial EDFAs. In such a glass, the power amplifier threshold would be reduced by a similar amount.

Many potential network applications call for amplified outputs well in excess of 100 mW; to obtain an amplified signal power of 200 mW, for example, will require a pump power close to 500 mW, for both optimized PDFFAs and EDFAs. This amount of pump power is certainly available from the Nd:YLF system, as described above, while recent advances in diode laser technology have resulted very recently in the commercial availability of integrated semiconductor oscillator-amplifiers with output powers at the 1 W level [27]. Use of these devices would make true diode-pumped operation of a PDFFA power amplifier a reality.

Although the discussion above indicates that efficient small-signal amplifiers and power amplifiers will become available in the near future, there still remains the issue of overall reliability of the amplifier as a network component. It is clear that, before large-scale deployment can be considered, reliability assessment must be carried out on the fibre itself, on the pump sources, and on the interface between the doped fibre and standard fibre. Work is already under way in this area, in a European collaborative 'RACE' project, 'GAIN' (glass amplifiers for integrated networks), which has a particular interest in the interface issues, and pump laser performance.

12.6 CONCLUSIONS

The results presented in this paper, combined with those of other workers [21, 22] represent a compelling body of evidence that PDFFAs could be an attractive option for networks requiring all-optical amplification within the second telecommunications window. It is now clear that acceptable small-signal efficiencies are possible, while in the saturated regime the power amplifier efficiency is extremely good. Furthermore, the spectral characteristics of these devices are ideally suited to second window applications, giving coverage over 50 nm from 1280 nm to 1330 nm.

However, many important issues have as yet to be resolved. Perhaps primary amongst these are questions relating to the noise and crosstalk performance and the question of a practical pump source. Theoretical considerations indicate that noise and crosstalk should not pose any problem to the operation of the PDFFA — near-ideal performance being expected in both cases. Furthermore, the preliminary systems experiments presented in this chapter offer encouraging evidence to this effect. With regard to the question of a viable 'real world' pump source, BTL's results and those of NTT [23] indicate that practical options do already exist, which, coupled with the current rapid development of semiconductor pump lasers around 1 μm, means that this issue should not ultimately deny the PDFFA's practicality.

The likely explosion of feature-rich broadband and narrowband services which is anticipated over the next decade means that radical changes within the transmission and switching fabric of current networks are unavoidable. Clearly the decisions made now regarding the most appropriate network upgrade strategy will be of critical importance to the ability of telecommunications companies to meet customer service demands in the next century. As discussed in Chapter 3, optical networks freed from line-rate and power budget constraints provide a uniquely flexible means of future-proofing the network while minimizing the network management overhead. A key question is therefore the means by which existing networks, with their mix of electrical switches, copper local access and 1.3 μm fibre transmission, can gracefully evolve into this high-capacity and highly flexible optical network. The praseodymium-doped fluoride fibre amplifier has the capability to play a central role in this evolutionary route, as it maximizes the potential of existing optical fibre networks while forming an essential building block for networks of the future.

REFERENCES

1. Millar C A: 'Fibre amplifiers in optical telecommunications — a perspective', BT Technol J, 9, No 4, pp 5-11 (1991).

2. Desurvire E: 'Erbium-doped fibre amplifiers for new generations of optical communications systems', Opt Photon News, 2, pp 6-11 (1991).

3. Urquhart P: 'Review of rare-earth doped fibre lasers and amplifiers', IEE Proc J (Optoelectron), 135, pp 385-407 (1988).

4. Mears R J, Reekie L, Morkel P R and Payne D N: 'Low-noise erbium doped fibre amplifier operating at 1.54 μm', Electron Lett, 23, p 1026 (1987).

5. Giles C R and Desurvire E: 'Modelling erbium-doped fibre amplifiers', IEEE J Lightwave Technol, 9, No 2, pp 271-283 (1991).

6. Bergano N S et al: '9000 km, 5 Gb/s NRZ transmission experiment using 274 erbium-doped fibre amplifiers', Proc OSA OAA92, Santa Fe, New Mexico (1992).

7. Forrester D S, Hill A M, Lobbett R A, Wyatt R and Carter S F: '39.81 Gb, 43.8 million way WDM broadcast network with 527 km range', Electron Lett, 22, pp 1882-1884 (1990).

8. Payne D B: 'Passive optical networks', Proc ECOC 92 (1992).

9. Ohishi Y, Kanamori T, Kitagawa T, Takahashi S, Snitzer E and Sigel G H: 'Pr^{3+}-doped fluoride fibre amplifier operating at 1.31 μm', Opt Lett, 16, No 22, pp 1747-1749 (1991).

10. Carter S F, Szebesta D, Davey S T, Wyatt R, Brierley M C and France P W: 'Amplification at 1.3 μm in a Pr^{3+}-doped single-mode fluorozirconate fibre', Electron Lett, 27, No 8, pp 628-629 (1991).

11. Durteste Y, Monerie M, Allain J Y and Poignant H: 'Amplification and lasing at 1.3 μm in praseodymium-doped fluorozirconate fibres', Electron Lett, 27, No 8, pp 628-629 (1991).

12. Davey S T and France P W: 'Rare earth doped fluorozirconate glasses for fibre devices', BT Technol J, 7, No 1, pp 58-68 (1989).

13. Whitley T J, Wyatt R, Szebesta D and Davey S: 'High output power from an efficient praseodymium doped fluoride fibre amplifier', Proc OSA, 17, (OAA92), Santa Fe, New Mexico, USA (1992).

14. Whitley T J, Wyatt R, Szebesta D and Davey S: 'Quarter watt output from a praseodymium doped fluoride fibre amplifier pumped with a diode-pumped Nd:YLF laser', Proc of OSA, 17, (OAA92), Santa Fe, New Mexico, USA (1992).

15. 'Handbook of Laser Science and Technology', p 23, CRC Press (1987).

16. Shinn M D, Sibley W A, Drexhage M G and Brown R N: 'Optical transitions of Er^{3+}-ions in fluorozirconate glass', Phys Rev B, 27, No 11 (1983).

17. France P W: 'Fluoride glass optical fibres', Ch 10 in: 'Applications', Blackie (1990).

18. Midwinter J: 'Optical Fibres for Transmission', Wiley (1979).

19. Miyajima Y, Sugawa T and Fukasaku Y: '38.2 dB amplification at 1.31 μm and possibility of 0.98 μm pumping in Pr^{3+}-doped fluoride fibre', Electron Lett, 27, pp 1706-1708 (1991).

20. Whitley T J and Wyatt R: 'Analytic expression for gain in an idealised 4-level doped fibre amplifier', Proc OSA, 17 (OAA92), Santa Fe, New Mexico, USA (1992).

21. Ohishi Y, Kanamori T, Nishi T and Takahashi S: 'A high gain, high saturation power Pr^{3+}-doped fluoride fibre amplifier operating at 1.3 μm', IEEE Photon Technol Lett, 3, No 8, pp 715-717 (1991).

22. Sugawa T and Miyajima Y: 'Gain and output-saturation limit in Pr^{3+}-doped fluoride fibre amplifier operating in the 1.3 μm band', IEEE Photon Technol Lett, 3, No 7, pp 616-618 (July 1991).

23. Yamada M, Shimizu M, Ohishi Y, Temmyo J, Wada M, Kanamori T, Horiguchi M and Takahashi S: '15.1 dB gain in a Pr^{3+}-doped fluoride fibre amplifier pumped by high power laser-diode modules', IEEE Photon Technol Lett, 4, No 9, pp 994-996 (1992).

24. Spectra-Physics TFR104C (1990).

25. Lobbett R, Wyatt R, Eardley P, Whitley T J, Smyth P, Szebesta D, Davey S T, Millar C A and Brierley M C: 'System characterisation of high gain and high saturated output power, Pr^{3+}-doped fluorozirconate fibre amplifier at 1.3 μm', Electron Lett, 27, No 16, pp 1473-1474 (1991).

26. Becker P C, Broer M M, Lambrecht V G, Bruce A J and Nykolak G: 'Pr^{3+}:La-Ga-S glass: a promising material for 1.3 μm fibre amplification', Proc OSA, 17 (OAA92), Santa Fe, New Mexico, USA (1992).

27. Spectra Diode Labs: '1 Watt cw single mode MOPA laser diode', SDL-5760 series.

13

VISIBLE FIBRE LASERS

M C Brierley, J F Massicott, T J Whitley, C A Millar, R Wyatt, S T Davey and D Szebesta

13.1 INTRODUCTION

A glimpse of the not-too-distant future reveals a world of personal full-colour visual communicators, visible in bright sunlight, probably with high-definition colour document transmission facilities. The home and office will use high-definition, three-dimensional colour images, generated on computers or received off-air or via cable, which can be projected anywhere under all light conditions. The home will have on-demand information services instantly available, and video entertainment services will be accessed on demand. A convergence of computer and communications technologies will have taken place. Home working with instant access to high-speed, high-capacity computers and visual services will be the norm. Paper mail and publications will have disappeared, replaced by electronic mail and publishing. Much of the enabling technology to transport such services has already been demonstrated, e.g. ultra-high bandwidth (10 Gbit/s) transmission for virtually unlimited distances over fibres similar to those already installed in BT's core network. Optical amplifiers, which enable both long repeaterless links and almost infinite distribution potential, are commercially available for 1.55 μm and are being developed for 1.3 μm (see Chapter 12). Indeed it is not so hard to visualize the simultaneous transmission of many hundreds of films, on an on-demand basis, to every home in the UK, using current technologies and simple extensions to the optical network [1]. Currently, telecommunications operating companies are excluded from these markets in most countries (except the USA) by regulation. This will be as influential as the available technologies in determining future developments. However, regulations can change much more rapidly than new technologies can be developed, so network operators must be ready to face the new opportunities.

The application of optics in telecommunications has so far tackled the problems of information transport. So much has been achieved using 1.3 μm and 1.55 μm infra-red signals over optical fibres that the network transport capability is assured for the foreseeable future. To fully utilize the information capacity of the network it is therefore necessary to address the human interface. The only communications media using exclusively aural senses are radio and telephony. Much more information can be rapidly assimilated by the human brain if that information is in full-colour pictorial form. Even in face-to-face conversation between two people visual contact plays a major role. Obviously infra-red optics can play little part here, but some of the technologies employed in the optical network can be extended or developed for visible operation.

13.2 APPLICATIONS FOR COMPACT VISIBLE LASERS

There are two obvious problems which cannot be readily solved by currently available technologies. Firstly, displays such as liquid crystal or CRT TV screens cannot be viewed in daylight. The best prospect here is reflective mode LCD, which at the present state of the art is restricted to monochrome, although research is under way to extend this to colour. Secondly, storage of the huge amounts of data required, both for information and entertainment services, and for control software of enhanced networks, is difficult. Visible lasers in the red, green and blue are at least one enabling technology towards a solution to display problems. Clearly this is only one component, but scanning systems are already available and widely used in laser light shows, and fast modulators are available, so the development of a full display technology could be rapid. Current videodisc storage technology, using 800 nm infra-red diode lasers, could be improved by up to a factor of ten in capacity if short-wavelength compact blue lasers were available. This is derived in part from the fact that the diameter of the spot to which a beam of light can be focused is directly proportional to the wavelength of that light. Hence a focused spot of 800 nm light will be twice the diameter, four times the area, of a 400 nm spot. This gives an immediate four-fold potential increase in storage capacity. Further space utilization can be gained by using lasers with better beam quality than those currently used, allowing higher packing density. It is also conceivable that the blue playback signal could be used to directly modulate transmission lasers, thereby bypassing an optics/electronics/optics interface. Coupled with high bit rate transmission, this technology could be used to download information or entertainment packages for use at the customer's leisure, freeing up the network for further traffic.

For read-only storage media, lasers around 400 nm in the violet with powers up to 1 mW would be adequate, whereas write-once applications would require a few milliwatts at the same wavelength. For small hand-held displays red lasers around 630 nm, green lasers around 540 nm, and blue lasers around 470 nm totalling less than 1 mW optical power will probably be sufficient, though for larger displays, such as large video walls, several watts of optical power may be required — consequently, 'wall-plug' efficiency will need to be high. In all of these applications the lasers will be required to operate at room temperature and use as little electrical power as possible, particularly for display applications in remote locations. Small physical size and rugged construction will be of great importance. As in all things, low item-cost will be needed to keep the cost of the user equipment at an acceptable level.

Efficient visible lasers will also find application in other beneficial areas, for example in the field of medicine where the last two decades have seen revolutionary changes to diagnostic techniques, treatment and surgery due to the widespread application of lasers [2]. Almost all known human life-threatening diseases can be diagnosed by cytology, the study of the properties of single cells, using tuneable visible lasers of a few milliwatts power. What prevents some of these techniques appearing in the general practitioner's office is the size, cost, and maintenance requirements of the currently available laser technologies (argon ion, neodymium YAG and CO_2 lasers are most common). The development of low-cost, reliable, compact and efficient visible lasers for communications applications could have considerable impact in medicine.

High-resolution colour printing on to film currently uses red and green helium neon lasers and air-cooled argon ion lasers. Major improvements in equipment size, electrical efficiency and cost could be made by using compact, electrically efficient devices.

In short, the development of compact, low-cost visible lasers represents an enabling technology, possibly the last required, for the current view of the home and office of the future. The remainder of this chapter is a review of alternative technologies which could lead to the rapid development of suitable lasers. As the title of the chapter suggests, emphasis is placed on developments in visible fibre lasers. Some key specialist terms are defined in the Appendix.

13.3 TECHNOLOGY ALTERNATIVES

13.3.1 Frequency doubling

It is imperative, then, to examine the technologies currently under consideration in research laboratories world-wide which will ultimately lead to such devices. One of the earliest approaches to this problem was to use nonlinear effects in crystals such as KTP to double or triple the frequency of infra-red lasers. A testament to this work is the commercially availability of green (532 nm), frequency-doubled, diode-pumped, neodymium YAG lasers with reasonable powers from several suppliers. Using similar techniques to frequency-double GaAlAs diode lasers using $KNbO_3$ crystals [3], blue light can be produced with 12% overall electrical to blue light power efficiency. Complex resonant enhancements and feedback mechanisms are required, however, to approach this efficiency.

An alternative approach has been taken by Marshall et al [4], who have produced a pulsed laser which is tuneable from the UV (380 nm) to green (520 nm). This was achieved by pumping an optical parametric oscillator (OPO) by a frequency-doubled Q-switched Nd:YAG laser (532 nm) and including a frequency doubler in the OPO cavity. The OPO produced output tuneable from 760 nm to 1040 nm with 30% conversion efficiency, which when frequency doubled (40% efficiency) gave the 380-520 nm tuning range. The 'wallplug efficiency' of the whole device was 0.4%. However, the output is pulsed and the device is quite large, employing a 50 cm laser cavity for the pump. Although it may be possible to reduce the size in the future, there are too many frequency-converting stages to envisage a truly small and efficient device.

13.3.2 Semiconductors

It would be desirable to generate blue light from a semiconductor source directly. Indeed much effort is being expended in this quest. ZnSe-based II-VI semiconductors have suitable band-gaps for blue-green operation, and work on this subject began in Japan in the early 1980s [5]. However, laser light generation was not achieved in this material until April 1991 [6]. Recent developments using ZnCdSe/ZnSe/ZnMgSSe structures have produced pulsed room temperature operation at 498 nm (blue-green) [7] and continuous wave (CW) room temperature operation at 523.5 nm in the green [8]. Here the laser threshold was 45 mA current with 2 mW output obtained for 55 mA current. These results represent a significant step forward in this technology.

UV-blue emission has been observed in gallium nitride and gallium aluminium nitride devices [9], though laser operation has not yet been observed. It is likely that it will take a number of years before these technologies reach maturity, although ultimately they will probably take the largest market share due to size, cost, and electrical efficiency considerations. Red semiconductor lasers are already commercially available at relatively low cost, e.g. modern battery-operated laser pointers. In research laboratories 3 W of CW red light [10] has been obtained from a diode array and 60 W of quasi-CW power [11]. A limitation of semiconductor lasers is the output mode shape and the astigmatism of the beam.

13.3.3 Up-conversion lasers

An alternative method of generating visible light is frequency up-conversion in either crystals or fibres. This differs from frequency doubling described above in that the output light is not restricted to harmonics of the pump light, and the lasers are generally tuneable independently of the infra-red pump laser. The target is the generation of visible laser light from infra-red diode pump sources. Triply ionized elements from the lanthanide series (rare earths), e.g. erbium or thulium, are employed as active ions in a host medium. These ions can be optically excited to several discrete energy states, each of which has a finite lifetime before decay occurs to a lower energy state. These finite lifetimes can be usefully employed to allow further interactions to occur, such as energy transfer or the absorption of a second photon (excited state absorption (ESA)). These are the two principal up-conversion mechanisms. Energy transfer, or co-operative up-conversion, occurs between two, or sometimes three, ions which are physically close to each other. In the simplest case of two ions, one ion will transfer all its energy to the adjacent ion, the first ion returning to a lower state with the second ion elevated to a higher state. All energy is conserved, and quantized, so for this process to occur requires energy states equally spaced above and below the excited state in question (Fig. 13.1). High dopant concentrations are also required to ensure that ions are sufficiently close together, typically $>1\%$ by weight. In lasers which employ sensitizers, the energy transfer mechanism is more complex and will not be described here. Energy transfer is the predominant mechanism used in up-conversion crystal lasers. ESA is simply the absorption of a further photon by an excited ion, elevating the ion to a higher excited state (Fig. 13.2). For practical visible lasers it is preferable that this photon is at the same wavelength as the original excitation, so only one pump source will be required; however, there is the flexibility to use more than one wavelength, as will be illustrated later. ESA is the predominant mechanism used in fibre

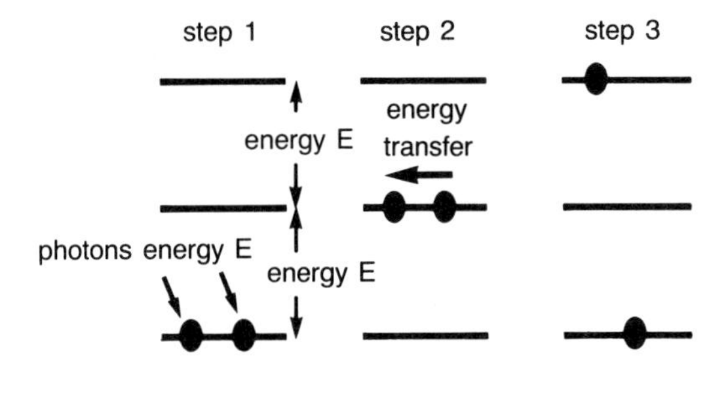

Fig. 13.1 Energy transfer.

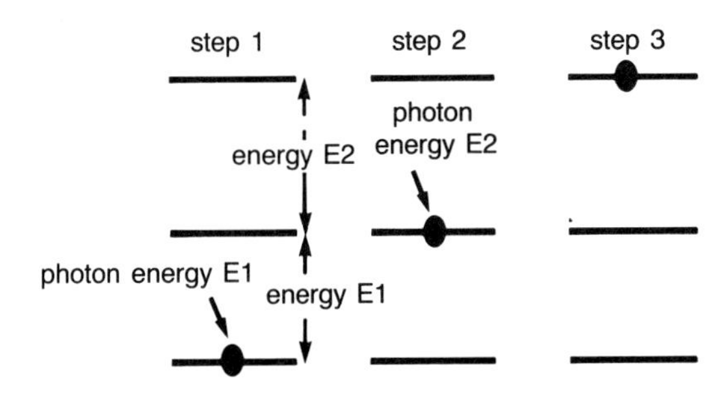

Fig. 13.2 Excited state absorption.

lasers with low dopant concentrations. A simple description of the energy states in rare earth ions is given by energy level diagrams, the most widely used being that of Dieke [12], and use will be made of these in the following sections.

13.4 UP-CONVERSION CRYSTAL LASERS

In reality the energy states of an ion are broadened by crystal field interactions to a greater or lesser extent, producing an energy band. This helps the up-conversion process in that the initial ground state absorption (GSA) can occur over a range of wavelengths and the energy level symmetry need not be so exact. However, unless all interactions are between the strongest parts of the bands, the efficiency can suffer. In crystal hosts the regular structure dictates that each dopant ion experiences similar fields, and therefore the bands remain relatively narrow. One consequence of this is that it can be difficult to find a suitable pumping laser which exactly matches the GSA, and another is that having found a pump laser, the probability that it will exactly match the ESA steps is low. The narrow bands also lead to a lower probability of energy gap overlaps, making energy transfer less efficient. This is reflected in the relatively low optical conversion efficiencies where the highest efficiency laser using ESA as the pump mechanism was 7% [13], and energy transfer 13% [14], both operating below liquid nitrogen temperatures.

Ytterbium is often used as a sensitizer for other rare earth lasers because it has two useful properties — it has a very broad energy level which absorbs strongly in the region covered by high power GaAs laser diodes and it has only one energy level above the ground state. Consequently it suffers neither from unwanted up-conversion due to energy transfer within its own system, whatever the concentration, nor ESA. The broad energy state overlaps with a range of energy levels in other rare earths, where direct diode pumping is not possible. This, coupled with the long lifetime of its single excited state, makes it a good sensitizer for transferring energy to nearly matched levels in other ions.

Crystal based up-conversion lasers have been demonstrated at many wavelengths covering the visible part of the electromagnetic spectrum, but the vast majority operate only at cryogenic temperatures, rendering them impractical for telecommunications devices. Blue and green up-conversion lasing has been reported in erbium-doped $YLiF_4$ crystals [15] and lasers at wavelengths as short as 380 nm (deep violet) have been demonstrated [16]. Recently Stephens and McFarlane [17] have demonstrated a 100 mW output power, diode-pumped erbium-doped $YLiF_4$ up-conversion crystal laser operating at 551 nm in the green. To achieve this the 3W 797 nm diode array pump source was line-narrowed to 0.1 nm linewidth using self-injection seeding in order to match the laser to the erbium absorption spectrum, the resulting power delivered to the crystal being reduced to 1.75 W. An IR-to-green conversion efficiency of 5% overall (8% slope) was achieved with the crystal held at 48 K. MacFarlane et al [18] have demonstrated blue, green, and yellow lasing in an erbium-doped $YLiF_4$ crystal containing 5% erbium. Pumping was into the lowest excited state, $^4I_{13/2}$, at 1.5 μm, with the proposed mechanism for exciting the higher levels being up to five or more sequential pairwise cross-relaxations, requiring eight pump photons. However, it is likely that a number of excitation routes occur simultaneously. This is the highest ratio of output to pump photon energy so far reported in any up-conversion laser. Despite the number of excitation photons required, the thresholds obtained for all three transitions are within the range of semiconductor pump lasers at 1.5 μm. The drawback with all these devices is the requirement to maintain the crystals at cryogenic temperatures.

Employing a complex energy transfer scheme, involving ytterbium as a sensitizer for the excitation of thulium ions, Thrash and Johnson [19] have recently reported chopped pump blue, green and red room-temperature laser operation in a BaYF:YbTm crystal. Diode-pumped and also CW operation was demonstrated for the red transition. A further complication in this laser is that for the blue and green lines to operate, the red line also needs to be lasing, acting as an extra pump source for ESA to take ions to the required $^1D_2$ upper state. This work demonstrates that simultaneous red, green and blue lasing may be possible in the same compact room-temperature device, which could be extremely useful in compact portable video displays for example.

As an alternative to rare-earth-doped crystals, Yang et al [20] used a nanosecond pulsed 830-890 nm laser to pump ZnSe and ZnSSe crystals to generate blue lasing. Again, cryogenic temperatures were required.

Another approach to the construction of compact visible lasers has been taken by Mukherjee [21], using DCM dye doped into a PMMA waveguide fabricated on a silica-on-silicon substrate. The waveguide was 12 μm $\times$ 7 μm $\times$ 2.5 mm. The pump source was a mode-locked picosecond dye laser giving ~4 ps pulses at 736 nm with 82 MHz repetition rate, synchronously pumped by a neodymium:YAG laser. Using the two cleaved-end facets of the guide to form a cavity with approximately 4% reflectivity at each end, lasing was achieved around 570 nm. However, for 2 mW average input power only 1 μW average output was achieved, the inefficiency being attributable to poor facet quality and a number of other parameters. Although this work is in its infancy, waveguide lasers offer a number of potential advantages in the quest for compact lasers, most notably the confinement of power. The most developed form of waveguide laser is the fibre laser, and progress on the visible fibre laser front is discussed in the next section.

13.5 VISIBLE FIBRE LASER TECHNOLOGIES

Fibre lasers date back to the early 1960s [22], and it is perhaps only chance that this was not the very first laser ever demonstrated. Other reference works on the subject of fibre lasers are available [23, 24]. Optical fibres offer two tremendous advantages over other laser technologies. Firstly, the small waveguide size allows enormous power densities to be achieved, around 1.5×10^8 W/m^2, for only 1 mW input power. Secondly, the optical quality of fibres developed for telecommunications allows laser cavity lengths to be chosen for optimum performance, rather than being restricted to crystal growth size. In addition, single-mode fibre produces Gaussian output beams with well-defined divergence and no astigmatism. The combination of power density with long interaction lengths allowed continuous room-temperature operation of the three-level 1.54 μm transition in erbium-doped silica fibre [25], and the development of the now commercially available erbium-doped fibre amplifier (EDFA) for the third optical telecommunications window. Unfortunately the silica-based glass system used for these amplifiers has high phonon energies, the effect of which is to 'deactivate', by non-radiative quenching, energy levels whose next lower level is closer than about 4600 cm^{-1} (see Chapter 12) [23]. This phenomenon deactivates almost all of the energy levels in the rare earths, including most of those from which visible emission would otherwise be possible. However, a fibre technology has been developed using glasses with low phonon energy. These glasses are composed mainly of the fluorides of zirconium, barium, lanthanum, aluminium, and sodium, and are commonly referred to as ZBLAN [26]. The glasses are of interest for ultra-low-loss transmission since they have an intrinsic loss one order of magnitude lower than that of silica. However, the requirement for such a medium has diminished with the advent of EDFAs. In ZBLAN glasses an energy gap of only 3100 cm^{-1} results in the predominant decay mechanism being radiative, and a gap of 2700 cm^{-1} can give a lifetime of 100 μs. This results in many more of the available levels having comparatively long radiative lifetimes. One consequence of this is that extremely high optical conversion efficiencies can be achieved, e.g. in a diode-pumped neodymium-doped ZBLAN fibre laser, quantum efficiencies (above threshold) of 97% have been measured [27]. Over 100 possible visible transitions which could occur in this glass have been identified, although it is not expected that all of these will produce lasing. However, the majority of fibre laser reports to date, and all but one of the visible fibre laser reports, have been in the ZBLAN glass system. The borderline between a level being active, where the predominant energy decay mechanism is fluorescence, and deactivated by non-radiative multiphonon decay, is a very strong function of energy gap; however, efficient operation of stimulated processes in borderline cases where the radiative efficiency is only a few percent is admirably demonstrated by the praseodymium amplifier for 1.3 μm (see Chapter 12).

The first visible up-conversion laser in a ZBLAN fibre was demonstrated by Allain et al [28] of CNET in France, using a thulium-doped fibre cooled to 77 K, pumped by two red lines, 647.1 nm and 676.4 nm, from krypton ion lasers. Lasing was observed at 455 nm and 480 nm. The second [29] used a single line from a krypton ion laser, 647.1 nm, to pump holmium-doped fibre at room temperature, producing tuneable lasing from 540 nm to 553 nm.

13.5.1 Blue, green, orange and red praseodymium-doped fibre lasers

The same group at CNET have produced tuneable lasing in the orange, red, and near-infra-red by pumping a praseodymium fibre, similar to those used for early 1.3 μm amplifier work, at 476.5 nm, directly into the upper energy level for these transitions [30]. This demonstrates that, as predicted from energy gap considerations, some of the higher energy levels in praseodymium-doped ZBLAN fibre are indeed active. It is not possible to pump these high levels in praseodymium by a single infra-red pump source using up-conversion because the energy levels are not suitably placed. However, Smart et al of Southampton University, in collaboration with BT Laboratories [31], used a two-wavelength up-conversion pumping scheme (Fig. 13.3) in a similar praseodymium-doped fibre to produce blue, green, and red lasing. The first pump wavelength, 1.01 μm, excites ions from the $^3H_4$ ground state energy level to the $^1G_4$ level which is the upper level for the 1.3 μm amplifier (see Chapter 12). The measured lifetime of this level is ~ 100 μs, which is sufficient time for the second wavelength, 835 nm, to further excite the majority of these ions to the $^3P_1$ level. This effectively switches off the 1.3 μm emission. The radiative lifetime of $^3P_1$ is sufficiently long to allow a population inversion between $^3P_1$ and $^3H_5$ to be established (this is a necessary condition for lasing to occur), allowing the 520 nm lasing. However, the majority of

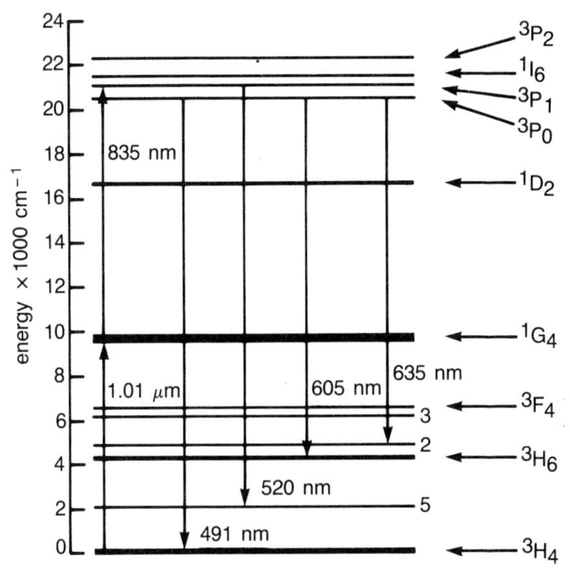

Fig. 13.3 Energy level diagram for praseodymium in ZBLAN showing pump scheme and visible laser transitions (after Smart et al [31]).

the population in $^3P_1$ will decay non-radiatively to $^3P_0$. The pump transitions are strong enough to establish a population inversion between $^3P_0$ and the lowest four energy levels, including the ground state, allowing lasing at 491, 605 and 635 nm. The four transitions have not been operated simultaneously, but pairs of laser lines have. At the time of these experiments, diode lasers at the first pump wavelength of 1.01 μm were not available, but these are now being developed by several commercial companies for 1.3 μm amplifiers. With further development one can envisage a single device pumped by two different diodes, operating in the red, green and blue, for use in displays. In Allain et al [32], the practicality of this laser system has been taken one stage further by including ytterbium as a sensitizer. In this case the ytterbium concentration is a relatively high 2% to ensure a high probability of close proximity to the 0.1% of praseodymium, facilitating energy transfer. The ytterbium is pumped at 810-860 nm, where praseodymium does not absorb, transferring energy to the praseodymium $^1G_4$ level. From there absorption of a further pump photon takes an ion up to the $^3P_1$ level (Fig. 13.4). So far this approach has only yielded room-temperature lasing at 635 nm in the red, but could be a step towards a single-diode-pumped red, green, and blue laser.

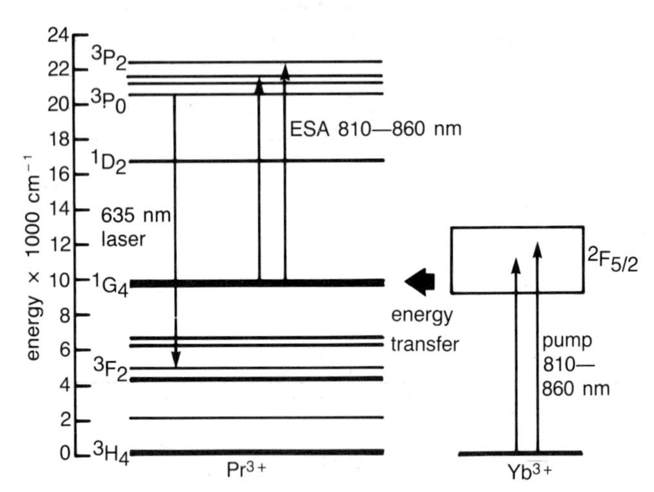

Fig. 13.4 Excitation route of $^3P_0$ level for 635 nm laser (after Allain et al [32]).

Recently Piehler et al [33] have operated the red 635 nm and green 521 nm transitions separately in a ytterbium:praseodymium co-doped fluoride fibre using two diode pump lasers. The first at either 1010 nm or 980 nm populates the $^1G_4$ level via energy transfer from ytterbium, while the second at 833 nm raises population from there to $^1I_6$ via ESA. 0.7 mW of green and 6 mW of red have been produced. Progress within the Southampton

University/BT collaboration [34] has resulted in a device producing blue, green and red lasing (separately) from a single pump source. Here a Ti:sapphire laser operating at 840 nm was used to pump a 1020 nm ytterbium-doped silica fibre laser, whose cavity was formed by in-fibre gratings. The output from this laser and the unabsorbed 840 nm pump were used to pump the praseodymium-doped fluoride fibre. In this work improvements in the fibre loss and design have led to significant improvements in laser thresholds. It should be possible to diode pump the ytterbium laser to produce a more practical system.

13.5.2 Blue thulium-doped laser

Modifications have recently been made to the early work [28] on blue lasers in thulium-doped ZBLAN fibre by Tohmon et al [35]. Here europium was added to the fibre to depopulate the lower level of the 455 nm transition, forcing this to be the dominant lasing route. Wavelengths of 780 nm and 643 nm, both achievable by diodes, were used for pumping, but temperatures of 77 K were still required.

Grubb et al [36] of Amoco Technology have recently published a blue laser using a thulium-doped fibre supplied by Le Verre Fluoré, pumped by a neodymium:YAG laser operating around 1120 nm. The pump laser had to be developed to produce this wavelength. The pumping scheme for this device is shown in Fig. 13.5. This produced a blue laser at 480 nm with 18%

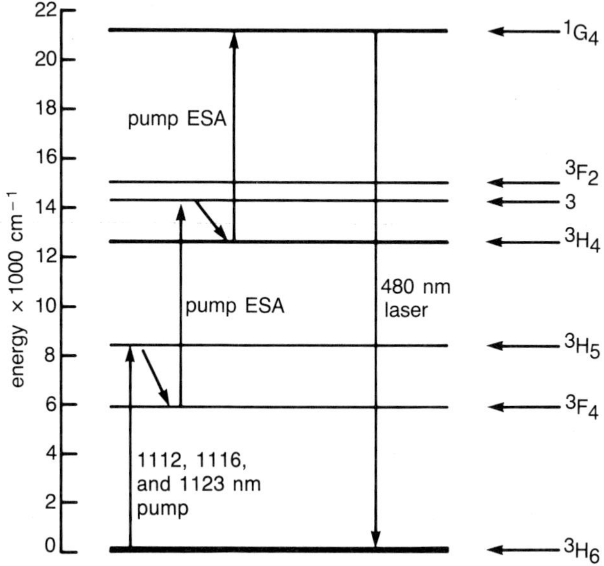

Fig. 13.5 Excitation route for the 480 nm blue thulium laser (after Grubb et al [36]).

slope efficiency, a threshold of around 30 mW, and an output power of 57 mW. This is probably the most practical blue laser source so far demonstrated.

13.5.3 Green erbium-doped lasers

At BT Laboratories an 801 nm ESA up-conversion pumping scheme has previously been used to facilitate efficient lasing [37] and amplification [38] on an 850 nm transition in erbium, which cannot be directly pumped at diode wavelengths. During this work a strong green fluorescence was observed, which originates from the same $^4S_{3/2}$ upper level (Fig. 13.6). Optimizing the cavity for green operation, but still using the fibre designed for 850 nm operation, 23 mW of 546 nm green lasing was obtained [39]. The pumping scheme of this laser employs ESA from both $^4I_{13/2}$ and $^4I_{11/2}$ (Fig. 13.6).

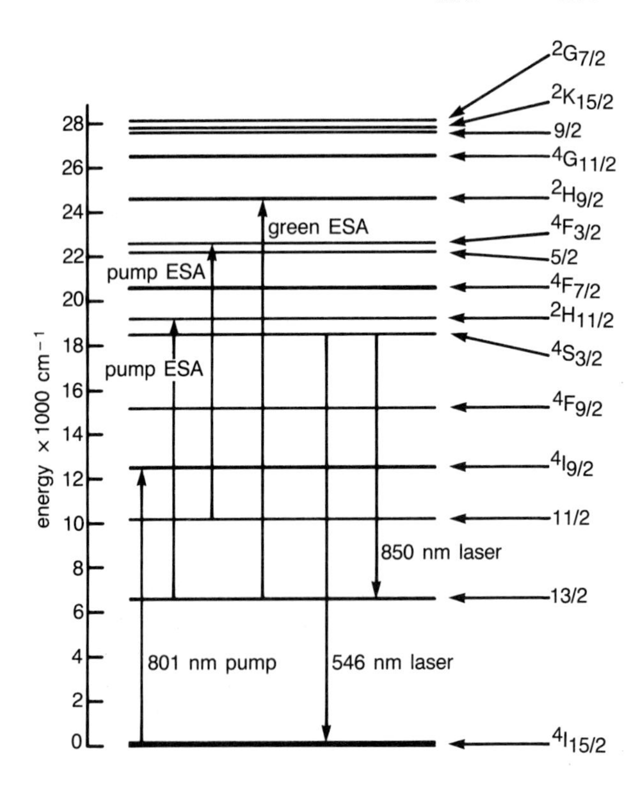

Fig. 13.6 Excitation route for 546 nm (green) and 850 nm erbium lasers (after Whitley et al [39]).

Although only that from $^4I_{13/2}$ to $^4S_{3/2}$ is necessary, the second, and a possible third ESA are unavoidable if population is present in the appropriate states. This is due to the multiplicity of energy levels spaced by the equivalent of 800 nm. This fact has proved detrimental to 800 nm pumping of the EDFA [40], but, as is seen here, can be usefully employed for up-conversion. The power characteristic curve of this green laser (Fig. 13.7) shows some saturation at high pump powers. This was originally attributed to competitive lasing on the 850 nm transition; however, subsequent work on similar fibres has shown similar effects with no lasing at 850 nm. The first ESA step reduces

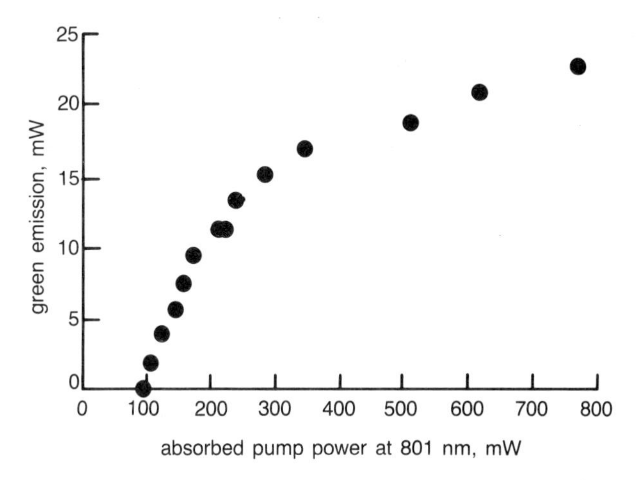

Fig. 13.7 Evolution of lasing power at 546 nm wavelength with absorbed pump power at 801 nm (after Whitley et al [39]).

the population in the $^4I_{13/2}$ lower state, allowing a population inversion to be maintained. The laser threshold was 100 mW, with a maximum slope efficiency of 11%. Allain et al [41] pumped the same transition at 970 nm with ground-state absorption elevating ions to $^4I_{11/2}$ followed by ESA to $^4F_{7/2}$. Fast non-radiative decay then populates the $^4S_{3/2}$ level (Fig. 13.8). This pumping scheme reduces the competition from 850 nm, but the authors also include a tuning prism in the laser cavity, and this prevents lasing at wavelengths other than that desired. The tuning element adds loss to the cavity, increasing the threshold to 200 mW pump power, but nevertheless a slope efficiency of 15% was obtained, with output powers of 50 mW. No saturation was observed. A second mirror was included in this experiment to force lasing at 1.55 μm to depopulate the $^4I_{13/2}$ level and so prevent ions trapped in this long-lived state from reducing efficiency. Both authors report an ESA of green light from $^4I_{13/2}$, which is not believed to be very detrimental to the laser performance.

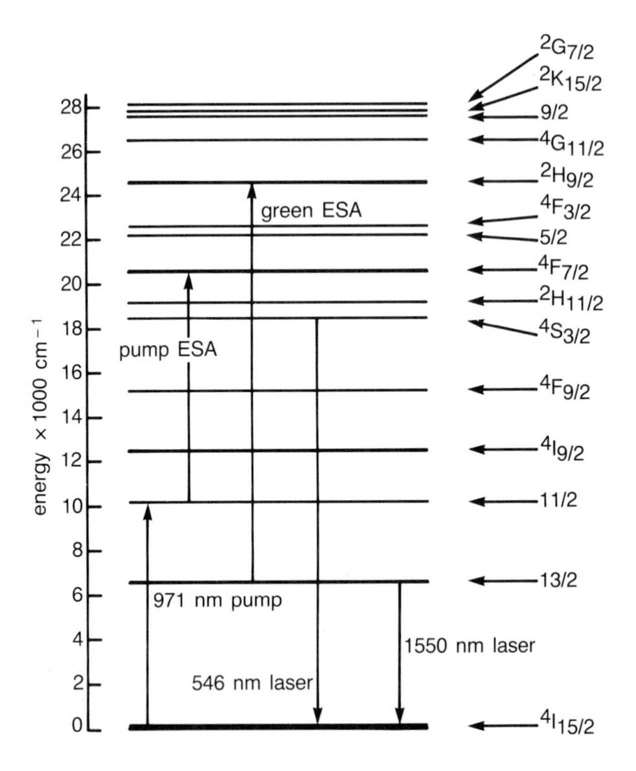

Fig. 13.8 Excitation route for 546 nm (green) erbium laser, 971 nm pumping (after Allain et al [41]).

More recent work at BT Laboratories has used this same 971 nm pumping scheme in a cavity with no tuning element. This decreases the cavity loss but also frustrates the lasing at 1.55 μm, which in Allain's work prevented population trapping. Figure 13.9 shows the laser characteristic obtained. The threshold is reduced to around 40 mW, with 25 mW output obtained for 470 mW of pump. The initial slope efficiency was 11%, as in previous work, though an efficiency roll-off is still apparent.

Piehler et al [42] have produced 2.5 mW of green (544 nm) using the 40 mW diode laser pump module from an EDFA which operated slightly below 980 nm. The same fibre pumped by a Ti:sapphire laser exhibited a 51% slope efficiency, and gave 14 mW of output for 59 mW of pump. The

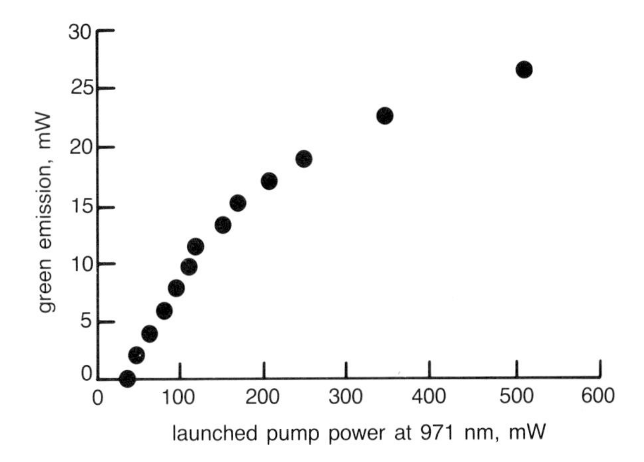

Fig. 13.9 Evolution of green laser power with 971 nm pump power.

latest results from BT Laboratories [43] used a commercial 801 nm semiconductor laser diode from which up to 40 mW could be coupled into an improved fibre design having a 1.1 μm diameter core and 0.4 numerical aperture. The lowest threshold obtained was 7 mW of coupled power, with over 3 mW maximum output power.

Taken together, these recent results represent a large step forward towards a practical green source which could take some of the market now enjoyed by green helium-neon lasers.

13.6 CONCLUSIONS

There are several options in the quest for efficient, low-cost, reliable visible lasers, the most promising of which seems to be that of visible fibre lasers based on fluoride glasses. These lasers already offer better room-temperature performance in the blue than any of the other technologies. There are clearly many steps to be taken before visible fibre lasers appear in the market-place, but this development can capitalize on the work already done for the closely related fibre amplifiers. Further development steps are required to generate display or storage devices, but these again can capitalize on existing products. The mechanics of optical disc technologies are well-established so that only the optical implications of changing from infra-red to blue light need be addressed. It is a larger step to generate a two-dimensional display from a single three-colour laser light source or three separate lasers, and an even larger step to generate a real-time three-dimensional display. However, fast-scanning systems do exist, as do modulators, so the display development is reduced to the problems of control, screen material (if any) and the persistence of the eye.

Visible lasers are a fundamental building block and enabling technology towards visual and information services of the future. Several technologies exist, but fibre lasers are believed to provide the best short and medium-term solution.

APPENDIX

Glossary

Active level: often referred to as metastable, an active level is an energy level where radiative energy loss dominates over non-radiative.

Deactivated level: an energy level from which non-radiative energy loss mechanisms dominate.

Frequency doubling: generation of light at twice the frequency (half the wavelength) of the original signal by nonlinear interaction.

Lifetime: the time for the number of ions in a given energy level (the population) to decay to $1/e$ of that number in the absence of pump power.

Population: the number of ions in a given excited state.

Population inversion: where the population of a higher excited state is greater than the population of a lower energy level.

Slope efficiency: the ratio of output power to pump power, in excess of the threshold power.

Threshold: the minimum pump power required to produce lasing, i.e. the pump power required to produce gain equivalent to the cavity losses.

Up-conversion: generation of light at higher frequencies than the source (pump) frequency.

REFERENCES

1. Hill A M: 'Network implications of optical amplifiers', Optical Society of America 1992 Technical Digest Series, 5, Opt Fibre Commun, pp 181-120 (February 1992).
2. Opt and Photon News, 3, No 10, Optical Society of America (October 1992).
3. Koslovski W J: 'Frequency-doubled diode lasers', Paper FA2, Optical Society of America 1992 Technical Digest Series, 6, Compact Blue-Green Lasers (1992).
4. Marshall L R et al: 'Continuously tuneable diode-pumped uv-blue laser source', Opt Lett, 18, No 10, pp 817-819 (1993).
5. Kukimoto H: 'Overview — blue-green semiconductor LED/laser work in Japan', Paper ThC2, Optical Society of America, Technical Digest Series, 6, 'Compact blue-green lasers' (1992).
6. Haase M A et al: 'Blue-green laser diodes', Appl Phys Lett, 59, No 11, pp 1172-1174 (September 1991).
7. Itoh S et al: 'Room temperature pulsed operation of 498 nm laser with ZnMgSSe cladding layers', Electron Lett, 29, No 10, pp 766-767 (1993).
8. Nakayama N et al: 'Room temperature continuous operation of blue-green laser diodes', Electron Lett, 29, No 16, pp 1488-1489 (1993).
9. Amano H et al: 'Perspective of GaN/GaAlN based ultra-violet/blue lasers', paper ThC4, Optical Society of America, Technical Digest Series, 6, Compact blue-green lasers (1992).
10. Geels R A et al: '3 W CW laser diodes operating at 633 nm', Electron Lett, 28, No 11, pp 1043-1044 (May 1992).
11. Haden J M et al: 'High power, 60 W quasi-CW, visible laser diode arrays', Electron Lett, 28, No 5, pp 451-452 (February 1992).
12. Dieke G H: 'Spectra and energy levels of rare-earth ions in crystals', John Wiley and Sons, New York (1968).
13. Macfarlane R M et al: 'Up-conversion laser action at 450.2 and 483.0 nm in Tm:$YLiF_4$', Conference on Lasers and ElectroOptics (CLEO'90), Anaheim, pp 318-319 (May 1990).
14. McFarlane R A: 'High-power visible up-conversion laser', Opt Lett, 16, No 18, pp 1397-1399 (September 1991).
15. Herbert T et al: 'Blue and green CW up-conversion lasing in Er:$YLiF_4$', Appl Phys Lett, 57, No 17, pp 1727-1729 (October 1990).
16. Macfarlane R M et al: 'Violet CW neodymium up-conversion laser', Appl Phys Lett, 52, No 16, pp 1300-1302 (April 1988).

17. Stephens R R and McFarlane R A: 'Diode-pumped up-conversion laser with 100 mW output power', Opt Lett, 18 , No 1, pp 34-36 (1993).

18. Macfarlane R M et al: 'Blue, green and yellow up-conversion lasing in Er:$YLiF_4$ using 1.5 μm pumping', Electron Lett, 28 , No 23, pp 2136-2138 (1992).

19. Thrash R J and Johnson L F: 'Tm^{3+} room temperature up-conversion laser', paper ThB3, Optical Society of America 1992 Technical Digest Series, 6 , Compact Blue-Green Lasers (1992).

20. Yang X H et al: 'Two-photon pumped blue lasing in bulk ZnSe and ZnSSe', Appl Phys Lett, 62 , No 10, pp 1071-1073 (1993).

21. Mukherjee A: 'Two-photon pumped up-converted lasing in dye doped polymer waveguides', Appl Phys Lett, 62 , No 26, pp 3423-3425 (1993).

22. Snitzer E: 'Optical maser action of Nd^{3+} in a barium crown glass', Phys Rev Lett, 7 , pp 444-446 (1961).

23. France P W (Ed): 'Optical Fibre Lasers and Amplifiers', Blackie (1991).

24. Digonnet M J F (Ed): 'Rare Earth Doped Fiber Lasers and Amplifiers', Marcel Dekker (1993).

25. Mears et al: 'Low-threshold tuneable CW and Q-switched fibre laser operating at 1.55 μm', Electron Lett, 22 , pp 159-160 (1986).

26. France et al: 'Progress in fluoride fibres for optical communications', BT Technol J, 5 , No 2, pp 28-44 (April 1987).

27. Brierley M C and Hunt M H: 'Efficient semiconductor pumped fluoride fibre lasers', SPIE 1171 , Fibre Laser Sources and Amplifiers, pp 157-159, Boston (September 1989).

28. Allain J Y et al: 'Blue up-conversion fluorozirconate fibre laser', Electron Lett, 26 , No 3, pp 166-168 (February 1990).

29. Allain J Y et al: 'Room temperature CW tuneable green up-conversion holmium fibre laser', Electron Lett, 26 , No 4, pp 261-263 (February 1990).

30. Allain J Y et al: 'Tuneable CW lasing around 610, 635, 695, 715, 885 and 910 nm in praseodymium-doped fluorozirconate fibre', Electron Lett, 27 , No 2, pp 189-191 (January 1991).

31. Smart R G et al: 'CW room temperature up-conversion lasing at blue, green and red wavelengths in infra-red pumped Pr^{3+}-doped fluoride fibre', Electron Lett, 27 , No 14, pp 1307-1309 (July 1991).

32. Allain J Y et al: 'Red up-conversion Yb-sensitized Pr fluoride fibre laser pumped in 0.8 μm region', Electron Lett, 27 , No 13, pp 1156-1157 (June 1991).

33. Piehler D et al: 'Laser diode-pumped visible up-conversion fiber laser', Digest of CLEO 1993, OSA Technical Digest Series, 11 , p 406 (1993).

34. Pask H M et al: 'Up-conversion laser action in Pr^{3+}-doped ZBLAN fibre pumpd by an Yb-doped silica fibre laser', OSA Topical Meeting on Advanced Solid State Lasers, Salt Lake City (February 1994).

35. Tohmon G et al: 'Up-conversion lasing at 455 nm in Tm^{3+}:Eu^{3+} co-doped fluorozirconate fiber', Digest of CLEO 1993, OSA Technical Digest Series, 11 , p 406 (1993).

36. Grubb S G et al: 'CW room-temperature blue up-conversion fibre laser', Electron Lett, 28 , No 13, pp 1243-1244 (June 1992).

37. Millar C A et al: 'Efficient up-conversion pumping at 800 nm of an erbium-doped fluoride fibre laser operating at 850 nm', Electron Lett, 26 , No 22, pp 1871-1873 (October 1990).

38. Whitley T J et al: '23 dB gain up-conversion pumped erbium-doped fibre amplifier operating at 850 nm', Electron Lett, 27 , No 2, pp 185-186 (January 1991).

39. Whitley T J et al: 'Up-conversion pumped green lasing in erbium-doped fluorozirconate fibre', Electron Lett, 27 , No 20, pp 1785-1786 (September 1991).

40. Wyatt R: Ch 4 in France P W (Ed): 'Optical Fibre Lasers and Amplifiers', Blackie (1991).

41. Allain J Y et al: 'Tuneable green up-conversion erbium fibre laser', Electron Lett, 28 , No 2, pp 111-113 (January 1992).

42. Piehler D et al: 'Up-conversion process creates compact blue/green lasers', Laser Focus World (November 1993), also OSA Topical Meeting on Compact Blue/Green Lasers, Salt Lake City, Utah (February 1994).

43. Massicott J F et al: 'Low threshold, diode pumped operation of a green, Er^{3+}-doped fluoride fibre laser', Electron Lett, 29 , No 24 (November 1993).

14

WAVELENGTH AND TIME-MULTIPLEXING DEVICES ON PLANAR SILICA MOTHERBOARDS

S A Cassidy, F MacKenzie, C J Beaumont, G D Maxwell, B J Ainslie, M Nield, C A Jones, J D Rush and A Thurlow

14.1 INTRODUCTION

Optical technology has been widely adopted in systems since it is a very convenient provider of cheap bandwidth. Equally, however, its place in the network is earned through its greater use of passive components. This has given such benefits as long unrepeated spans, passive distribution, less power feeding and low failure rates. The networks are therefore easier to run, manage and maintain.

These passive components not only enable effective point-to-point links, but also can give the network greater function and upgradability. For example, wavelength multiplexing enables the addition of further connections or services entailing only a change in the terminal equipment. As examples, wavelength switching can reduce the switching complexity in the network, while optical processing in the time domain has shown header recognition at high speed with a minimum of active components. Network capacity upgrades and automatic self-routeing can therefore be done with minimum complexity of equipment.

In many cases, any one function can be realized in a choice of technological approaches. In optics the range of functions required and hence the range of technologies is likely to be diverse. However, minimizing the number of separate technologies will give economies of scale. The question is — how do we make the correct choice of the optimum subset of technologies? The set used must maintain the robustness and as far as possible the passive properties of optics, as well as the low costs which are all intrinsic to optical technology.

The cost of components and subsystems is dominated by the assembly techniques. This argues for the maximum amount of integration to be used, subject to attaining sensible yields and performance. It is plain to see the effect of these forces in the form of electronic technology today. As there is a more diverse technology set in optics, our device design needs to take into account the assembly methods capable of integrating the final subsystem. The interfaces need to be as simple and the structure as modular as possible. What is needed, therefore, is a technology which can realize the device and subsystem functions which will be required, as well as providing an integration method which enables the subsystem function to be built from modules of different technologies in a cost-effective way.

Planar silica is a widely studied technology which can be used to realize a wide range of passive optical functions. The ability to fabricate low-loss guides means that complex interconnect patterns between its own passive functions and active optoelectronic components is possible. The choice of a technology based on micromachineable silicon wafers offers a simple uniform approach to the interface to active semiconductor devices as well as the interface to the optical fibre in the network. In principle, therefore, silica on silicon can be used as a single fabric to interface between the network fibre and the electronic layer of the system, while performing optical functions on the way.

Simple devices are already commercially available, providing the power-splitting function for passive optical networks. In the laboratory it has been shown that it is possible to combine guides, wavelength-selective gratings and couplers, Fourier plane diffractive elements, lenses, time-delay circuits and highly complex interconnect patterns on the one chip. Its basis of silicon technology has enabled accurately aligned interfaces to fibre arrays and active devices, greatly simplifying the interfaces which integrate the technologies into a functional component. This concept has been termed the 'optical motherboard' [1].

This chapter describes examples of wavelength and time domain devices in planar silica technology and also recent progress in the ability of this technology to form standard interfaces between sources and detectors and the fibre of the system. Future directions are outlined.

Four sections describing the devices and technology follow — a description of a 16-channel wavelength-multiplexing device, a time-delay loop for use as the basis of optical packet header recognition, some fundamental results on guide-packing density, and progress on the optical motherboard principle interfaces. The conclusion describes the overall status and impact of this area.

14.2 WAVELENGTH MULTIPLEXER

This section describes an example of a wavelength-multiplexing device in planar silica. Its 16 channels each have a bandwidth of 1.3 nm, separated by 2.7 nm.

A number of technological approaches have been taken to wavelength-multiplexing devices, falling into the broad categories of bulk optics [2], or waveguide optics. Bulk optics currently give the best outright performance. Waveguide approaches, however, offer compatibility with fibres and ability to interface directly with optoelectronics, small size, and are suitable for mass production. It is to be expected that loss and crosstalk performance will approach that of bulk optic devices in the next few years as the technology develops. Using the waveguide approach, a number of devices have been proposed, such as cascaded Mach-Zehnders [3], arrayed-waveguide gratings [4, 5], photosensitive gratings (see Chapter 15) and focusing grating mirrors [6-8]. The device discussed in this chapter is of the last type.

A 16-channel wavelength-multiplexing device has been fabricated in planar silica in which the input and output guides are aligned relative to the grating mirror as part of the lithographic process, so needing no post-alignment. Special processing techniques have enabled the fan of input/output guides to be integrated with the Fourier plane in which the dispersion takes place. The guides fan in from a spacing typical of fibre arrays or transmitter/receiver arrays down to a size which allows the whole device to occupy only 0.5×3 cm of chip. The device can be interfaced to such arrays in one alignment operation. Any error in array positioning does not result in an error in the selected wavelengths.

14.2.1 Device design

The device in detail consists of a fan array of waveguides integrated with a slab-guiding region (Fourier plane), into which is etched a focusing echelle grating, operating in the Littrow configuration. The operation is therefore very similar to a 'slice' through a bulk optical device — resulting in a very compact form. The grating is made by etching a vertical step through the Fourier plane, which is then metallized to give high reflection. The circular grating element is made up of a number of confocal and concentric circular segments, joined by small radial steps to form a staircase pattern (see Fig. 14.1).

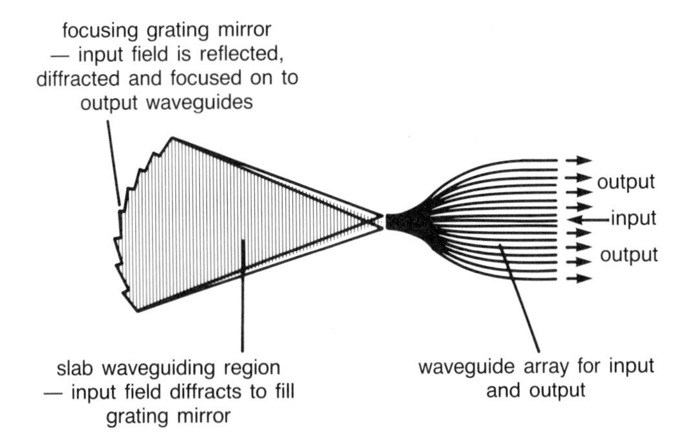

Fig. 14.1 WDM chip design showing shaped waveguide array for input and output and circular grating element.

One further feature of the design is that by increasing the radii of the circular elements, the staircase grating is moved further from the origin, and the individual feature sizes of each step are expanded. This has no effect on the basic optical behaviour. This then allows feature sizes to be chosen which are capable of being patterned on a wafer using standard optical lithography.

Detailed discussion of the device design has been published elsewhere [9], and will not be repeated here. Broadly speaking, it can be shown that the resolving power of the grating is proportional to the overall optical depth of the grating, with the free spectral range being inversely proportional to the step depth, t. The linear dispersion depends on the ratio of t/D, with D being the step width. The main parameters of the device design are summarized in Table 14.1. With these values, the bandwidth should be 0.85 nm, with the channel separation being 2.63 nm, very close to those obtained. These parameters were chosen as approximate fits, to speed the mask design and test the device principle.

Table 14.1 Summary of device design parameters.

Central wavelength	1.55 μm
Waveguide cross-section	6.5×6.5 μm
Waveguide spacing	20 μm
Δn	0.005
Focal length of mirror	28.6 mm
t/D ratio	0.205
Grating mode order	9

14.2.2 Description of the technology

The waveguide layers used to make the device were formed using flame hydrolysis deposition [10, 11]. This is basically the same technique that is used for fibre preform manufacture, but modified for planar geometries. The planar core waveguide is patterned and etched using standard lithographic and reactive ion etching techniques. For this device, the process is slightly more complicated in that there are two patterning and etching steps. The first step is used to pattern and etch the input and output waveguides. These are then given a planarizing cladding. The second step is then to pattern and etch the focusing grating mirror through the cladding. The mirror facet is then coated with a reflecting layer of titanium and gold. The wafer is then diced and the waveguides polished for assessment.

14.2.3 Results

Spectral response was measured using a low power broadband light-emitting diode source (1.4-1.6 μm), with a single-mode fibre array for input and output. The output was fed into a spectrum analyser. A montage of traces for each channel is recorded in Fig. 14.2. All 16 channels can clearly be seen. Insertion loss and crosstalk were measured for a number of individual channels employing a tuneable narrowband long external cavity (LEC) laser source, used for its higher spectral power density. Discrete point spectra obtained in this way for channels 13 to 17 are shown in Fig. 14.3. The number of measurable channels was limited by the sweeping range of the LEC. Each trace is a line drawing based on 80 measured points. A wavelength meter was used to check each wavelength point used. As a reference, the LEC power spectrum is included. Channel spacing is ~2.67 nm with the −3 dB bandwidth being ~1.3 nm. The wavelength spacing is within 0.04 nm from that expected from the design parameters, whereas the bandwidth of each channel is slightly higher. This is likely to be due to a slight oversizing of the capturing guides, and some variations on the etched positions of the

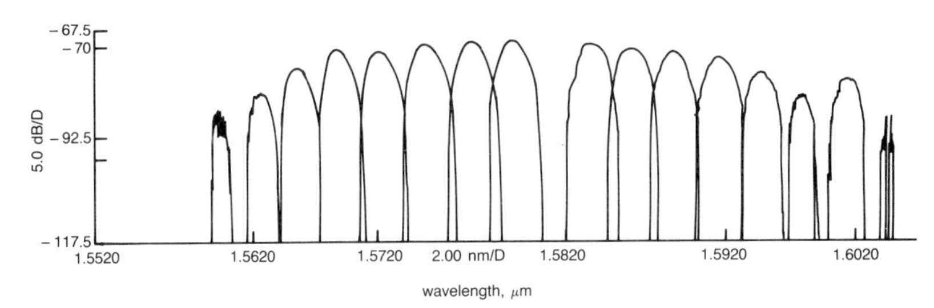

Fig. 14.2 Output spectra from broadband source input.

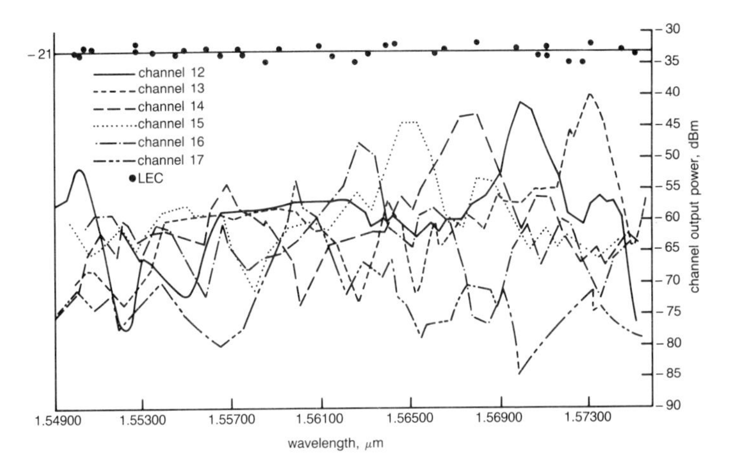

Fig. 14.3 Output from LEC input.

mirror surfaces. An insertion loss of ~19 dB was measured for channel 12 by tuning the LEC to the peak wavelength of that channel. This is dominated by scatter from the mirror and some residual non-verticality. It is to be expected that progress in etching techniques will reduce this loss well below the first result. Roll-off of the channel peak power is observed in the outer channels. This is a result of the guides terminating on a straight line, as opposed to following the curvature of the focal plane. A secondary result of this roll-off is that the best crosstalk of about −16 dB between channels 12 and 13, deteriorates to ~ −4 dB between channels 8 and 10. Future iterations of the device will terminate the guides on the curved focal plane, or 'Rowland circle' [12], which will considerably reduce this effect (see Fig. 14.4).

The very close agreement between the expected and measured wavelength spacings is a clear demonstration of the advantage of being able to design

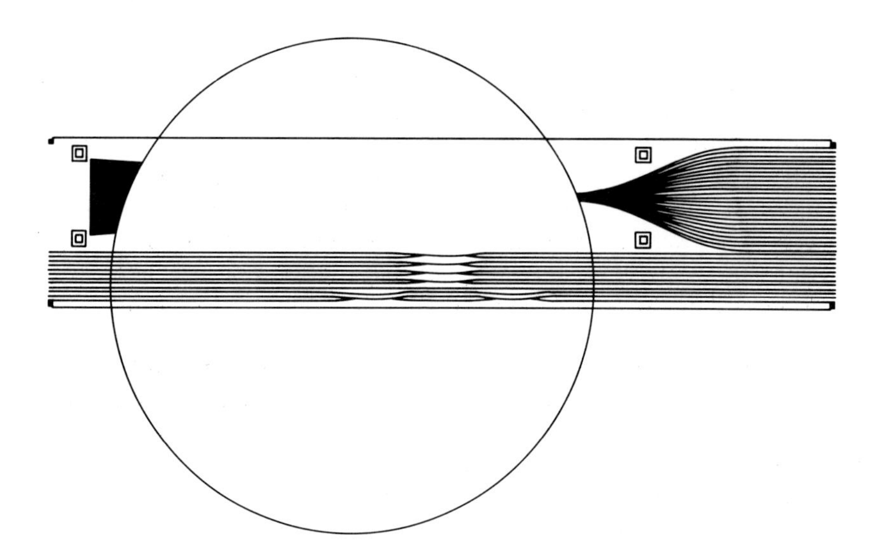

Fig. 14.4 Schematic diagram of second-generation device, showing shaped waveguide array set on a Rowland circle.

and fabricate a WDM device to a specification in a single lithographic alignment. The absolute wavelength positioning is governed by the mask alignment accuracy between the grating and guide array. The mask position accuracy is likely to be only a small proportion of the guide width. The wavelength accuracy is therefore likely to be small compared with the channel bandwidth. Refractive index control of the material is important in this respect. The variations achieved in practice indicate that absolute wavelength positioning accuracies below 1 nm are now possible. In a commercial, regularly run process, this is obviously going to be controlled to an even higher degree. Thermal expansion rates of the materials are very low and for most purposes it is unlikely that sophisticated temperature stabilization will be necessary. Further work on the etch processes is needed to tie down the loss and, as a result, the crosstalk performance across the whole bandwidth.

A wavelength-multiplexing device, whose specification is entirely predefined and stable, and which can be interfaced in a single alignment to fibre or device arrays, is a real possibility.

14.3 TIME-DELAY STRUCTURES FOR PROCESSING

Direct processing of bit streams for pulse pattern recognition, and filtering communication bandwidths below 1 GHz, have so far only seen convenient practical realization in the electronic domain. Performing these functions directly in the optical domain has advantages for simplified switching structures at very high speeds expected in the future main network (see Chapter 4). To carry out the necessary system functions requires long path-length structures in a realistic and convenient technology. Arsenic-doped planar silica allows very tight radii of curvature and hence very compact stable devices.

Long delay-line structures, in integrated optical materials with a small core-cladding refractive index difference (Δn), are limited in the practical size reductions and bend radii that can be fabricated [13]. This restriction can cause problems where a high level of optical integration is required. High Δn technologies have tended to imply high losses. By using a silica-based technology which is capable of large Δn, small guide dimensions can be used [14, 15], significantly reducing the bend radii while keeping the losses low. Arsenosilicate glass (ASG) is such a material, with a Δn of ≈ 0.03. Single-mode guides operating at 1550 nm with diameters $\approx 3\ \mu$m can be fabricated. The bend radii made possible by use of this technology allow centimetres of delay line to be produced with only a few millimetres total diameter.

14.3.1 Device design

A set of practical constraints was incorporated into the design for a family of delay line devices. These were:

- minimization of curvature loss;
- reduction of curvature discontinuity loss;
- easy interconnection with other optical elements for use in a composite design.

Designs which do not incorporate a crossover require a reverse bend to escape from the centre of the device (see Fig. 14.5(b)). This 'double back' design generates excessive curvature discontinuity losses, which escalate rapidly with decreasing bend radius, limiting the smallest device dimensions. However, in-plane guide crossovers at angles of 90° cause only minimal loss.

Consideration of this loss balance led to the design shown in Fig. 14.5(a), where the guide exits the centre of the device by crossing all the encircling loops at 90°. The bend loss is negligible and the curvature discontinuities are the same for single- and multiple-loop devices. Therefore the maximum practical delay length is governed only by the intrinsic material loss and crossover loss. The details of curvature and crossover losses can be found in section 14.4.

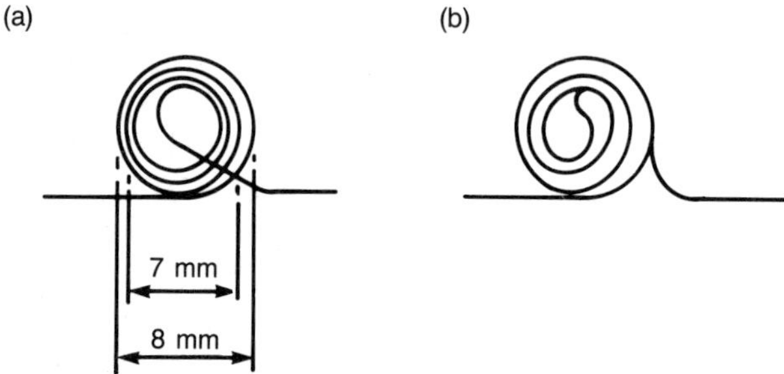

Fig. 14.5 Two ways of exiting a time delay loop.

14.3.2 Results

A 'single-turn' loop (equating to about 2 cm of linear guide length or 100 ps of delay) was used to test the initial idea. A nine-turn 'loop' (or 20 cm long guide, implying a 1 ns delay) can be fabricated within the same boundaries as a single-turn device using this geometry, although the loss would be rather high at the present state of development.

The main coils have an outer and inner diameter of 8 mm and 7 mm respectively. The device can also be directly inserted into an in-line composite design without the need for extra fitting. The spiralling guide was designed to be 3 μm $\times$ 3 μm with a centre guide separation of $\approx$40 μm between adjacent loops. As can be seen from the scanning electron microscope (SEM) photograph (Fig. 14.6), the guide crossovers are orthogonal and show little sign of crossover deformities as a result of the fabrication process.

Two lensed fibres were aligned with input and output guides and laser-welded to the walls of a surrounding package. After this process the fibre-to-fibre loss was measured as 11 dB. Just under half of this loss is accounted for by the guide losses, currently around 1 dB/cm. The remainder comes from the lensed fibre interfaces, which were not optimized for these particular guides.

The device was used to produce pulse pairs from a mode-locked DFB laser with 20 ps pulsed output. The two pulses were separated by 98.4 ps, which is within the measurement error of the 100 ps of the design.

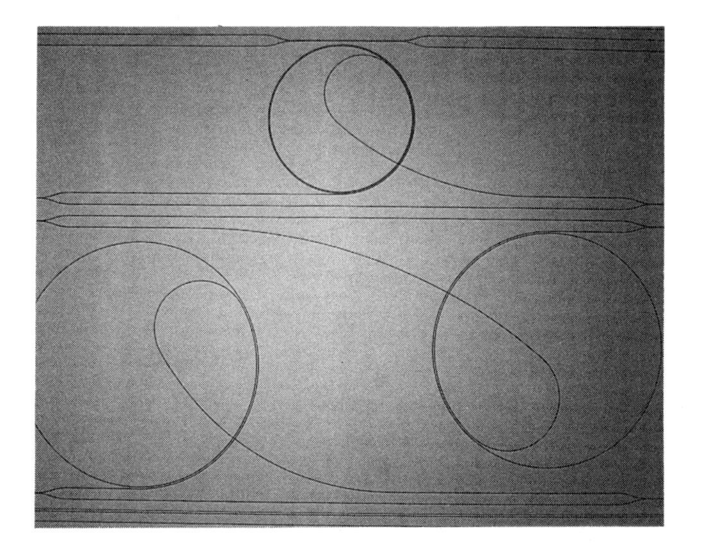

Fig. 14.6 Photographs of guide crossovers.

14.3.3 Discussion

Delay lines fabricated in planar silica technology are extremely stable on the length scale of the pulses being processed. Temperature variations for example are likely to change the path lengths by micrometres without external stabilization. While this path difference variation can be used to switch Mach-Zehnder structures which rely on interferometric path lengths, in this application pulse overlap is the criterion. Typical pulse lengths are on the millimetre-centimetre scale. The 2 cm delay demonstrated, corresponding to a 10 Gbit/s system, would need to be made in multiples of this length, in conjunction with nonlinear optical devices, to construct a full header-recognition circuit. Longer delays are possible, limited only by the material loss. In the future, rising transmission speeds imply that shorter rather than longer delays will be necessary. Since shorter delays are technologically easier, the passive part of pulse pattern generation or recognition circuits in this technology can easily keep pace with the increasing bit rates processed by future networks.

14.4 PACKING DENSITY AND GUIDE CROSSOVERS

The size of a device has an effect on the final cost as it governs the number of devices which can be fabricated on each wafer. Packing density is governed in turn by the ability to turn sharp corners and to form waveguide cross-overs with low loss. It is an ability inherent to passive optics to form these crossovers without signal interference. There is a small scattering loss penalty which must be quantified. In principle, highly complex interconnect patterns can be achieved in very small areas. This section presents theoretical and experimental results which quantify these effects. The guide sizes and indices chosen correspond to those of ASG technology.

14.4.1 Curvature losses

Finite element methods were used to derive guide parameters which were used to calculate curvature (see Fig. 14.7) and curvature transition loss (see Fig. 14.8). The loss resulting from bend radii down to 1 mm is extremely small. It is to be expected therefore that very intricate and densely packed interconnecting guides are possible in this technology.

Interestingly, it can be seen from Fig. 14.7 that the loss resulting from a continuous bend is very small indeed for curves as small as 1 mm radius. The main loss accrues from sudden changes in curvature. Figure 14.8 shows the point loss resulting from an abrupt change from a 1 mm curve to an effectively straight guide at the right of the graph. As the second curvature approaches 1 mm, the point loss obviously disappears for a curve in the same sense as the first, whereas it becomes prohibitively high for a curvature

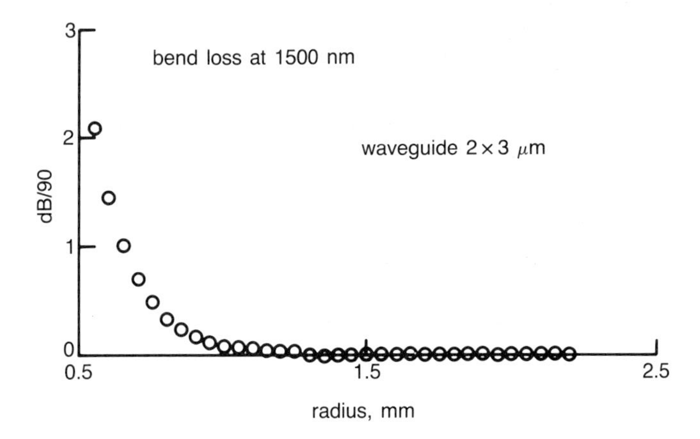

Fig. 14.7 Modelled excess loss for a 90° bend.

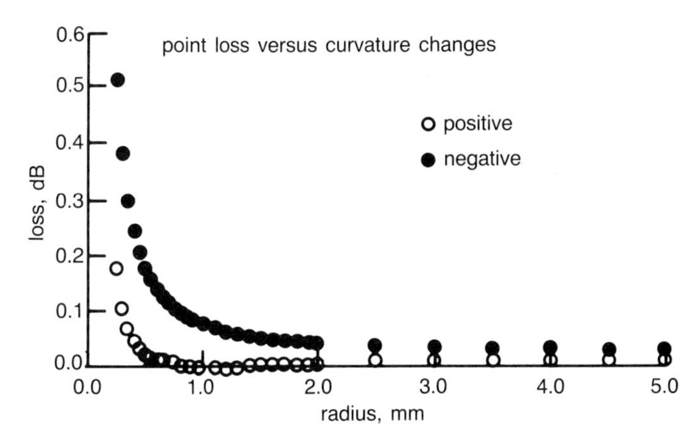

Fig. 14.8 Point loss versus radius of curvature transferring from a bend radius of 1 mm.

reversal. It is possible to reduce these losses by using sinusoidal curves rather than abrupt ones, at the expense of slightly more chip area.

14.4.2 Crossovers

Test structures have been made in order to measure the crossover loss directly. The chips included straight waveguides of size 2 μm × 2 μm, with a varying number of 90° crossovers, ranging from 0 to 100. The through power of each guide was measured at 1.3 and 1.55 μm and compared with the 0° crossover case. The results are summarized in Fig. 14.9.

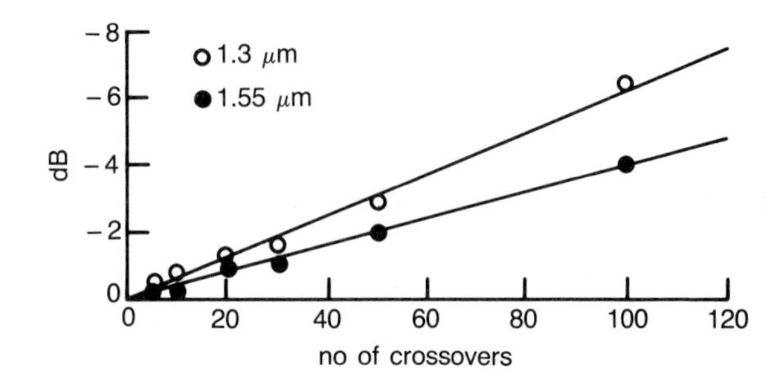

Fig. 14.9 Guide-crossing loss for large numbers of crossings.

It can be seen that the average loss per crossover is 0.06 dB (1.4%) at 1300 nm and drops to 0.04 dB (0.9%) at 1550 nm. Similar measurements on 3 μm guides indicate loss per crossover of ~0.07 dB (1.6%) for both 1300 nm and 1550 nm cases.

A detailed theoretical treatment can be found in MacKenzie et al [16]. A simple analysis of the crossover structure can be used to derive an analytical expression for the loss [17]. This model assumes diffraction from the waveguide over the crossover region, and a partial re-excitation of the fundamental waveguide mode thereafter, some of the power having been lost to radiation.

The simplest analytical approximation uses the 'equivalent fibre', where the fibre radius, a, is chosen such that the core cross-section is equal to the guide cross-section ($\pi a^2 = 4r^2$). In practice, this will be a reasonable approximation as reflow of the planar guide results in a near circular cross-section. The spot-size can then be approximated [18] by the expression:

$$w/a = 0.65 + 1.619/V^{3/2} + 2.879/V^6$$

where V is the 'equivalent fibre' V-value and the approximate loss is then given by the analytical expression for loss derived in Love and Ladouceur [17]:

$$\text{fractional loss} \sim \Delta/V^2(r/s)^4$$

where, following the nomenclature of Love and Ladouceur [17], $V = (2\pi/1)rn_{co}\surd(2\Delta)$, r is now the equivalent fibre radius, s is the spot-size in the nomenclature of the reference, $\Delta \sim (n_{co} - n_{cl})/n_{co}$ and n_{co}, n_{cl} are the core and cladding indices respectively, with $2w^2 = s^2$. It can be seen that the theoretical loss increases with increasing V-value.

Measured loss values (Fig. 14.9) for 2 μm and 3 μm guide crossovers are higher than the theoretical value, by ~0.3 dB, but support the V-value trend predicted by the diffraction model (see Fig. 14.10). This discrepancy can be used to estimate the physical details of the crossovers. An explanation for the higher measured loss could be broadening of the guide at the cross-over region, due to a 'rounding' of the corners to the tune of 0.5 μm (see Fig. 14.11). This would effectively cause the structure to behave as if the width of the crossing guide were greater.

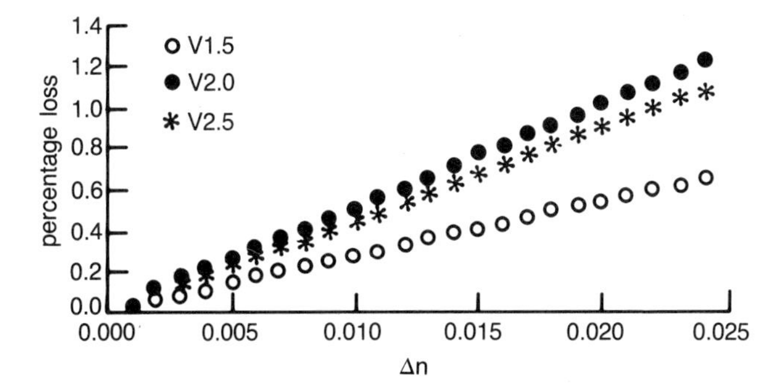

Fig. 14.10 Crossover loss as a function of guide parameters.

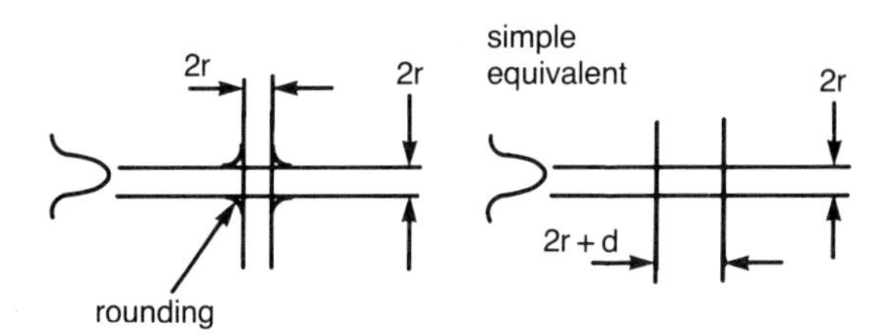

Fig. 14.11 Simple crossover loss model.

The theoretical loss values for 2 μm and 3 μm guides were recalculated, using the simple model in Fig. 14.11. The spot-sizes were retained, while the crossing guide was increased in width from $2r$ to $2r + d$, the chosen value of d being 0.5 μm for both guide widths.

It can be seen from Table 14.2 that the revised values are in close agreement with the measured loss.

Table 14.2 Comparison of loss values.

	Guide width	
	2 μm	3 μm
Wavelength = 1300 nm		
V-value	1.49	2.24
Measured loss (dB)	0.06	0.07
Theoretical loss including rounding	0.06	0.07
Wavelength = 1550 nm		
V-value	1.25	1.88
Measured loss (dB)	0.04	0.07
Theoretical loss including rounding	0.03	0.08

14.4.3 Discussion

The loss associated with five crossovers is equivalent to that associated with a single loop-back structure of 1 mm curve radius (i.e. 0.2 dB) and is sufficiently small to allow optical waveguide circuits of some complexity to be formed using crossover structures without prohibitive loss penalties. Combining this ability to curve sharply without incurring excessive loss enables complex interconnect patterns to be fabricated. Not only does this mean that compact devices can be made, but it becomes a practical proposition to include guides which simply have the function of aiding power alignment of input/output guides or active components. Close guide-packing therefore has an impact upon both the cost-governing factors in passive optical circuits — the chip area used, and the cost of assembly techniques.

14.5 INTERFACES — THE OPTICAL MOTHERBOARD

This section looks in more detail at the costs implied by the assembly techniques. The technology of silica waveguides on a silicon wafer means that a whole range of techniques which have been developed in detail in mature silicon technology can be enhanced to provide alignment and assembly aids.

14.5.1 Principles

Several types of interface are needed for a completely modular optical system. The optical circuit needs to interface to the outside world in the form of a fibre array. The circuit itself may well contain interfaces to active components — lasers, detectors, optical integrated circuits in semiconductors, and ultimately hybrid integrations to some localized electronic circuitry. This is the concept of the 'optical motherboard'. The main optical interfaces of this approach fall into three categories — waveguide to fibre, waveguide to detector and waveguide to laser, and hence to optoelectronic integrated circuits in semiconductors.

Silicon microengineering plays a major role in realizing these interfaces. Fabrication of planar silica waveguides on silicon substrates, suitably etched, allows hybrid integration of the optical components on to one motherboard. Minute physical features etched into the surface of the components to be assembled enable them to lock together in the correct orientation automatically. For example, the inclusion of precision etched V-grooves also allows fibre-to-waveguide coupling on to and off the motherboard and helps provide a low-cost interconnection technology. Etched features on laser chips can locate with corresponding features in the silicon motherboard to give physical alignment of active components. The details of the work described in the following sections refer to work carried out at BT Laboratories (BTL), but much progress has been achieved by others, e.g. Grant et al [19], and Armiento [20].

14.5.2 Technology

The planar waveguides are made from arsenosilicate glass. This has a high refractive index difference which gives strong guiding properties and allows relatively thin (6 μm) oxide layers to be used for the buffer and cladding layers; the waveguide layer is 2 μm thick. This is an important factor in fabricating the motherboard chip. Thinner layers can be deposited and etched more accurately; also, height control between different regions is easier to achieve. In addition, the ASG of the core is deposited by normal chemical vapour deposition, and the cladding layers by a plasma-enhanced chemical vapour deposition technique, which means that all processing temperatures can be kept below 1000 °C. Temperatures greater than 1000 °C can introduce crystal damage and seriously affect the dimensional control of the V-groove etching.

By careful design of the process sequence both the alignment features and waveguides can be defined in one photolithographic operation. This approach enables precise alignment between components to be built into the motherboard and is particularly important for achieving the submicrometre tolerances required for some of the interfaces.

Figure 14.12 shows a schematic of a hybrid optical motherboard, which includes examples of the interfaces needed. Some of the work being done at BTL in connection with these interfaces will be described in the following sections.

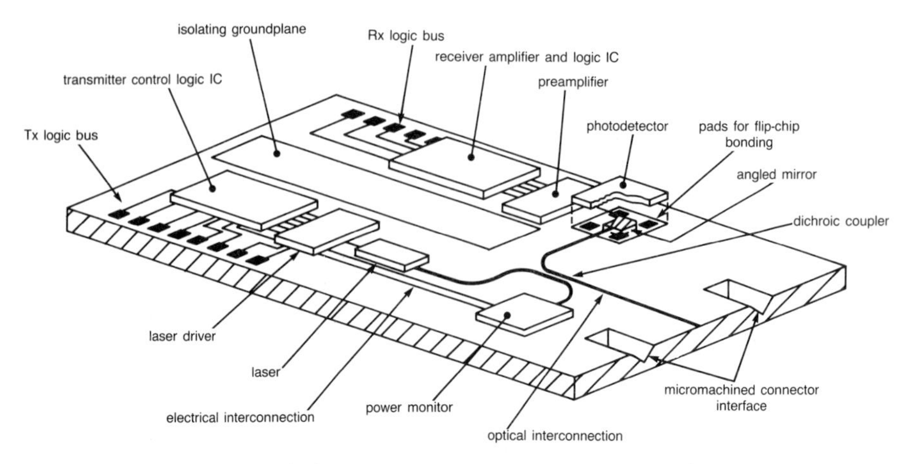

Fig. 14.12 Hybrid optical motherboard transceiver.

14.5.3 Principal interfaces

14.5.3.1 Waveguide to fibre

The aim of this work is to demonstrate the possibility of low-loss, low-cost connections between fibre arrays and silica guide arrays on the motherboard. Initial work has centred on the fabrication of fibre-to-fibre array connectors as representative of the same problem. A schematic of the array is shown in Fig. 14.13 as the component on the left-hand side of the diagram. These have been used successfully to demonstrate the V-groove technology by connecting two arrays of eight fibres together. This gave insertion losses between each of the eight pairs of fibres of typically 1 dB. Recently commercial fibre-to-fibre connectors have become readily available (e.g. the 'MT-8'™). The precision metal pins used in the alignment can interface directly to the silicon V-grooves of the motherboard. The optical circuits can then be connected to the system in a simple demountable way, using commercial components.

Measurements of waveguide-to-fibre array coupling have also been carried out to evaluate the mode-matching aspects of the connection. A schematic of the complete connector is shown in Fig. 14.13.

Figure 14.14 shows a photograph of a complete assembly with a fibre array connected to each end. Optical measurements were obtained using commercial fibre arrays using active alignment rather than locating pins. These results are presented in Fig. 14.15. Two sets of results are shown — the lower loss results are for each guide individually aligned to the fibre, the upper curve results from simultaneously optimizing the outer two guides only.

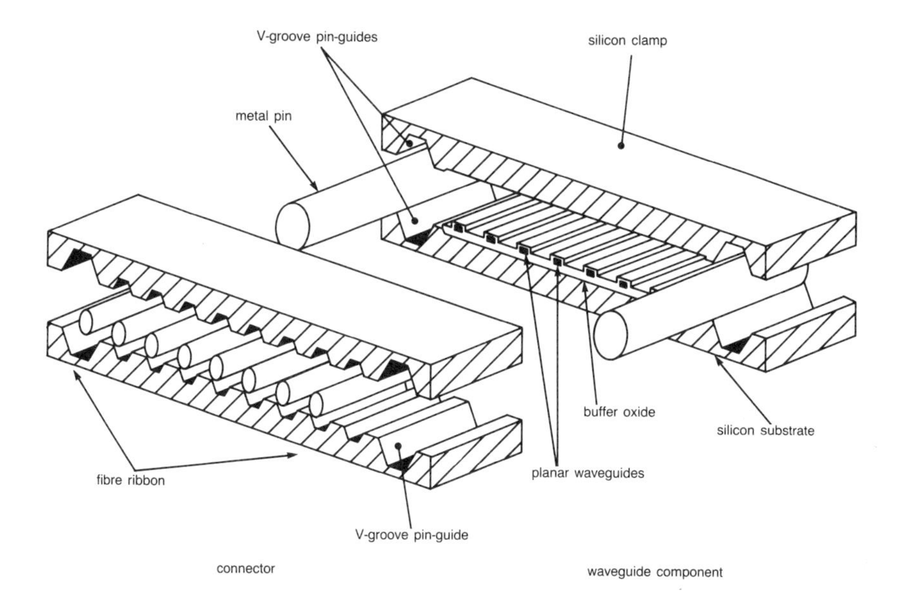

Fig. 14.13 Plug connection of multiple fibres to waveguides.

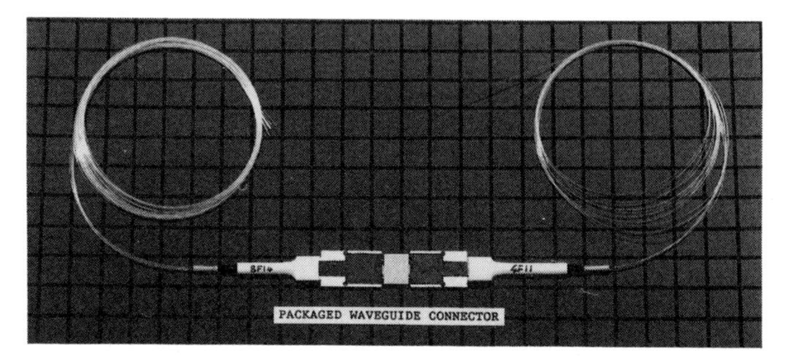

Fig. 14.14 Packaged waveguide connector with a fibre array connected to each end.

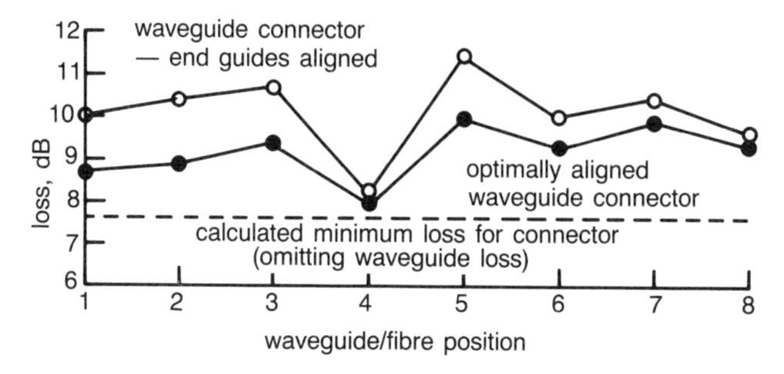

Fig. 14.15 Waveguide connector loss measurements.

The coupling levels are within about 2 dB of the calculated value for perfect alignment. The power coupling for perfect alignment is limited by the fundamental mode mismatch between fibre and the high refractive index guides to 3.8 dB. This limit can be reduced by adding mode transformers at the ends of the guides. One way of doing this is to fabricate waveguide tapers which spread the optical field to match that of the fibre. Recently it has been shown that this mode mismatch loss can be halved by tapering the guide ends down to 1 μm at the chip perimeter. Theoretical studies suggest that this can be reduced still further. These results demonstrate successful integration of waveguides and V-grooves on a single substrate. Assessment of a demountable connection between a commercial fibre array connector and motherboard is shortly to be established.

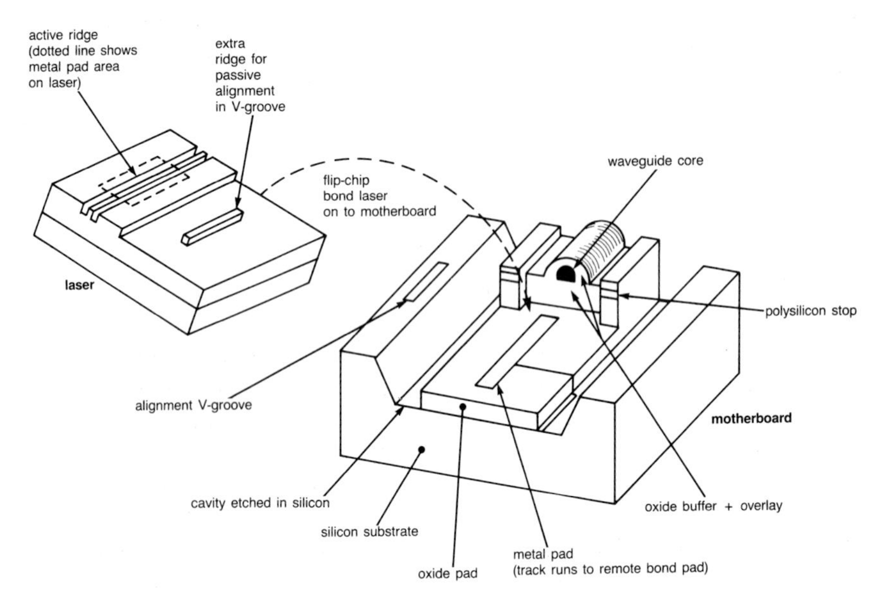

Fig. 14.16 Schematic showing how laser is mounted on hybrid optical motherboard.

14.5.3.2 Laser to waveguide

Figure 14.16 shows a method for achieving automatic precision alignment of a laser to a waveguide in three dimensions. The vertical alignment is achieved by precise control of the thickness of the deposited layers. The spacing between the laser facet and the end of the waveguide is determined by pushing the laser chip against accurately located physical stops on the motherboard. Lateral positioning is controlled by locking a ridge on the laser chip into an etched V-groove on the motherboard.

For maximum coupling efficiency the laser and waveguide should exactly butt together, in which case the coupling efficiency is limited by the mode size mismatch; for the structures of current interest, this results in a theoretical minimum coupling loss of 3.8 dB. Silicon microlens arrays, diffractive optics or mode spreading in the laser are all promising techniques for reducing this mode mismatch loss. In practice, a small separation between the laser and waveguide is preferred in order to reduce the axial alignment sensitivity, or to include lens elements. For many applications, however, a relatively large laser-to-waveguide coupling loss of up to 10 dB is acceptable in applications where cost rather than power efficiency is the issue. It has been calculated that a separation of 6 μm is the optimum in order to give the lowest sensitivity to axial misalignment; this gives horizontal and vertical alignment tolerances of 1.8 μm and 1.6 μm respectively, before the target loss of 10 dB is exceeded. The work on the fibre-aligning V-grooves has shown that submicrometre tolerances are achievable in this technology. The task remaining is to incorporate the microetching of features with the other motherboard processes.

14.5.3.3 Waveguide to detector

The waveguide to detector interface uses an etched and metallized V-groove sidewall as a mirror to deflect light from the waveguide into a substrate entry photodiode. This is shown in Fig. 14.17. Positioning tolerances required for this are much less severe than for the laser, due to the large active area of the photodiode. It has been shown by using a fibre and V-groove mirror that nearly 100% of the light can be coupled into the detector; even with 25 μm misplacement 70% efficiency was achieved. The accuracies attained by some pick-and-place machines would be adequate for this operation.

14.5.4 Discussion

The development of processing for a hybrid integrated optical motherboard is now at an advanced stage, and many of the critical process steps have been demonstrated. Devices which include waveguides, lasers and detectors on a single substrate are currently being processed, and waveguide-to-fibre array connectors have already been shown. The extension of the processing to combine all of these device structures on a single substrate will enable the full hybrid optical motherboard to be fabricated.

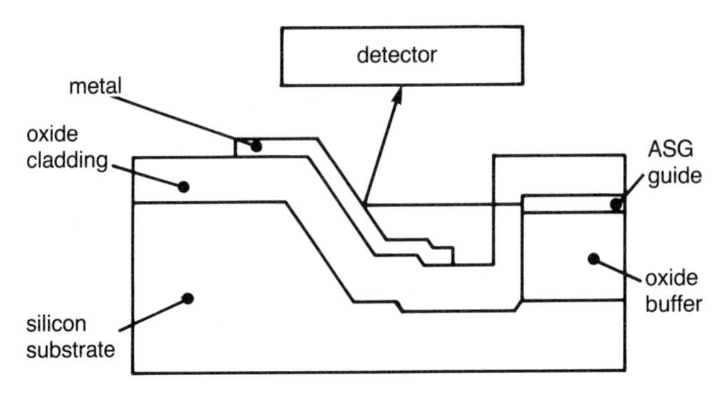

Fig. 14.17 Schematic of detector-to-waveguide interface.

14.6 CONCLUSIONS

The activity world-wide on planar silica components ranges from simple commercial devices for use in passive networks, to advanced materials work on integrated amplifier structures. The examples of passive devices presented in this chapter are those which show that important system functions can be built in as a fundamental part of manufacture, and result in stable devices. Passive optics is capable of great packing density. Guides can turn corners in small areas and cross others without signal interference. Of great interest is the ability to combine the technology of silica with that of silicon on which the silica guides are fabricated. This opens a whole range of silicon micromachining techniques allowing automatic alignment of component parts to great accuracy. This affects the single most important cost factor of practical device fabrication — assembly. No longer do precision active alignments need to be carried out as part of the construction. The devices will 'click' together with lasers, fibre arrays and active optical integrated circuits in semiconductors. All of the key interfaces have been demonstrated in principle. Painstaking work now needs to be carried out to ensure that the whole process is capable of being integrated into one. The next year or two should see significant progress in this area in laboratories around the world.

REFERENCES

1. Godfrey D, Bailie S, Cooper K, Nield M, Hill J and Welbourn D: 'Fully integrated silica based optical motherboard', IEEE Multichip Module Conference, pp 146-149, Santa Cruz (March 1992).

2. Wisley D R: '32 channel WDM multiplexer with 1 nm channel spacing and 0.7 nm bandwidth', Electron Lett, 27, No 6, pp 520-521 (14 March 1991).

3. Toba H, Oda K, Takato N and Nosu K: '5 GHz-spaced, eight channel, guided wave tunable multi/demultiplexer for optical FDM transmission systems', Electron Lett, 23, No 15, pp 788-789 (16 July 1987).

4. Dragone C, Edwards C A and Kistler R C: 'Integrated optics N × N multiplexer on silicon', IEEE Photon Technol Lett, 3, No 10, pp 896-899 (October 1991).

5. Takahashi H and Hibino Y: 'Arrayed-waveguide grating wavelength multiplexers fabricated with flame hydrolysis deposition', Fourth Optoelectronics Conference (OEC '92) Technical Digest, pp 306-307 (July 1992).

6. Fujii Y and Minowa J: 'Optical demultiplexer using a silicon concave diffraction grating', Appl Opt, 22, No 7, pp 974-978 (April 1983).

7. Grand G, Jadot J P, Valette S, Dennis H, Fournier A and Grouillet A M: 'Fiber pigtailed wavelength multiplexer/demultiplexer at 1.55 μm integrated on silicon substrate', Eight Annual European Fibre Optic Communications Conference (EFOC '90), pp 108-113 (June 1990).

8. Soole J B D, Scherer A, Leblanc H P, Andreadakis N C, Bhat R and Koza M A: 'Monolithic InP-based grating spectrometer for wavelength-division multiplexed systems at 1.5 μm', Electron Lett, 27, No 2, pp 132-134 (17 January 1991).

9. Maxwell G D, Cassidy S A, Beaumont C J, Barbarossa G and Powling R: 'Monolithic 16-channel Fourier optic WDM in planar silica', SPIE Fibres '92, Boston, Proc SPIE J, 1792, pp 78-88 (1993).

10. Kawachi M: 'Silica waveguides on silicon and their application to integrated-optic components', Opt Quantum Electron, 22, pp 391-416 (1990).

11. Kashyap R, Ainslie B J and Maxwell G D: 'Second harmonic generation in GeO_2 ridge waveguide', Electron Lett, 25, No 3, pp 206-208 (February 1989).

12. Richardson D: 'Diffraction gratings', in Kingslake R (Ed): 'Applied Optics and Optical Engineering', 5, pp 17-46 Academic Press, New York (1969).

13. Nishikido J, Okuno M and Himeno A: '4.65 Gbit/s optical four-bit pattern matching using silica based waveguide circuit', Electron Lett, 26, No 21, pp 1766-1767 (1990).

14. Welbourn D, Beaumont C and Nield M: 'Directional couplers in a high index silica waveguide', International Photonics Research Conference, Monterey (April 1991).

15. Beaumont C J, Cassidy S A, Welbourn D, Nield M and Thurlow A: 'Integrated silica optical delay line', Proc 17th European Conference on Optical Communication, Paris, France, pp 421-422 (1991).

16. MacKenzie F, Beaumont C J, Nield M and Cassidy S A: 'Measurement of excess loss in planar silica X-junctions', Electron Lett, 28 , No 20, pp 1919-1920 (1992).

17. Love J D and Ladouceur F: 'Excess loss in singlemode right-angle X-junctions', Electron Lett, 28 , pp 221-222 (1992).

18. Marcuse D: 'Loss analysis of single-mode fibre splices', Bell System Tech J, 56 , pp 703-711 (1977).

19. Grant M et al: 'Low loss coupling of ribbon fibres to silica on silicon integrated optics using preferentially etched V-grooves', Integrated Photonics Research (April 1992).

20. Armiento C: 'Hybrid optoelectronic integration on silicon', IEEE Symposium on High Density Integration and Computer Systems, Boston (1991).

15

LIGHT-SENSITIVE OPTICAL FIBRES AND PLANAR WAVEGUIDES

R Kashyap, J R Armitage, R J Campbell, D L Williams, G D Maxwell, B J Ainslie and C A Millar

15.1 INTRODUCTION

Since the proposal by Kao and Hockham [1] in 1966 of low-loss transmission in glass-fibre waveguides, research into optical fibres has yielded rich dividends. It is evident from the current network in the UK and elsewhere that optical fibres have come of age within an extremely short time period. With its large data capacity (Tbits/s), optical fibre can easily meet the bandwidth requirements for the foreseeable growth in communications. However, the transmission of data is only part of future communications systems. The challenge is to develop data handling and processing techniques which will allow new customer services to be implemented cost-effectively. Services of the future, such as HDTV and user interactive facilities [2], cannot be processed at the likely data rates by conventional electronic methods economically. The optimum solution is to exploit the fast optical-processing potential of optical technology along with well-developed electronics. These advanced services of the future are likely to need not only high bandwidth but also a large variety of systems and subsystems comprising units capable of fulfilling very basic yet crucial optical functions.

It is with these processing applications in mind that research into light-sensitive optical fibres and waveguides is being performed. This work has already revealed highly cost-effective and elegant solutions to a number of problems in future networks. It is also likely that options available in the future may revolutionize the design philosophy of networks. This exciting technology has grown rapidly and caused tremendous interest around the world.

15.2 WHAT ARE PERMANENT LIGHT-SENSITIVE CHANGES IN OPTICAL FIBRES?

15.2.1 Light-induced reflection gratings

Since 1978 transmission characteristics of optical fibres at certain visible wavelengths have been known to be dependent on the light power propagating in the fibre. This dependence takes the form of permanent changes in the loss at visible wavelengths and also in the local refractive index — of order 0.01-1% of the index difference between the core and cladding glasses. Fortunately, for optical fibre communications, these effects are restricted to wavelengths shorter than 600 nm and have been recently shown to peak around 240 nm [3]. Hill et al [4] first reported that blue-green light propagating in the fibre interferes with counter-propagating light from a small far-end reflection, causing the fibre to change its transmission characteristics. A periodic refractive index change in sympathy with the interference pattern was created in the fibre. This light-induced spatial modulation of refractive index acted as a grating which reflected light at the same wavelength as the illumination. The grating satisfied the Bragg conditions for a guided-mode reflection at the writing wavelength. This Bragg reflection grating was shown to grow in strength with time of illumination. A schematic of the grating is shown in Fig. 15.1.

Theoretical analysis of this observation was provided by Bures et al [5] using coupled mode equations linking the forward- and backward-propagating guided modes, similar to ones describing coupled pendulums. These gratings had weak coupling coefficients with narrow bandwidths (500 MHz), being several tens of centimetres in length. The wavelength of light used in these experiments was either single-frequency 488 nm or 514.5 nm from an argon ion laser, and required careful observation since these long fibre gratings tended to be highly temperature-sensitive. Since the Bragg condition was satisfied at the writing wavelength alone, the applications were considered somewhat restrictive. Thus, not much attention was paid to the

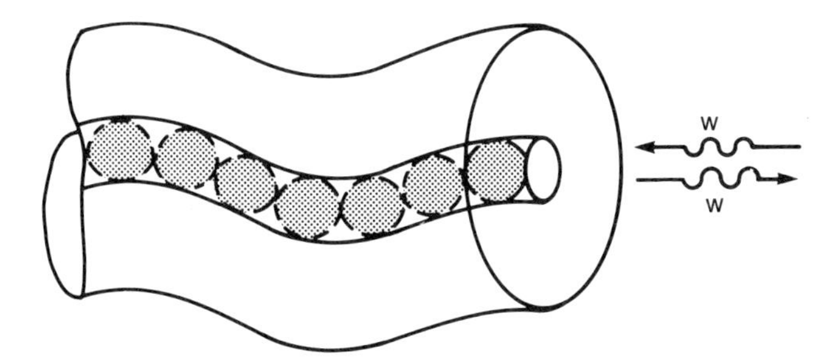

Fig. 15.1 A periodic refractive index modulation induced by the standing wave intensity causes the light at frequency, ω, to be reflected back (dotted region shows higher index regions after the Bragg grating is formed — light is thus reflected back).

phenomenon apart from being a curiosity of academic interest, while Hill's original fibre remained somewhat of an unknown quantity, being the only photosensitive fibre for several years.

15.2.2 Frequency doubling in optical fibres

The next major development was as a result of a curious effect observed in optical fibres in 1986. High optical intensities at 1064 nm from an Nd:YAG had been used since almost the beginning of research on optical fibres with predictable nonlinear effects. So it came as a surprise to the scientific community when Österberg and Margulis reported that an ordinary single-mode optical fibre which had been accidentally illuminated for several hours with invisible high-power pulsed Nd:YAG light at a wavelength of 1064 nm, was emitting green light at exactly the frequency-doubled wavelength of 532 nm [6]. It was soon realized that the fibre had undergone a physical metamorphosis allowing a phenomenon normally forbidden in material such as glass. A clue came from the knowledge that application of electric fields can allow frequency doubling in glass; could it be that charges were being released in the glass, under influence of the light in the core, which were being self-organized to create an internal field in a way that allowed the frequency-doubled light to grow? Since the refractive index at 532 nm is different from that at 1064 nm, exchange in energy from the infra-red to the green is only possible if dispersion can be compensated by periodic readjustment of the phase difference between the two wavelengths as they propagate in the fibre. This gives rise to the concept of the internally self-organized grating comprising a spatially periodic electric field [7-10].

This may be better appreciated in Fig. 15.2, where the internally generated electric field undergoes a change in direction in exactly the correct distance to cancel the effect of dispersion. Remarkably, highest reported power conversion from the infra-red to the green in optical fibres stands at 13% [11].

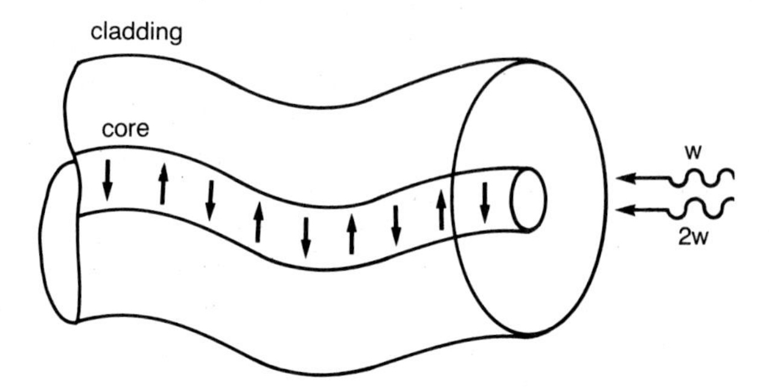

Fig. 15.2 An alternating internal electric field of the correct period for phase matching is generated by nonlinear interaction of the infra-red and its frequency-doubled field.

15.2.3 Self-organization

A self-organized grating was recognized as the reason for 'phase-matching', as it had been in the self-induced Bragg reflection grating case. Interest was immediately renewed in 'photosensitivity' of optical fibres. New models (e.g. Anderson et al [12]) appeared and the debate continued; however, it was recognized that a particular defect associated with core-dopant germanium which has a characteristic absorption in the ultraviolet (UV) at 240 nm was the probable source of the electric charge and also of the observed refractive index changes. Added to the discovery of frequency doubling in optical fibres was the demonstration by Meltz and Morey [13] of using the interference of two beams of ultraviolet radiation at around 250 nm to form reflection gratings for use at a wavelength of 647 nm. This technique was extended [14] to reflection gratings for use in the important 1500 nm telecommunications window. The latter demonstration stimulated activity around the world in the field of photosensitivity, since it was recognized that highly reflective all-fibre gratings used as filters can have an enormous impact on communications in all of the useful windows of transparency of optical fibres.

15.3 MATERIAL CONSIDERATIONS

In order to fully exploit the potential of photosensitivity of optical fibres, it is important to understand the origins of the effect and to be able to optimize the response of the waveguide materials. To this end, studies have been carried out by extending techniques such as UV emission/absorption spectroscopy to fibres and the measurement of the maximum induced change in index, as a function of processing and fabrication procedure. This approach has fundamentally influenced the understanding of the phenomenon.

It is known that a common defect in germanium-doped silica, thought to be responsible for the changes in the material, has a characteristic absorption at 240 nm [15]. This absorption has been identified with the oxygen-deficient defect, a schematic of which is shown in Fig. 15.3. There are several other defects possible in Ge-doped silica; however, this defect is suspected to be the prime mover in observation of the effect. Germanium is normally co-ordinated with four oxygen atoms. Here, one oxygen has been replaced by Si (or Ge) and has an extra donor electron associated with it. This bond may be broken by photoexcitation and the transmission spectrum altered permanently. The presence of defects has been used to good effect by Krol et al [16]. By incorporating other defect-rich dopant ions like cerium, frequency doubling was reported in germanium-free optical fibres. Index changes have also been observed in Ge-free fibres containing ions like aluminium and europium [17, 18].

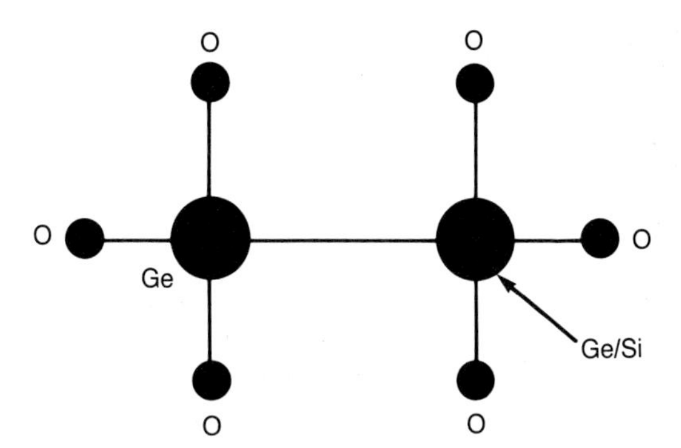

Fig. 15.3 The oxygen-deficient Ge defect thought to be responsible for photosensitivity of optical fibres. An electron may be ejected by breaking the Ge-Ge/Si bond by illumination with radiation at 240 nm.

It is also known that major structural changes can occur in other amorphous/glass systems such as in chalcogenides. In these materials, amorphous-to-crystalline changes occur at modest optical illumination power levels with index changes that are orders of magnitude larger than in Ge-doped silica. However, Ge-doped silica fibre technology is far advanced and well controlled in comparison, and far outweighs the advantage of the large index changes. These materials may be useful as the next-generation material systems.

15.3.1 Microscopic effects

The breakage in the 240 nm defect bond on photo-excitation releases charges which can move through the glass network via electron pathways [19] and are trapped at hole defect sites. Since defects are a rich source of charges (electron and holes) it is possible to investigate light-assisted charge transport by the measurement of electric currents.

Charge generation has been investigated as a function of UV illumination power in experiments on planar waveguides with gold electrodes [20]. Measurement allowed the calculation of total charge released and hence the defect population in the waveguide. Early samples showed a low defect population of order 0.023% of the Ge population, indicating a correspondingly low photosensitive response, and indeed attempts to induce index changes in standard Ge-SiO_2 planar waveguides using 240 nm radiation gave poor results. Estimates of the induced index change in optical fibres by Russell et al [21] show that the defect population has to be a large fraction of the Ge ion concentration before substantial index changes can occur.

15.3.2 Spectroscopic studies — enhancement of light sensitivity

Studies of UV absorption spectroscopy of thin samples of Ge-doped silica to investigate the effect of fabrication conditions and the influence of material systems have also been carried out successfully. These have included measuring the transmission of thin samples of preforms and fibres before and after irradiation with UV light [22]. Figure 15.4 shows the transmission spectrum before and after UV irradiation. The changes in the absorption may be related to the changes in index of refraction via the Kramers-Kronig relationship [22]. These calculations performed on optical fibres and planar films of photo-enhanced Ge-doped silica on silica show that the index change is an order of magnitude lower than that measured in waveguides. Hence,

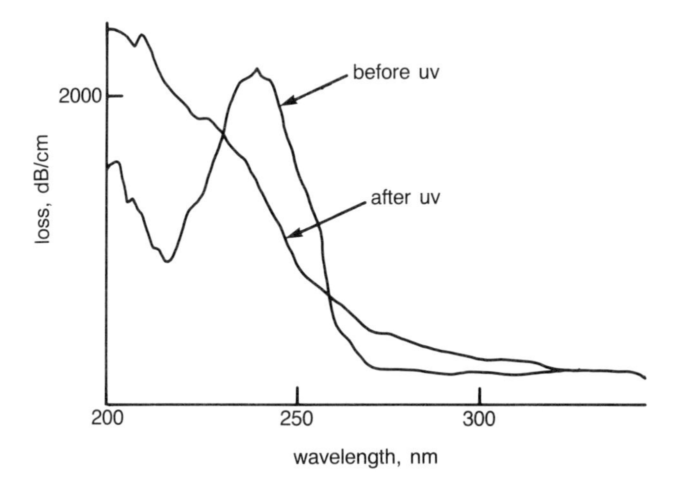

Fig. 15.4 The transmission spectrum of 15 μm thick preform before and after irradiation with 244 nm wavelength light.

it has been suspected that conformational changes, such as a change in density, may also be occurring to enhance the effect [21].

Many fibres manufactured at BT Laboratories have shown a photosensitive response. With a view to understanding the important material parameters influencing the optically induced index change, boron/germanium co-doped fibres were fabricated, since boron has been shown to undergo large volumetric changes on annealing [23]. These fibres showed high photosensitivity. Inclusion of hydrogen at elevated temperatures has also improved the sensitivity not only in optical fibres but also in planar films and waveguides. The absorption spectrum in the UV with hydrogen treatment is shown in Fig. 15.5. The enhanced 240 nm peak shows that the number of oxygen-deficient germanium defects has increased. Indeed, irradiation with UV light reduces the absorption at 240 nm dramatically, as shown in Fig. 15.6, giving rise to a refractive index change as a consequence of the Kramers-Kronig relationship.

Diffusion of hydrogen into optical fibres at room temperatures and high pressures has also been used to enhance the photosensitivity of optical fibres [24]. Hydrogen at pressures up to 800 atmospheres has been used to activate almost every germanium site in the core on UV irradiation so that large index changes (0.01) have been achieved. Using this scheme it has been possible to directly write channel waveguides into Ge-doped planar silica films. This technique allows any Ge-doped fibre to be photosensitized, but does increase OH formation, and therefore loss, at 1328 nm. Unless stored at low temperatures, unactivated hydrogen eventually diffuses out, returning the

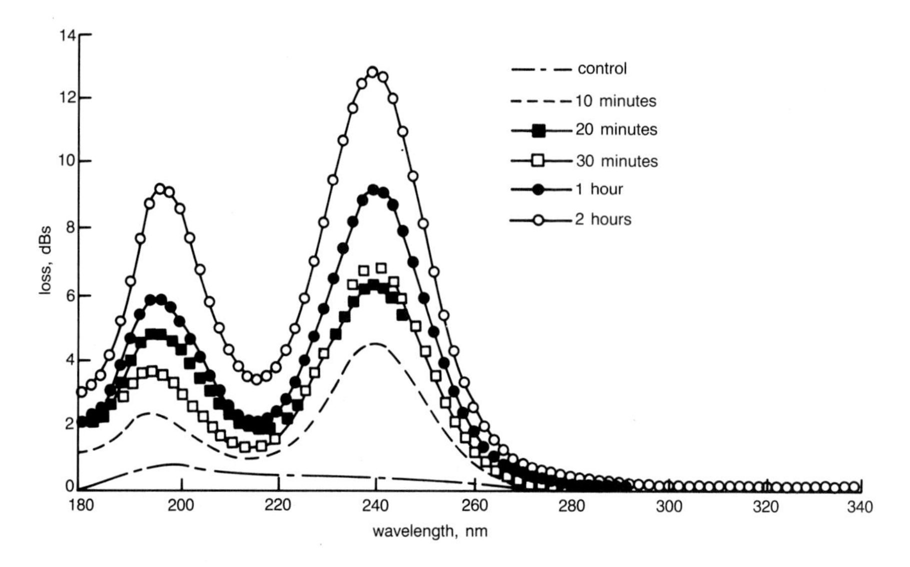

Fig. 15.5 The effect of increasing hydrogenation treatment in the UV spectrum of planar films of Ge-doped silica.

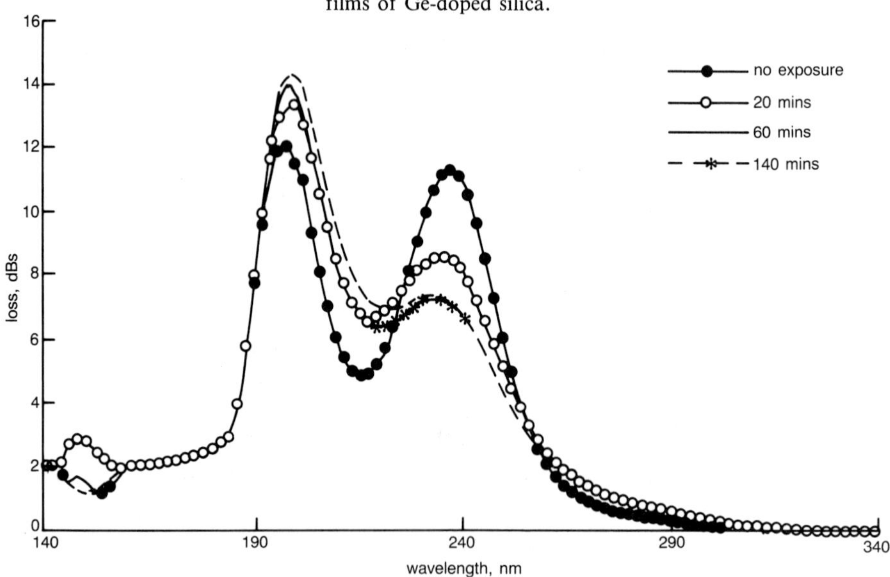

Fig. 15.6 The effect of increasing UV exposure on hydrogenated sample of Ge-doped silica. Note the reduction of the 240 nm absorption and the increase in the 213 nm peak.

fibre to its original low photosensitive state. Once activated by UV exposure, however, the index change is permanent. The nature of the chemical reactions is not fully understood.

Another method used for photosensitizing optical fibres and planar waveguides is by flame-brushing with a hydrogen-rich burner. At high flame temperatures, hydrogen is able to diffuse into the core so that the treated waveguide has an enhanced photosensitivity [25].

Gratings written at high optical fluences show characteristics of material damage [26]. These gratings are formed by physically damaging the core to cause periodic relief to gratings that are very robust at high temperatures. A disadvantage of this method is the high short-wavelength loss resulting from coupling of the guided mode to the radiation field (see section 15.6.1).

It is important to draw a distinction between the photorefractive and the photosensitive effects. The former effect is apparent in ferroelectric materials and uses the electrooptic effect. The charge released on photoillumination migrates through the crystal and on retrapping gives rise to an internal field. This field modulates the refractive index via the electrooptic effect, forming a phase hologram. In the photosensitive effect, the charges give rise to large internal fields, but since the quadratic electrooptic coefficient is several orders of magnitude smaller than the electrooptic coefficient in photorefractive crystals, the resulting index modulation is correspondingly small. Thus it can be deduced that a major component to the index change in fibres illuminated by UV radiation is through a change in the UV absorption, and possibly via a change in the density [22, 23].

15.4 TECHNIQUES FOR WRITING BRAGG REFLECTION GRATINGS IN OPTICAL FIBRES

Figure 15.7 shows the basic set-up for writing gratings optically into waveguides either in fibre or planar form. Two UV beams at 244 nm, derived from an intra-cavity frequency doubled argon ion laser, are allowed to interfere at some adjustable angle, θ, at the fibre or waveguide core [14]. Light from an edge-emitting LED (ELED) operating at around 1550 nm is coupled to a 50:50 splitting ratio fibre X-coupler. One output end of the coupler is used to launch light into the photosensitive fibre, while the second input arm can be used for monitoring reflections on a spectrum analyser as the grating is being written. The output from the fibre or waveguide in which the reflection grating is written can also be coupled to the spectrum analyser to monitor the transmission spectrum. A single-prism interferometer is used to write the gratings, as shown in Fig. 15.8. The two beams are generated by aiming the centre of a single beam at the apex of the prism so that half of the beam folds on to itself to interface at the fibre placed at the back face. This interferometer is highly stable and very easy to align. Another

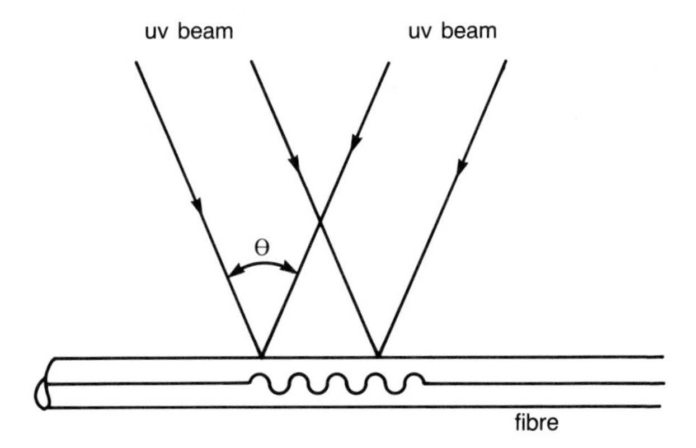

Fig. 15.7 Two UV beams incident at an angle, θ, interfere at the fibre to form a refractive index grating in the core.

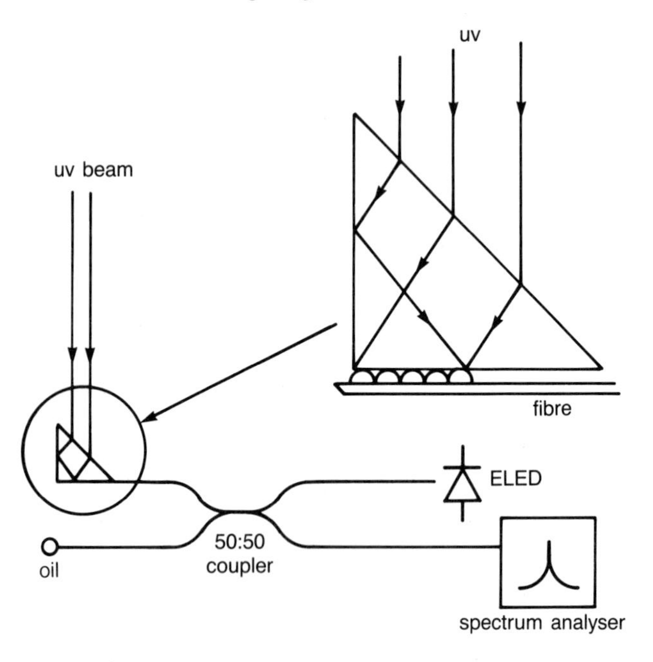

Fig. 15.8 The prism interferometer set-up used for forming photosensitive gratings in fibre cores. The inset shows a close-up of the interferometer.

interferometer used at BT Laboratories since 1991 is based on holographic replication of phase-mask directly into the core of a fibre. A computer-generated hologram [27] is fabricated in UV transmitting silica using e-beam technology. The hologram is then replicated in the core of a photosensitive fibre using the interferometer configuration shown in Fig. 15.9. The two mirrors indicated in the figure are two sides of a rectilinear silica block at which total internal reflection takes place to combine the diffracted beams at the fibre. This technique has been used to generate many of the gratings and devices shown in this paper. A cylindrical lens is used to create a line focus at the fibre held in a jig so that accurate alignment is made possible.

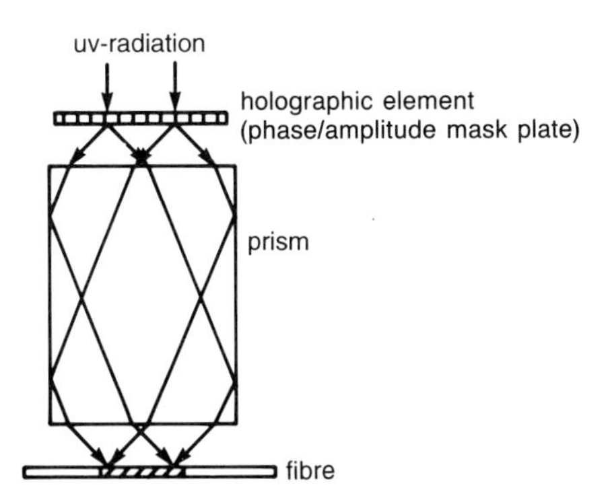

Fig. 15.9 A transmission diffraction grating used at BT Laboratories for writing a grating into the core of a fibre. A phase diffraction grating in silica is used as the mask.

15.4.1 Refractive index changes in fibres and waveguides

Recent measurements show that the best UV-exposed induced index change is larger than 10^{-3}, about as large as the core-cladding index difference, Δn [28-30]. Photosensitivity-enhanced planar single mode waveguides have shown around half of this index change [31]. These remarkably large index changes in optical waveguides with Δn values of between 0.4 and 0.3×10^{-3} can be used to solve many difficult problems in communications systems.

15.4.2 The generic component — the band-stop filter

The most important and also one of the first applications of photosensitivity of optical fibres in telecommunications was the reflection grating filter [14] in the 1500 nm window. Figure 15.10 shows the reflection/transmission spectrum of an optical fibre in which several gratings have been written at physical locations separated by a few millimetres. These reflection gratings fulfil a basic requirement for a band-stop filter — all wavelengths within a wide spectrum are transmitted other than those which meet the Bragg condition defined by the gratings. These gratings find applications mainly in areas where a narrowband reflection is required rather than stopping those wavelengths from being transmitted. However, by writing multiple narrowband gratings, it is possible to build specially shaped filters, which will be discussed in the next section.

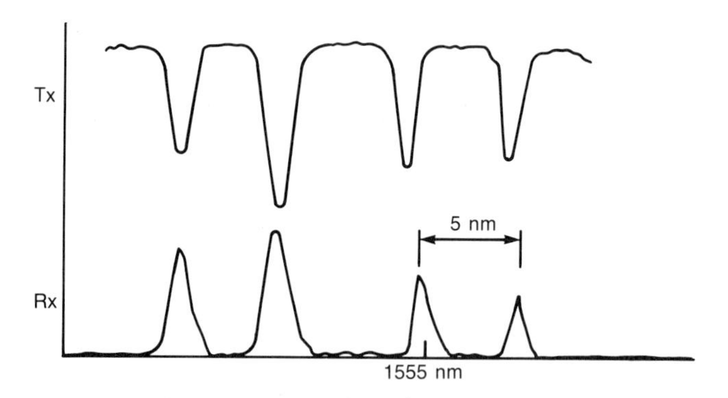

Fig. 15.10 Transmission and reflection spectrum of several gratings written into the core of a fibre at slightly different physical locations.

15.5 APPLICATIONS

15.5.1 The all-fibre grating mirror

The easiest, and perhaps the simplest, application of the fibre grating in telecommunications is as a reflector. Since fibre gratings can be fabricated with reflections of up to ~100% reflection with bandwidths of several nanometres to ~0.01 nm and grating lengths of a few millimetres, they have found applications in communications as narrowband reflectors for fibre and semiconductor lasers [32-35].

15.5.2 The semiconductor external fibre-cavity laser

An external fibre-cavity laser has been demonstrated using 40-60% reflectors. A schematic of this laser is shown in Fig. 15.11. A lensed-fibre end is offered to an antireflection-coated semiconductor laser. A fibre grating is then spliced directly with the pigtailed fibre or written directly into the pigtail, defining the lasing wavelength after the packaging process has been completed. This technique will be extremely useful in mass production since all the lensed-fibre semiconductor packages may be produced without specifying the exact lasing wavelength.

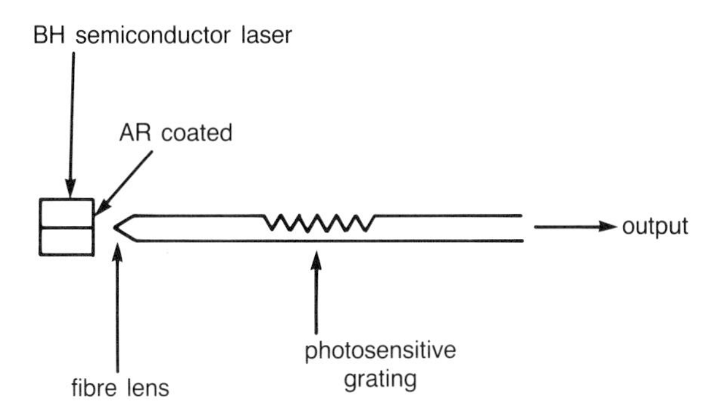

Fig. 15.11 Fibre grating external cavity semiconductor laser using an antireflection-coated facet and fibre lens [31].

The semiconductor external fibre grating cavity laser has been shown to have an extremely low chirp (<0.5 MHz) at a modulation frequency of 1.1 GHz [32]. This type of laser should find application in long-haul systems in which dispersion is a critical parameter limiting performance.

These lasers can be used not only as sources for communications but also in defining the semiconductor pump laser wavelength, e.g. 1480 nm/980 nm pump for the erbium amplifier at 1550 nm, or 1020 nm pump for the praseodymium system operating at 1300 nm.

15.5.3 Fibre lasers

The first demonstration of a laser operating with a fibre grating was of an erbium fibre laser with a reflection of only 0.5% from the grating at the lasing wavelength [14]. In that laser only one grating was used while the other was a high reflecting mirror. All-fibre lasers have also been demonstrated recently [33-35]. Fibre lasers are extremely versatile since the pump semiconductor laser may be integrated by splicing an external fibre grating cavity directly to the all-fibre grating laser, making it an extremely compact and simple system.

The erbium fibre grating laser has also been demonstrated as a tuneable mode-locked fibre system. Up to 50 mW of laser power was available from a laser producing 10 ps pulses (wavelength tuneable by stretching the grating) at a repetition rate of over 1 GHz [35]. These lasers combined with planar-waveguide modulators form a powerful combination and may find applications in many high bit rate transmission systems.

15.5.4 Sensors

Another major application of the fibre grating is in the area of sensors, with the earliest applications [36, 37] aimed at sensing in the first telecommunications window at 850 nm. These gratings may be stretched [37], heated [36] or used for detecting acoustic vibrations [38]. The temperature coefficient of the gratings is 0.016 nm/°C and they have a strain coefficient of 1 nm/% ϵ at 1500 nm. These values are easily detectable and form ideal parameters for sensing.

Recently all-fibre sensor networks incorporating gratings have also been demonstrated [39].

15.5.5 Bandpass filters

A common requirement of filters is the bandpass filter, which allows only certain wavelengths to be transmitted while blocking all others.

This type of filter is difficult to fabricate since it generally requires two identical gratings to be written into a stable interferometer. Figure 15.12 shows the schematic of this filter, using a 50:50 split-ratio Michelson interferometer [40] made with fibres. The principle of operation is as follows. Gratings reflecting at identical wavelengths are written into each of the arms. If the reflected light from both arms is returned in phase at the coupler, then all the light appears in the launching port 1. However, if the two paths back to the coupler are $\pi/2$ out of phase, all of the reflected light appears at port 2. Unfortunately, fabrication using optical fibre makes the device inherently unstable since the path difference between the two arms has to be stablized to well within a wavelength of light. Any differential temperature change or vibration upsets this balance. Since it is difficult to write identical gratings at exactly the right location such that the phase differences are correct for bandpass operation, it is necessary to be able to adjust the interferometer so that it remains in balance.

Recently, this problem was overcome by fabricating a four-port bandpass Mach-Zehnder interferometer filter in a Ge-doped planar silica which had been made photosensitive [41]. The first demonstration showed how powerful the method is. The path difference could be easily trimmed by exposing part of one arm of the interferometer with UV light to rebalance it (by inducing an index change over 3 mm), optimizing the performance after

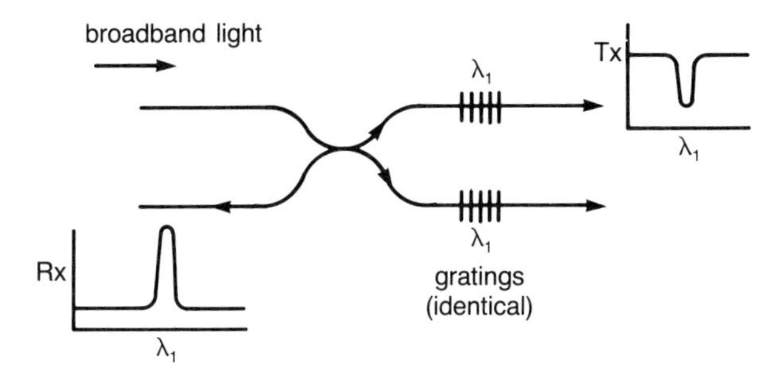

Fig. 15.12 A bandpass filter made using a fibre Michelson interferometer with identical gratings written into each arm [38].

the gratings had been written. Figure 15.13 shows the scheme used for fabricating the bandpass filter, including the location where trimming was done. Figure 15.13 also shows the bandpass transmission spectrum from port 2 of the Mach-Zehnder after the interferometer had been balanced. The transmission spectrum from port 3 is shown in Fig. 15.14. After optimization, 60% of the light at port 1 appeared at port 2 within the transmission bandpass, while 10% appeared at port 1.

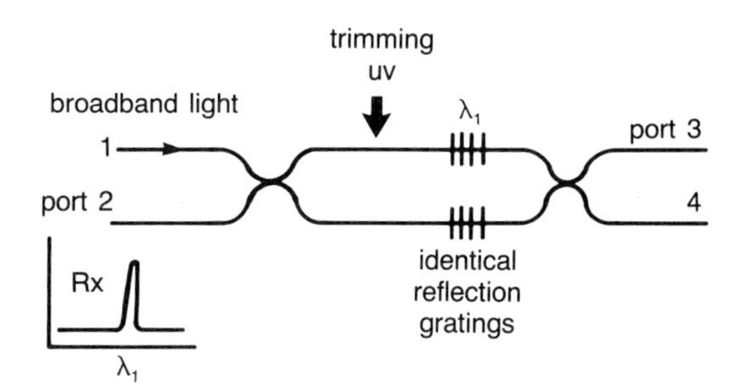

Fig. 15.13 A four-port bandpass filter fabricated in a Ge-doped photosensitive Mach-Zehnder interferometer. Balancing of the interferometer was carried out by exposing a 3 mm long section of one arm as shown in the figure.

Within the band-stop, the extinction at the output was in exces of 95% at port 3 (see Fig. 15.15). The fibre-to-fibre insertion loss was estimated to be 1.3 dB. Narrowband transmitting filters of this type are thus highly compatible with optical fibres.

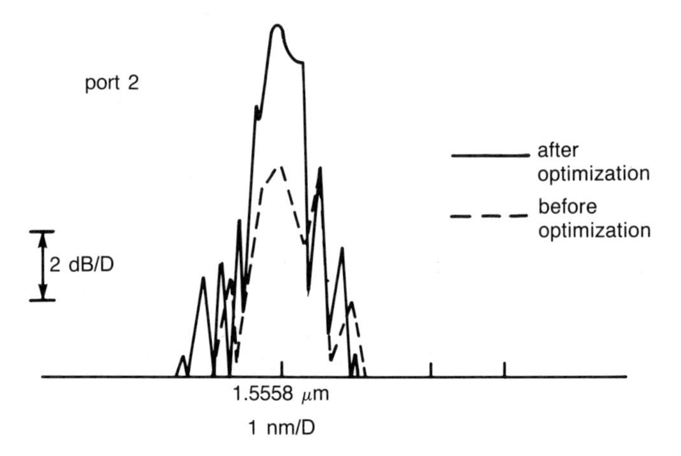

Fig. 15.14 Transmission spectrum of a band-stopping filter at port 2 before and after rebalancing of the interferometer.

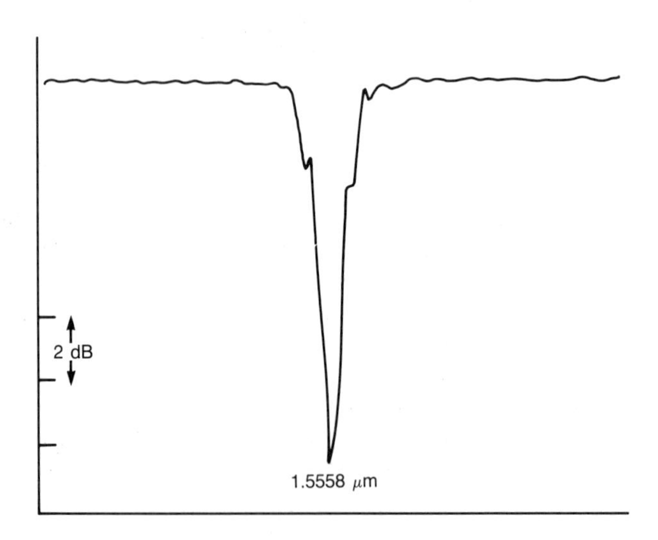

Fig. 15.15 Transmission spectrum of port 3 showing 97% rejection at the band-stop of the grating.

15.6 SPECIAL DEVICES

15.6.1 The radiation mode tap

If the normal to the grating vector is at an angle to the propagation direction of the guided mode, radiation loss may occur when a certain phase-matching condition is met [42]. Figure 15.16 shows a grating with its normal inclined at an angle α, written into the core of the fibre. If weak guidance is assumed, as in the case of most telecommunications fibres, shallow inclination angles are adequate to impart sufficient sideways momentum to the propagating mode to overcome the confinement due to guidance.

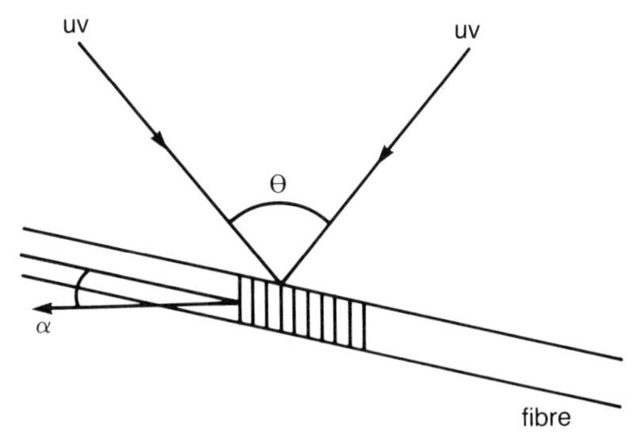

Fig. 15.16 An inclined grating formed in the core of the fibre for use as an optical fibre tap.

Figure 15.17 shows the transmission loss of one such side tap designed to flatten the gain spectrum of an erbium fibre amplifier. The length of the inclined grating was 3 mm while the centre wavelength at the peak of the amplified spontaneous emission spectrum (ASE) was 1531 nm. When this filter was spliced to a 3 m long length of erbium fibre, the peak in the ASE spectrum was eliminated.

Figure 15.18 shows the flattened spectrum. The ± 0.5 dB bandwidth is 35 nm, while 25 nm of the spectrum is within ± 0.1 dB. The gain as estimated from the ASE spectrum is around 15 dB with the 15 mW 980 nm diode pump. With some optimization, it should be possible to increase the bandwidth to over 40 nm with at least similar performance.

Side-tap gratings may be positioned anywhere within the transmission wavelength spectrum of optical fibres, by choosing the appropriate period of the phase mask. Figure 15.19 shows the loss spectrum of eight side-tap gratings written sequentially into a 100 mm length of fibre [43]. The gratings

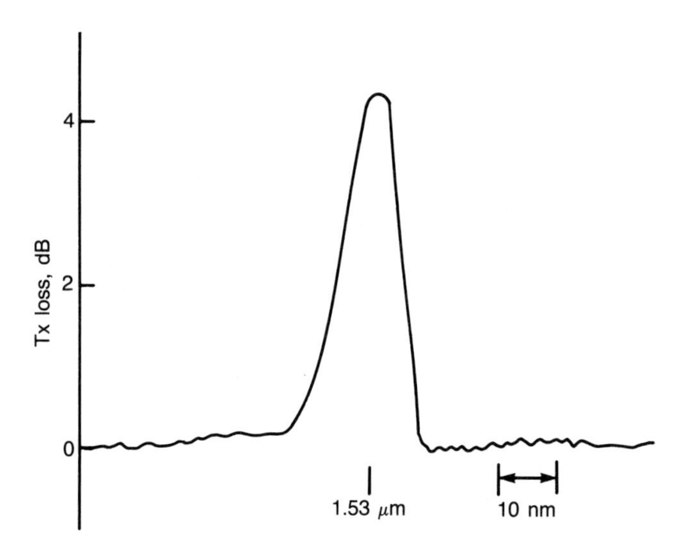

Fig. 15.17 Transmission loss of a side-tap grating as a function of wavelength. The peak loss is around 4 dB, with an FWHM bandwidth of ~7 nm.

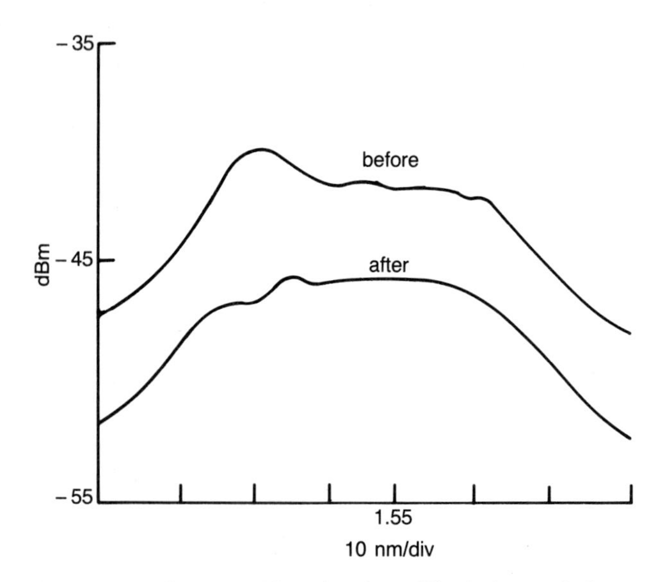

Fig. 15.18 ASE spectrum from an erbium-doped amplifier before and after wavelength flattening with a side-tap filter [42].

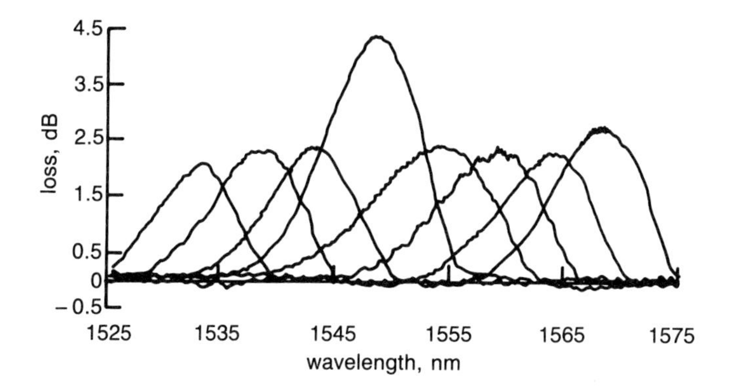

Fig. 15.19 Loss spectrum of eight individual side-tap gratings written into single-mode optical fibre. Each grating was approximately 4 mm long. The loss peak may be arranged to be at any wavelength by choosing the appropriate pitch of the phase mask grating.

were approximately 4 mm long. These were written as eight individual transmission phase gratings with Bragg wavelengths separated by 5 nm and fabricated on a single phase mask plate. It is possible to create wavelength blocking filters by writing multiple gratings spanning the 1300 nm or 1500 nm window. A peak loss of 19 dB with a blocking bandwidth of 10 nm has also been demonstrated using a single 8 mm long grating (see Fig. 15.20).

Side-tap gratings may be used to tap a small amount of light non-invasively from a fibre for monitoring and for providing multiplexed services in a ring/star network.

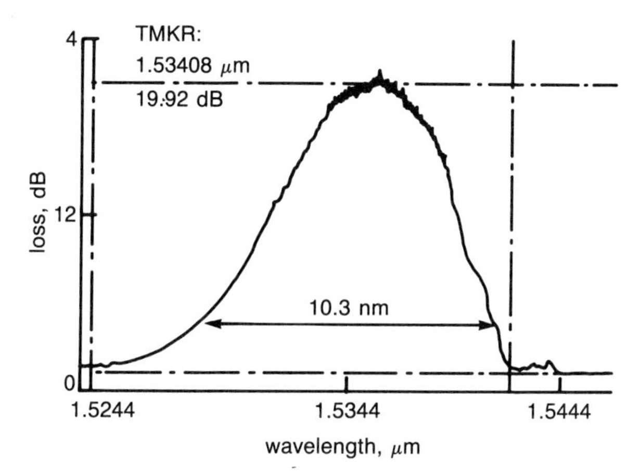

Fig. 15.20 Wavelength-blocking filter with a peak loss of 19 dB made with a single 8 mm long side-tap grating.

15.6.2 Shaped filters

Other types of shaped filters may be fabricated by writing several gratings closely separated in wavelength resulting in broadband reflectors. One such filter is shown in Fig. 15.21. The normal bandwidth of the single-reflection grating was 0.9 nm. However, the composite filter now has a bandwidth of around 1.4 nm and almost a flat top. Such broadband reflectors should find applications as laser mirrors and transmission filters when combined with the planar interferometer described earlier.

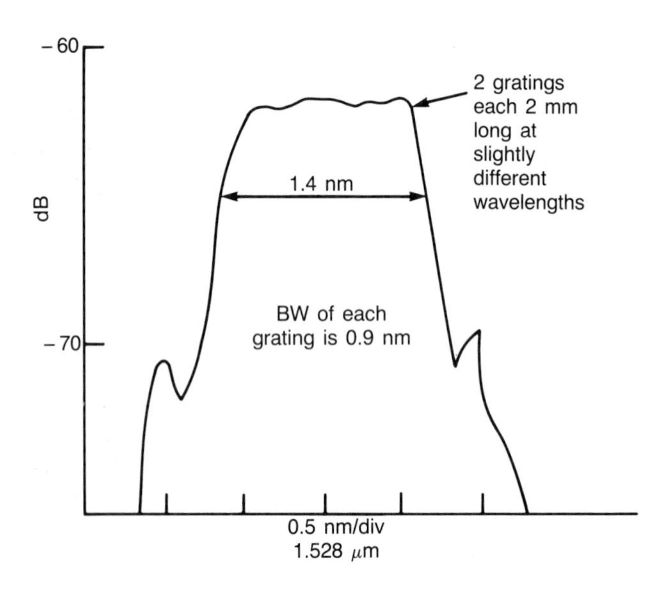

Fig. 15.21 Broadband reflection filters fabricated by writing two gratings at slightly different wavelengths. Each grating has a bandwidth of 0.9 nm while the composite bandwidth is 1.4 nm.

Apodized filters are easy to fabricate by tapering the intensity of the UV writing beams along their length, giving rise to a non-uniformly modulated grating. Other possibilities include the fabrication of chirped gratings for dispersion compensation [44] in long-haul submarine systems.

15.6.3 Oscillator systems

Recently, it has been shown that if two reflection gratings, slightly separated in wavelength, are used in an erbium fibre laser (semiconductor laser as well), the laser can be made to oscillate at those two wavelengths simultaneously [45]. The beating between the two lasing lines, f_1 and f_2, produces a modulated output signal given by the difference $f_1 - f_2$. Figure 15.22 shows the schematic of a dual-frequency erbium fibre laser source [45]. Two pairs of gratings define the operating wavelengths of two lasers. These lasers are separated by ~0.5 nm in central wavelengths and are concatenated such that single-end pumping is possible. The output of the lasers is a beat frequency defined by the difference in operating wavelengths. A novel feature of the lasers is the way in which each wavelength is locked into single-frequency operation by the residual reflection in the wings of the bandwidth of the two middle gratings. Each laser operated with a linewidth of 16 kHz. This sinusoidal modulation, which may be at several THz, can be transformed into a soliton pulse train by simply launching it into a chosen length of tapered dispersion fibre (linearly chirped dispersion) [46]. The laser was thus used in conjunction with specially selected dispersion-mixed fibres to suppress Brillouin scattering while simultaneously forcing the beat signal to form into solitons. A mark-to-space ratio of 1:7 was achieved at 59 GHz [47, 48].

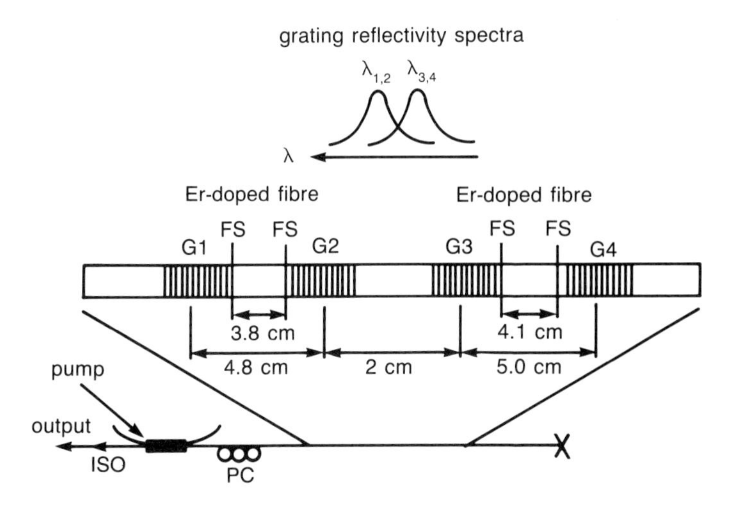

Fig. 15.22 Dual-frequency all-fibre grating erbium laser. This laser is single-end pumped and generates a beat frequency of 59.1 GHz. Each lasing wavelength has a measured linewidth of 16 kHz.

15.7 CONCLUSIONS

It has been shown in this chapter that photosensitive fibre may be exploited to provide solutions to a wide variety of problems in communications — from cheap narrow-bandwidth semiconductor/fibre laser sources, to applications in bandpass filtering and sensors, to name but a few. The future potential is enormous, giving rise to new ideas in communications networks. It is expected that the benefits of this technology will be felt in the near future and be far-reaching. This chapter has mainly reviewed recent advances at BT Laboratories — the authors' apologies to those whose work may not have been included here.

REFERENCES

1. Kao K C and Hockham G A: 'Dielectric-fibre surface waveguides for optical frequencies', IEE Proc, 133 , Pt J, No 3, pp 191-198 (1986).

2. Tobagi F A: 'Fast packet switch architectures for broadband integrated services digital networks', Proc IEEE, 78 , No 1, pp 133-167 (1990).

3. Simmons K D, LaRochelle S, Mizrahi V, Stegeman G and Griscom D L: 'Correlation of defect centres with a wavelength-dependent photosensitive response in germania-doped silica optical fibres', Opt Lett, 16 , pp 141-143 (1991).

4. Hill K O, Fujii Y, Johnson D C and Kawasaki B S: 'Photosensitivity in optical fibre waveguides: Application to reflection filter fabrication', Appl Phys Lett, 32 , No 10, pp 647-469 (1978).

5. Bures J, Lacroix S and Lapierre J: 'Bragg reflector induced by photosensitivity in an optical fibre: model of growth and frequency response', Appl Opt, 21 , No 19, pp 3502-3506 (1982).

6. Österberg U and Margulis W: 'Dye laser pumped by Nd YAG laser pulses frequency doubled in a glass optical fibre', Opt Lett, 11 , No 8, pp 516-518 (1986).

7. Stolen R H and Tom H W K: 'Self-organised phase-matched harmonic generation in optical fibres', Opt Lett, 12 , No 8, pp 585-587 (1987).

8. Baranova N B and Ya Zeldovich B: 'Extension of holography to multi-frequency fields', JETP Lett, 45 , No 12, pp 717-720 (1987).

9. Dianov E M, Kazansky P G, Yu Stepanov D and Sulimov V B: 'Photovoltaic mechanism of photo-induced second-harmonic generation in optical fibres', Conf on Integrated Photonics Research, Technical Digest Series, 5 (Optical Society of America, Washington DC), pp 46-50 (1990).

10. Kamal A, Stock M L, Szpak A, Thomas C H, Weinberger D A, Frenkel M Y, Nees J, Ozaki K, Valdmanis J, in Optical Society of America Annual Meeting Technical Digest, (OSA, Boston MA), paper PD24 (1990).

11. Farries M C, Fermann M E and St J Russell P, in Proceedings of the Topical Meeting on nonlinear guided wave phenomenon: Physics and Applications, Technical Digest Series, 2 (Optical Soc of America), pp 246-249 (1989).

12. Anderson D Z, Mizrahi V and Sipe J E: 'Model for second-harmonic generation in glass optical fibres based on asymmetric photoelectron emission from defect sites', Opt Lett, 16, No 11, pp 796-798 (1991).

13. Meltz G, Morey W W and Glenn W H: 'Formation of Bragg gratings in optical fibres by a transverse holographic method', Opt Lett, 14, No 15, pp 823-825 (1989).

14. Kashyap R, Armitage J R, Wyatt R, Davey S T and Williams D L: 'All-fibre narrowband reflection gratings at 1500 nm', Electron Lett, 26, No 11, pp 730-732 (1990).

15. Tsai T E and Griscom D L: 'Compositional effects on the radiation response of Ge-doped silica-core optical fibre waveguides', Appl Opt, 19, No 17, pp 2910-2916 (1980).

16. Krol D M and Simpson J R: 'Photo-induced second-harmonic generation in rare-earth-doped aluminosilicate optical fibres', Opt Lett, 16, No 21, pp 1650-1652 (1991).

17. Broer M M, Cone R L and Simpson J R: 'Ultraviolet-induced distributed-feedback gratings in Ce/sup 3+/-doped silica optical fibres', Opt Lett, 16, No 18, pp 1391-1393 (1991).

18. Hill K O, Malo B, Bilodeau F, Johnson D C, Morse T F, Kilian A, Reinhart L and Kyunghwan Oh: 'Photosensitivity in Eu^{2+}:Al_2O_3-doped-core fibre: preliminary results and application to mode converters', OFC'91, paper PD3-1, in Proceedings, pp 14-17 (1991).

19. Lampert M A and Mark P: 'Current Injection in Solids', Academic Press (1970).

20. Kashyap R, Maxwell G D and Williams D L: 'Photoconduction in germanium and phosphorus doped silica waveguides', Appl Phys Lett, 62, No 3, pp 214-216 (1993).

21. Russell P St J, Hand D P and Chow Y T: 'Optically-induced creation, transformation and organisation of defects and colour-centres in optical fibres', SPIE, 1516, International Workshop on Photo-induced Self-Organisation effects in Optical Fibres, pp 47-54 (1992).

22. Williams D L, Davey S T, Kashyap R, Armitage J R and Ainslie B J: 'Direct observation of uv induced bleaching of 240 nm absorption band in photosensitive germaniosilicate glass fibres', Electron Lett, 28, No 4, p 369-371 (1992).

23. Camlibel I, Pinnow D A and Dabby F W: 'Optical ageing characteristics of borosilicate clad fused silica core fibre optical waveguides', Appl Phys Lett, 26, No 4, pp 185-187 (1975).

24. Lemaire P J, Atkins R M, Mizrahi V and Reed W A: 'High pressure H_2 loading as a technique for achieving ultra-high uv photosensitivity and thermal sensitivity in GeO_2-doped optical fibres', Electron Lett, 29, No 13, pp 1191-1192 (1993).

25. Hibino Y, Abe M, Yamada H, Ohmori Y, Bilodeau F, Malo B and Hill K O: 'Increases in photosensitivity in silica-based optical waveguides on silicon', Electron Lett, 25, No 7, pp 621-622 (1993).

26. Archambault J L, Reekie L and Russell P St: '100% reflectivity Bragg reflectors produced in optical fibres by single excimer laser pulses', Electron Lett, 29 , No 5, pp 453-455 (1993).

27. McKee P, Wood D, Dames M and Dix C: 'Fabrication of multiphase optical elements for array spot generation', SPIE 1461, Electron Imaging Sci and Technol, pp 17-23 (1991).

28. Fertein E, Legoubin S, Douay M, Canon S, Bernage P, Niay P, Bayon F and Georges T: 'Shifts in resonance wavelengths of Bragg gratings during writing or bleaching experiments by UV illumination within germanosilicate optical fibre', Electron Lett, 27 , No 20, pp 1838-1839 (1991).

29. Duval Y, Kashyap R, Fleming S and Ouellette F: 'Correlation between ultraviolet-induced refractive index change and photoluminescence in Ge-doped fibre', Appl Phys Lett, 61 , No 25, pp 2955-2957 (1992).

30. Limberger H G, Fonjallaz P Y and Salathy R P: 'Spectral characterization of photo-induced high efficiency Bragg gratings in standard telecommunication fibres', Electron Lett, 29 , No 1, pp 47-49 (1993).

31. Maxwell G D, Kashyap R and Ainslie B J: 'UV written 1.5 μm reflection filters in single mode planar silica guides', Electron Lett, 28 , No 22, p 2106-2107 (1992).

32. Bird D M, Armitage J R, Kashyap R, Fatah R M A and Cameron K H: 'Narrow line semiconductor laser using fibre grating', Electron Lett, 27 , No 22, pp 1115-1116 (1991).

33. Ball G A, Morey W W and Waters J P: 'Nd^{3+} fibre laser utilizing intra-core Bragg reflectors', Electron Lett, 26 , No 21, pp 1829-1830 (1990).

34. Zyskind J L, Mizrahi V, DiGiovanni D J and Sulhoff J W: 'Short single frequency erbium-doped fibre laser', Electron Lett, 28 , No 15, pp 1385-1387 (1992).

35. Davey R P, Smith K, Kashyap R and Armitage J R: 'Mode-locked erbium fibre laser with wavelength selection by means of fibre Bragg grating reflector', Electron Lett, 27 , No 22, pp 2087-2088 (1991).

36. Morey W W, Meltz G and Glenn W H: 'Fibre optic Bragg grating sensors', SPIE 1169, Fibre Optics Sensors VII, pp 98-107 (1989).

37. Campbell R J, Kashyap R, Millar C A, Davey S T and Williams D L: 'Narrowband optical fibre grating sensors', Proc of OFS'90 (1990).

38. Melle S M, Kexing Liu and Measures R M: 'A passive wavelength demodulation system for guided wave Bragg grating sensors', IEEE Photon Technol Lett, 4 , No 5, pp 516-518 (1992).

39. Blair L and Cassidy S A: 'Wavelength division multiplexed sensor network using Bragg fibre reflection gratings', Electron Lett, 28 , No 18, pp 1734-1735 (1992).

40. Morey W W: 'Tuneable narrow-line bandpass filter using fibre gratings', OFC'91, paper PD20-1 (1991).

41. Kashyap R, Maxwell G D and Ainslie B J: 'Laser trimmed four-port bandpass filter fabricated in simple mode photosensitive fibre', Photon Technol Lett, 5, No 2, pp 191-194 (February 1993).

42. Kashyap R, Wyatt R and Campbell R J: 'Wideband gain flattened erbium fibre amplifier using a photosensitive fibre-blazed grating', Electron Lett, 29, No 2, pp 154-156 (1993).

43. Kashyap R, Wyatt R and McKee P F: 'Wavelength flattened saturated erbium amplifier using multiple side-tap Bragg gratings', Electron Lett, 29, No 11, pp 1025-1026 (1993).

44. Ouellette F: 'All-fibre filter for efficient dispersion compensation', Opt Lett, 16, No 5, pp 303-305 (1991).

45. Chernikov S, Kashyap R, Taylor J R and McKee P F: 'Dual frequency all-fibre grating laser source', Electron Lett, 29, No 12, pp 1089-1090 (1993).

46. Chernikov S V, Taylor J R, Mamyshev P V and Dianov E M: 'Generation of soliton pulse trains in optical fibre using 2 CW single mode diode lasers', Electron Lett, 28, No 10, pp 931-932 (1992).

47. Chernikov S V, Kashyap R and Taylor J R: 'Soliton pulse train generated by integrated all optical fibre technique using comb-like dispersion profiled fibre', Proceedings of Nonlinear Guided Phenomena, Cambridge, UK, Paper PD3-1 (1993).

48. Chernikov S V, Taylor J R and Kashyap R: 'Integrated all-optical fibre source of multi-gigahertz soliton pulse train', Electron Lett, 29 (1993).

16

NEW APPLICATIONS OF OPTICS FROM MODERN COMPUTER DESIGN METHODS

P F McKee, J R Towers, M R Wilkinson and D Wood

16.1 INTRODUCTION

There are a wide range of design techniques available within the engineering world, whose usage spans from kitchen design to processor chip layout. At its simplest level a designer might just draw out a plan of the object or system. The design process then consists of moving or changing elements within it until an arrangement is found that best suits the functions required of the system. Much more sophisticated techniques are also available for those design problems that can be analysed in terms of mathematical formulae. Many of these designs can be 'solved' using simple analytical or computational techniques — the length of a ladder can be determined using Pythagoras's formula for right-angled triangles, simple lens formulae can be used to determine the placement of lenses in optical imaging systems, resistor networks can be obtained from solving systems of simultaneous linear equations. A large body of mathematical expertise has been applied to such 'inverse' problems and a substantial part of the mathematics development since the days of Newton has been used towards solving such problems.

In more recent times a radically different approach to design has begun to emerge; this has been made possible by the phenomenal increase in the computer power that is now widely available. The roots of this new approach lie very close to the traditional methods of a skilled designer — juggling with the available elements until a best compromise is found between the constraints and the needs of the system. Such mathematical theory that exists to underpin this new approach comes from the field of statistical physics [1], rather than the existing theories on inverse problems. The earliest algorithm based on this new approach was given the name 'Simulated Annealing' by the workers at IBM who developed it [2]. It was applied with great success to a number of very concrete problems of the partitioning of circuits into discrete chips and other examples, all loosely based on the classic travelling salesman problem of route optimization. A brief description of this algorithm is given in section 16.2.

The reason why this development was so important was that it showed how a global best solution could be found, rather than a local maximum in the neighbourhood of the starting parameters. It also removed the need to make gross simplifications to the real problem in order to make the mathematics of the design process tractable. Because of this it became possible to include into the formulation the detailed constraints of the problem. Finally, it allowed the designer the scope to use quite sophisticated judgements of the measure of performance used to assess the value of a given solution. The disadvantage was that the demands on computer time were massively increased compared with the more conventional algorithms. Hence in the early days the algorithm's use on real problems was restricted to a few workers with access to fast computers. Since then a number of related techniques have been developed from these basic ideas and their use is now widespread over a vast range of design problems.

The work described in this chapter covers two main areas of the technology of diffractive optics. These areas are the free-space optics of surface relief holograms and longitudinal gratings in optical fibres and planar waveguides. The problems encountered in the design of diffractive elements are very different to those in network design and layout problems more usually associated with simulated annealing techniques. The algorithms developed at BT Laboratories (BTL) for the design of longitudinal waveguide gratings are the first, to our knowledge, to be used successfully to synthesize gratings with specific, complex properties. Each of these two topics have many potential applications, both for the technology currently used by BT and for the technology and concepts that may be needed for a telecommunications network of the future. The free-space diffractive optics can be exploited for interconnections — either from chip to chip [3], from a single source to an array of modulators or fibres [4, 5], or simply from a laser source to an optical fibre [6, 7], or as an optical antenna to distribute light safely and in a controlled manner over a predefined area in a room or a building [8].

The longitudinal gratings can be used as filters or wavelength selective mirrors in a number of systems [9] — for wavelength-division multiplexing, for reducing noise or flattening the gain spectrum in networks employing optical amplifiers, and as feedback elements in fibre or semiconductor lasers.

The next section gives a brief description of the simulated annealing algorithm in both its general form and for the problems encountered in designing free-space and waveguide-diffractive optics. Then follow two further sections that give more details of applications in free-space holograms and waveguide gratings and show some examples of the experimental results obtained. Each of the two areas of diffractive optics place great demands on the fabrication processes needed to realize any designs. Included in these two sections are brief descriptions of how these optical elements are made and the fabrication tolerances needed. Finally, the concluding section looks to the future of the technique as a means of programming optical signals in space and wavelength.

16.2 SIMULATED ANNEALING ALGORITHMS

To illustrate how simulated annealing algorithms work, we will begin by describing their application to the travelling salesman problem. The salesman is required to visit a number of cities and the aim of this problem is to find the optimum route for him to follow. In order to use any optimization algorithm it is first necessary to describe the system in terms of parameters which can be varied to explore the possible solutions. In this example the system can be described simply by the order in which the cities are visited.

The next step in the design process is to formulate some measure of how well a system with given parameters performs. This measure can be very simple or very sophisticated. In this example, we could use just the total route length as a measure or include complicated data on mountain heights, traffic statistics and local comforts. The algorithm then changes one of the parameters in the system at random and recalculates this measure to assess whether or not the change should be accepted. A typical change would be to reverse the order in which the cities are visited for part of the route. Changes of this kind are useful because the performance measure need only be recalculated where the route has been changed, which is much quicker than repeating the calculation for the whole route.

Changes which improve the performance measure, M, are always accepted. However, changes which reduce the performance are not always rejected — some are accepted according to a certain probability, P, such that:

$$P(\Delta M) = e^{\frac{\Delta M}{T}} \quad \text{if } \Delta M \leq 0$$

$$P(\Delta M) = 1 \quad \text{if } \Delta M > 0$$

where ΔM is the change of the performance measure and T is a notional 'temperature'. This process of making a random change, and deciding whether to keep it or not, is then repeated many times. As the design progresses, the 'temperature', T, is gradually reduced so that the probability of accepting the 'detrimental' changes decreases. At the start of the design process T is chosen so that almost all changes are accepted, and at the end only those changes that improve the performance, M, are accepted. In between these extremes, M fluctuates as the parameters change. More time is spent in parameter regions where M is high than where M is low, and this asymmetry increases as T decreases. It has been shown that if T reduces sufficiently slowly then the system evolves towards a solution close to the global maximum of performance. If the 'detrimental' changes were never accepted the system would get stuck at a solution with a poor local maximum of performance close to the starting conditions.

The advantage of simulated annealing over conventional design methods becomes apparent for systems with a large number of parameters where the solution space can be very complicated and the 'global' maxima may be well hidden in parameter space. For the travelling salesman problem the total number of possible solutions increases exponentially with the number of cities. This means that it becomes virtually impossible to find exact solutions to the problem with more than a few hundred cities because of the vast amount of computation required. Where it is possible to obtain an exact 'optimum' solution, the answers obtained by the simulated annealing algorithm are almost indistinguishable from such solutions. Excellent results for several thousand cities with very complex measures of performance can be obtained using simulated annealing algorithms [2], where exact optima are, in practice, unobtainable. There are, of course, many other applications for such optimization algorithms. In particular, simulated annealing methods have been used at BTL for training neural networks in both software applications [10] and for optical hardware [11, 12].

Many further developments to the basic ideas of simulated annealing have led in some cases to significant improvements, particularly in the speed of convergence. Most notable of these developments are those known as 'genetic algorithms' [13], in which ensembles or 'populations' of possible solutions are held. The iterations then proceed by 'mutations' (changes to one or more of the parameters) or by 'cross-fertilization' (interchange of selected parameter values between members of the populations). There are technical reasons why genetic algorithms are impractical to use on the two diffractive optics examples given here. In other situations, however, they have been used with great success [14].

Our own uses of simulated annealing in the design of diffractive optics have little in common physically with the travelling salesman example discussed above. A more detailed description of the two examples will be given in later sections; here, discussion will cover only those aspects connected with the optimization. Each of the two examples is concerned with the synthesis of a diffractive element that transforms the input light into some specified output pattern. In the free-space example the spatial distribution of the input light is transformed to a given output; in the waveguide example the wavelength response of the grating to the input wave is determined.

A common feature of these two different implementations is that a 'target' characteristic is specified and the performance measure determined by the correspondence between the actual output of the device and the 'target' output. This means that the exact requirements of the system can be used as an input to the design algorithm, without any of the limitations associated with solving the problem analytically. Also, it is possible to include details concerning which properties of the system are most important by giving them more 'weight' when evaluating the performance. How well the algorithm performs in trying to match these desired characteristics depends on the constraints imposed by the physical properties of the optical element.

Another advantage of this approach is the ability to take into account the real practical limits of the fabrication processes. For example, the size of the pixels in a surface relief hologram is determined, in practice, by the lithography and etching processes. Similarly, the total length of a grating and the smallest feature sizes are determined by the fabrication method. Each of these properties and constraints can be included in the algorithms used so that only designs that can be realized in practice are produced.

16.3 FREE-SPACE HOLOGRAMS

Over distances larger than about a wavelength, monochromatic light is described (for the purposes of this chapter) accurately as a classical wave phenomenon. Its behaviour can be thought of in terms of a complex-valued wave, whose intensity is proportional to the power density of the light. In free space, the phase of the wave is given by the distance travelled from the source. This phase rotates by an angle of 2π radians over a distance of one wavelength. When the light passes through a transparent material it is slowed down. The effective distance, or 'optical path', travelled is increased in proportion to the refractive index of the medium. A transparent medium retards the phase of the incident light wave, and hence a patterned depth profile printed on the surface of the transparent object introduces a corresponding phase pattern into the light. A phase pattern with the full dynamic range of angles, 0-2π, can be induced by a surface depth profile only a few wavelengths deep.

Figure 16.1 gives a schematic view of how surface relief diffractive elements (holograms) are used. The light passes from the source, through the hologram surface to the target area. The field on the target is given by the coherent sum of the phase contributions from every path from the source, through the hologram to the point in the target. The problem is to find the surface depth profile on the hologram that leads to a given intensity distribution on the target — or rather to an optimum fit to the desired distribution. Some simple diffraction problems have well-known solutions — for example, imaging from one point to another requires the surface profile to be one which keeps the phase of all the optical paths, from the object to the image point through the surface, equal. This is known as a Fresnel zone plate and represents simply the phase of the corresponding conventional lens taken modulo 2π. Other problems, including many of BTL's own

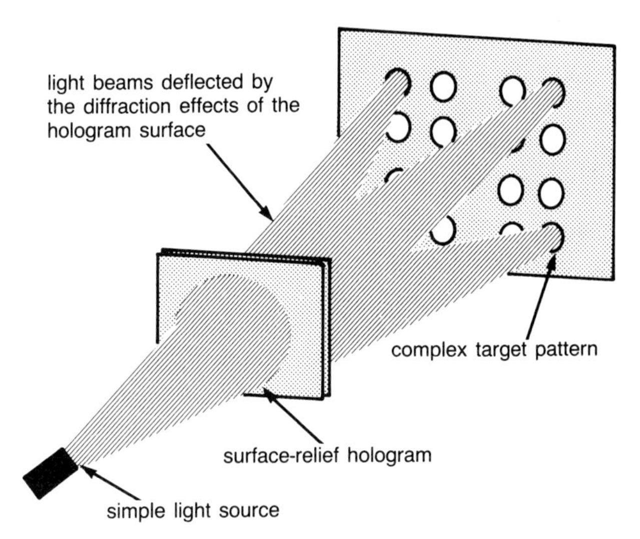

Fig. 16.1 Application of computer-designed holograms — they may be used for interconnections in micro-optics where the diffraction distances are comparable to the hologram size (a few millimetres), or as antenna devices where the target is a shaped pattern in the far field, in another part of the room or another building.

applications — such as multiple imaging or transforming the shape of the light beam — are much more complicated.

In order to define this design problem, the surface profile is divided into an array of pixels, whose values correspond to the etch depth. These pixels correspond to small squares, of a given phase depth, that cover the patterned area. At present these depths are restricted to two or four possible values — corresponding to 0, $\pi/2$, π and $3\pi/2$ phase retardations. The pixel array is initialized with random values, then the intensity on the target is calculated — and hence the measure, M, which is being used to determine the fit with the intended target. Pixels are then changed at random and the change to the measure, ΔM, is calculated; the simulated annealing algorithm is used to decide which changes are accepted and to evolve the pattern towards its optimum solution. This design procedure, in some cases, does stretch our computer resources to their limits. Changes to a single pixel can be calculated very rapidly but there may be a huge number of pixels — up to four million are used in some designs. However, once a design is complete, the calculations do not need to be repeated and the real limiting process lies in the fabrication rather than the design.

Using the design process outlined above, holograms can be produced for use in a wide variety of applications. Each hologram produced falls into one of two broad categories — holograms designed for use in the far field or the back focal plane of a lens, known as Fourier holograms, and near-field holograms which also perform the lensing function, known as Fresnel holograms. For the far-field holograms, the formulae for the optical path through the surface simplify into Fourier transforms and, in many cases, further symmetries in the problem enable the pattern to be specified as a two-dimensional repeating pattern of cells. Therefore, the surface relief pattern only needs to be designed over a small area of the hologram and hence the design calculations are greatly simplified. For near-field, Fresnel holograms no such simplifications or symmetries are available, so the computational problems are very much more difficult. In addition, for most applications, the feature sizes of the Fresnel hologram patterns have to be considerably smaller and at least four phase levels are needed to achieve reasonable efficiency.

Electron beam lithography is used to define a resist mask and the depth profile is imprinted into quartz substrates by reactive-ion etching through these resist masks. A typical feature size for these holograms is 1 μm, and etch depths of up to 2.5 μm are needed. In order to achieve such etch depths the resist stack includes metal layers to increase etch resistance. Multiple-phase levels have to be built up in layers, which have to be accurately aligned; any misalignment reduces device operating efficiency [15]. Accurate control of etch depth is also essential for efficient operation [16]; hence test structures have to be included in the patterns to aid this control. Etch depths, currently, can be controlled typically to within 30 nm.

Fourier holograms have been designed and fabricated at BT Laboratories which split an incoming collimated beam into a predetermined array of output beams, for example, a square array or a more complex pattern. The intensities of the output beams can also be controlled during the design stage, the intensities distributed according to a predetermined weight matrix. We can produce two-phase level, Fourier holograms routinely with measured output beam intensities accurate to within 1% of their target values [5]. Measured diffraction efficiencies, i.e. the ratio of power in the output pattern relative to the total incoming power, are also high and close to the theoretical values. For example, this hologram producing a 4×4 square output array has a measured efficiency of 74% with a (measurement limited) accuracy of better than 1%. Its theoretical efficiency was 78% and the theoretical variation of power between the beams was below 0.1%.

Fourier holograms which split an incoming beam into an array of output beams of different intensities are particularly relevant to neural network applications. Optoelectronic neural networks making use of spatial optics for interconnection have been proposed as potential building blocks for powerful parallel system controllers, necessary in the future when the capacity and versatility of telecommunications switches becomes too great for conventional serial processors to handle. At BT Laboratories we have constructed several optical neural network demonstrators over the last few years which include holograms designed by our simulated annealing technique [17]. Figure 16.2 shows the output of one of the holograms used in the latest of these. This hologram produces eight groups of four images with predefined weights within each group of 1, 2, 4 and 8 units. In this case the input image had also been generated by a hologram which produces a 4×4 array of beams on to an indium phosphide, multi-quantum well modulator array which provides high-speed data on to the beams. This beam array is then further split by the more complicated hologram into the eight groups of the four weighted images. The quality of the image shown in Fig. 16.2 is limited due to the video monitor system. In reality, measured intensities were within 2% of the target values and well within system specifications for the neural network.

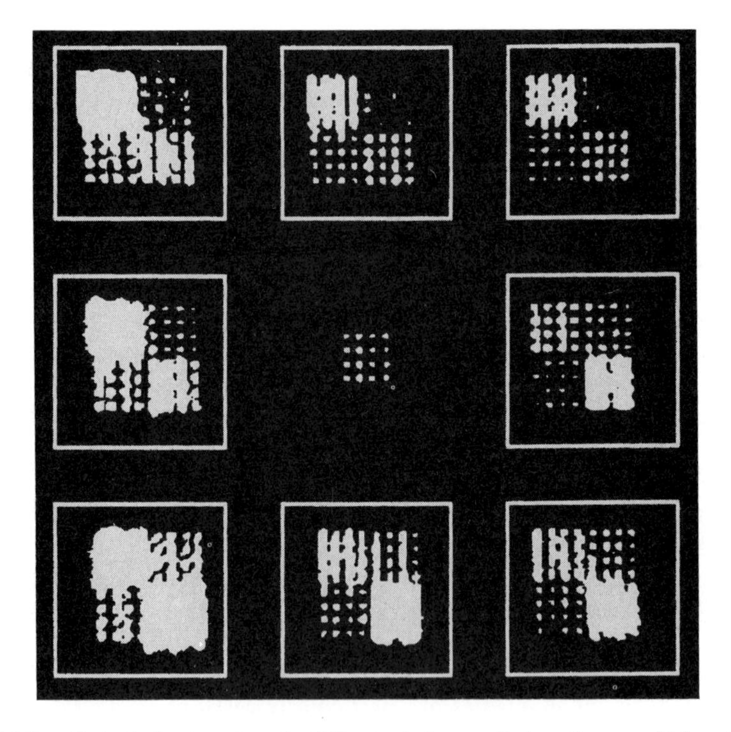

Fig. 16.2 Output of a two-phase level Fourier hologram that produces multiple weighted images of an input 4×4 array of spots. The output images are grouped into sets to image the input with weights of 1, 2, 4 and 8 units. Image quality here is limited by the video monitor.

Many potential applications might involve the use of additional lenses to focus the hologram output. For ease of alignment and to reduce the packaging complexity, and therefore the cost, of a future device, it would be beneficial if the lensing and beam-splitting functions could be integrated into a single holographic element. The design algorithms can cope with this requirement and the two functions — splitting and focusing — can be combined to produce a single hologram design. The resulting holograms are Fresnel holograms. We have designed and fabricated several different types of multiple imaging elements including some specifically for optical fibre applications, for operation at a wavelength of 1.55 μm. A scanning electron-microscope photograph of the four-phase level surface of a Fresnel hologram that performs 1-16 multiple imaging from optical fibres is shown in Fig. 16.3. Figure 16.4 shows the captured camera form of the image plane of one of these 1-16 focusing splitters as a three-dimensional intensity plot. Most differences between peak heights seen in Fig. 16.4 are due to the camera. The holographic element producing this output is 1 mm^2 in size and was

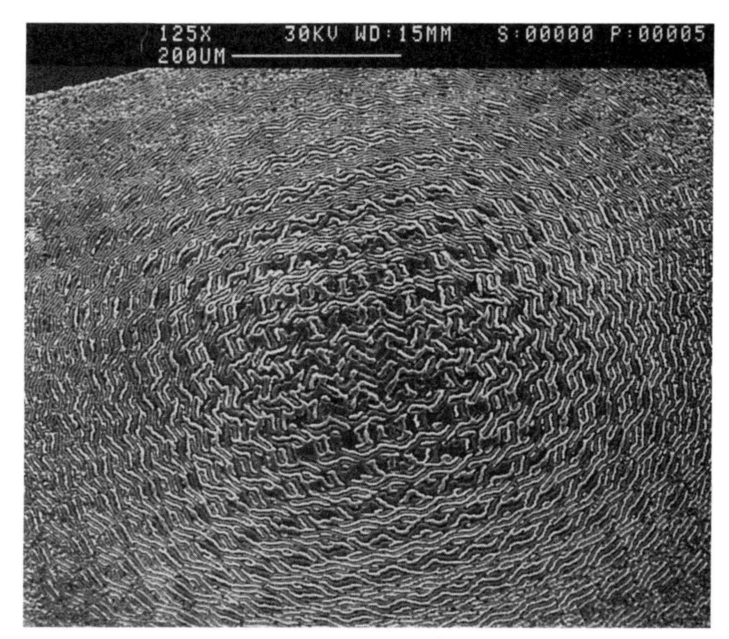

Fig. 16.3 An electron microscope picture of a four-phase-level, surface-relief Fresnel multiple-lens, f/2 hologram that produces 16 images of a single input spot.

designed to produce an array of 4×4 focused spots from a single collimated beam. Each output spot was diffraction limited, with the size determined by the diffraction limit from the entire 1 mm^2 of the hologram surface. The spot sizes are therefore much smaller than could have been achieved using a lens array, where the diffraction limit would have been by the lens diameters. The diffraction efficiency of this hologram was measured to be approximately 63%, with an additional 4% loss through reflections off the etched face that was not anti-reflection coated, and output spot powers were uniform to within 6%. These first results are very encouraging, as the elements are small enough to be integrated on to modulator array packages and the surface-relief profiles can easily be copied for low-cost, mass production.

The more recent work on Fourier holograms has concentrated on 'antenna' elements to distribute light safely and in a controlled manner over a predefined area in a room or building. These are designed as elements to produce a very large array of beams (up to 81×81 beams) with high diffraction angles. Holograms of this nature are important components of proposed optical wireless systems where safe distribution of high amounts of optical power, a potential hazard to the eye, is necessary [8]. More details about this application can be found in Chapter 6.

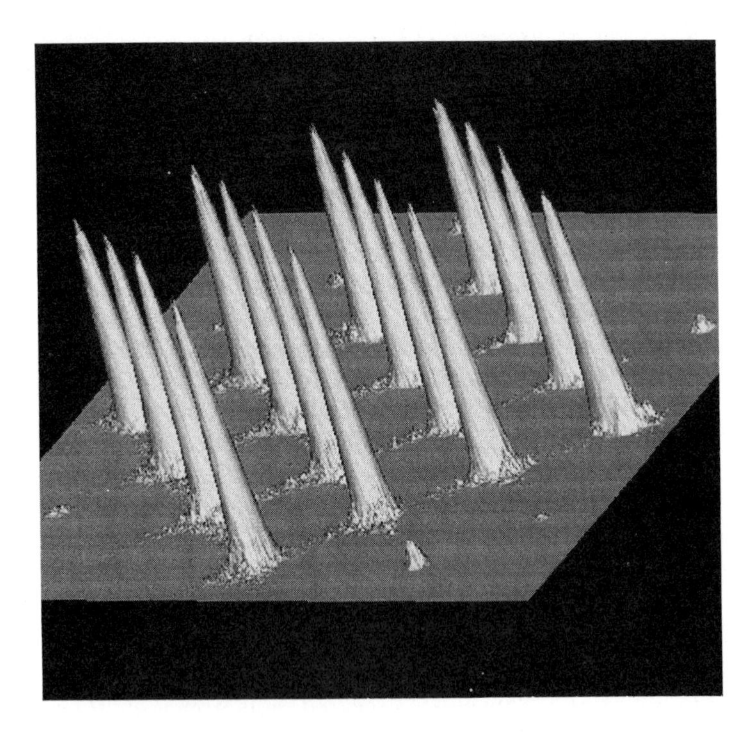

Fig. 16.4 Three-dimensional intensity plot of the captured camera output of the image plane from a 1-16 multiple-focusing, Fresnel lens-splitter. Most of the differences in height of the spots are due to camera distortion. Measured spot powers showed uniformity to within 6%.

16.4 LONGITUDINAL WAVEGUIDE GRATINGS

Longitudinal grating structures can be used in various waveguide technologies to produce narrowband filters and wavelength selective mirrors, which have many applications in optical communications [18,19]. This section will concentrate on results which have been obtained using simulated annealing algorithms to design overlay gratings for reflection filters. These are based on D-fibre, which is simply a D-shaped single-mode fibre with its core within $\sim 1\ \mu m$ of the flat surface. This provides a convenient technology for gaining access to light in the core of a fibre over lengths of several millimetres.

A schematic of a D-fibre grating filter is shown in Fig. 16.5. The grating is overlaid on to the flat surface of the fibre with its lines perpendicular to the fibre core. The grating itself has a surface relief profile which causes a pattern of changes in the effective refractive index of the fibre mode along the length of the device. The grating is typically a few millimetres long and consists of a sequence of lines with a spacing of $\sim 0.5\ \mu m$. This acts as a Bragg filter, reflecting incident light around a resonant wavelength back along the waveguide.

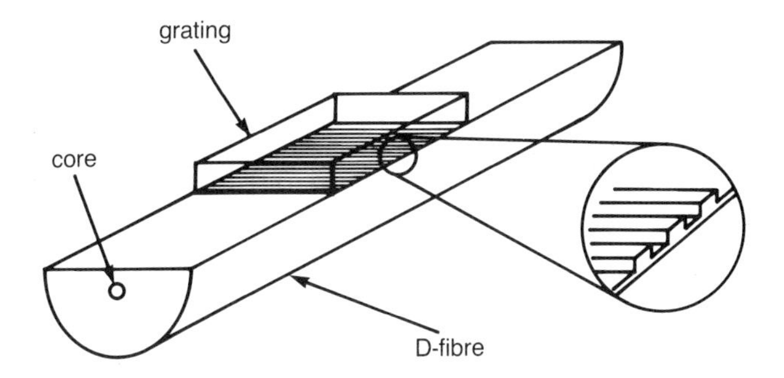

Fig. 16.5 A schematic diagram of a D-fibre grating overlay filter.

The design technique to produce particular filter shapes involves changing the sequence of lines and spaces in the grating pattern itself. It does not rely on altering the properties of the underlying waveguide structure, although these do, of course, form part of the input to the design process. Therefore this technique is not specific to this technology but is generally applicable to a wide variety of waveguide grating structures such as planar silica waveguides and semiconductor lasers.

It is the filter characteristics in the wavelength domain which are to be optimized using a simulated annealing algorithm. For a conventional, uniform grating the filter characteristics are determined by the total grating length and the strength of the interaction. For low reflectivities, the reflection has a characteristic $\sin(\lambda)/\lambda$ profile in the wavelength domain and the bandwidth is approximately inversely proportional to the length. Increasing the interaction strength increases the reflectivity and the bandwidth but also exaggerates the sidelobes. This makes it difficult to obtain a high reflectivity and a useful bandwidth together with low sidelobes. There are various changes that can be introduced into the grating to alter these characteristics. One technique is to vary the grating line spacing along its length, producing a 'chirped' grating. However, chirped gratings are difficult to fabricate and only give limited design flexibility. The approach taken here is to keep the line spacing constant but to change the sequence of lines and spaces in the grating. Unfortunately, it is impractical to consider changes to individual lines in the same way as the pixels in a hologram because there are $\sim 10\,000$ lines in a typical grating and computation of the filter response is quite lengthy. Although this is not a problem for a single calculation, there are a vast number of possible patterns ($\sim 2^{10\,000}$) to explore and the computing power to perform the required number of iterations is not available. Therefore, to simplify the computation of the filter response, the grating is divided into a number of sections, each containing many repetitions of a short sequence of lines and spaces (a 'word' pattern). Examples of typical 'words' are shown in Fig. 16.6. The lines and spaces are $\lambda/4$ wide (where λ is the resonant wavelength) and some 'words' are phase shifts of others to allow destructive interference within the grating, giving more flexibility in design.

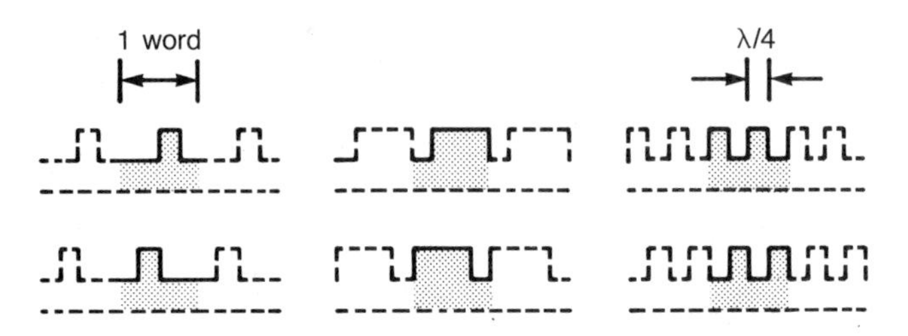

Fig. 16.6 Some example 'word' patterns which make up the grating.

The algorithm makes changes to the grating by choosing a random section and changing either the 'word' or the number of repetitions in that section. The filter response is then recalculated and compared with a target response to obtain the performance measure. The main advantage of dividing the grating into sections is that it is only necessary to recompute the filter response in the sections which have changed, thus reducing the computation time considerably.

Using this approach, grating filters with almost any arbitrary reflection profile can be designed. The only limitations are those imposed by the physical properties of the grating and the waveguide technology. For example, in a D-fibre device the maximum length of the grating is restricted to the length over which good optical contact is achievable, which is determined by the flatness of the grating and the substrate holding the fibre.

The gratings are fabricated by electron beam lithography and reactive-ion etching into quartz substrates in a similar way to the surface relief holograms. However, the challenges associated with these devices are quite different. The features which make up a grating are approximately 0.25 μm in size, the exact size being dependent on the operating wavelength of the filter. To image such high-resolution structures successfully, a thin resist layer is to be preferred but this will have a low etch resistance. So, for grating exposure, a bi-level resist process is used combining a thin imaging layer with a thicker etch-resistant layer. To attain the precise line spacings required the electron beam deflection field is reduced in size, thus increasing its pixel resolution. This means that most gratings have to be made using more than one deflection field 'stitched' together. Using the electron beam microfabricator's beam-deflection calibration system and the high-resolution laser stage beam, position errors within 20 to 25 nm have been demonstrated at BT Laboratories [20].

A photograph of a fabricated grating is shown in Fig. 16.7. This design is ~4 mm long and contains 64 sections. Obviously it is not possible to resolve individual grating lines or 'word' patterns in this picture, but some of the

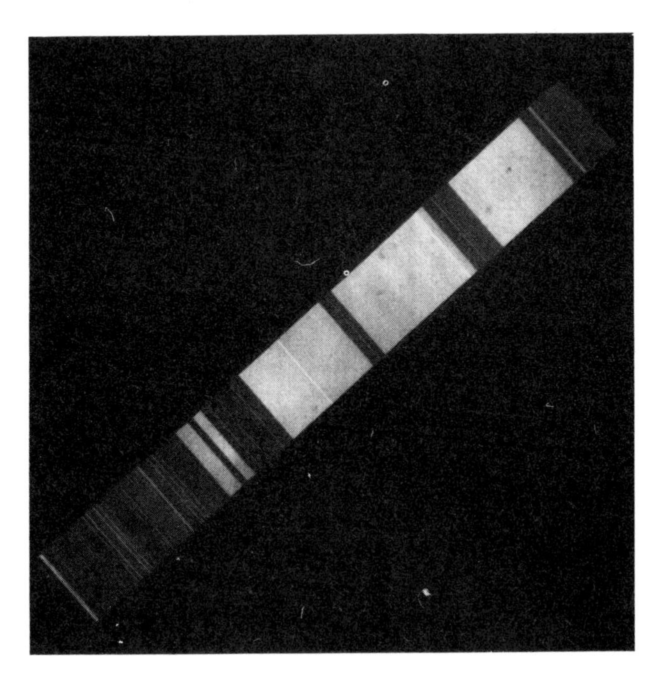

Fig. 16.7 A photograph of a 4 mm long grating overlay filter. The detailed structure of the grating is not resolved in this photograph but the transitions between different word patterns can clearly be seen.

sections can be clearly distinguished because of the different diffractive properties of the patterns within them.

Figure 16.8 shows the calculated reflection profile from a filter designed with two peaks separated by 4 nm in wavelength. Although it would be possible to obtain this type of response using two gratings with different line spacings, this grating uses only one fixed spacing, demonstrating the versatility of the design technique. Figure 16.9 is the experimental measurement of this filter, showing excellent agreement with the calculated profile. The small differences are due to either slight errors produced in the fabrication of the grating or non-uniform contact between the grating and the D-fibre.

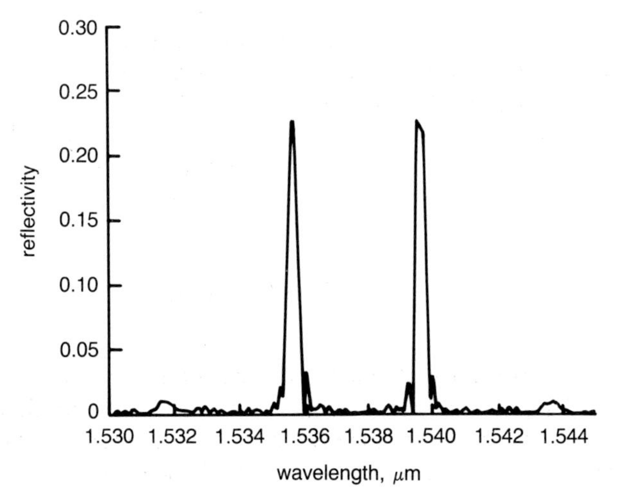

Fig. 16.8 Calculated reflectivity of a grating design showing two peaks separated by 4 nm.

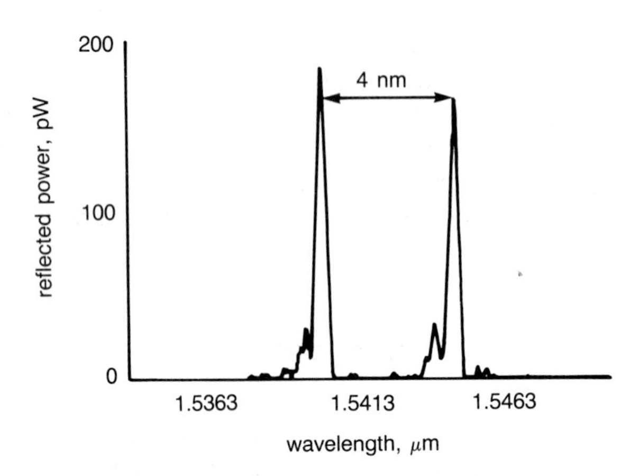

Fig. 16.9 Experimental measurement of reflected power spectrum from the D-fibre grating overlay filter.

Having demonstrated that the designed filter characteristics can be reproduced experimentally, it is now possible to design other filters by defining the desired wavelength response. This can be anything from a single reflection with a particular shape to a complicated profile with different features at a number of wavelengths. Recently, filter designs have been produced with relatively broad, flat-topped profiles for applications in wavelength multiplexed systems [21] and Gaussian transmission filters for noise suppression.

The transmission filters were obtained by reflecting the wavelengths on either side of a central channel. In this case the performance measure was chosen to have more 'weight' at the centre of the filter than at the edges, where the characteristics are less important. This is a good example of how the design process can take account of the critical aspects of the device without being constrained in other areas where the properties need not be so well-defined.

16.5 CONCLUSIONS

This chapter has introduced the idea of using modern computer techniques to produce diffractive optical elements with complicated characteristics designed to perform specific functions for systems of the future. The methods described allow the exact requirements of the element to be used as an input to the design process, without any restrictions concerning the analytical solutions to the problem. Ultimately, the limit of what can be achieved in practice is set by the fabrication processes and these constraints can be built into the design algorithms. As a result, it is possible to produce realistic designs for optical elements with properties that could not be produced using conventional methods. As computing power continues to increase and fabrication methods develop, this type of design method has great potential to produce sophisticated optical elements which are able to process signals in space and wavelength for many new applications in future communications networks.

REFERENCES

1. Metropolis N, Rosenbluth A, Rosenbluth M N, Teller A and Teller E: 'Equation of state calculations by fast computing machines', J Chem Phys, 21, p 1087 (1953).

2. Kirkpatrick S, Gellatt C D and Vecchi M P Jr: 'Optimization by simulated annealing', Science, 220, pp 671-680 (1983).

3. Bergman L A, Wu W H, Johnston A R, Nixon R, Esener S C, Guest C C, Yu P, Drabik T J, Feldman M and Lee S H: 'Holographic optical interconnects for VLSI', Opt Eng, 25, No 10, pp 1109-1118 (1986).

4. Feldman M R and Guest C G: 'Iterative encoding of high-efficiency holograms for generation of spot arrays', Opt Lett, 14, 479-481 (1989).

5. Dames M P, Dowling R J, McKee P and Wood D: 'Efficient optical elements to generate intensity weighted spot arrays: design and fabrication', Appl Opt, 30 , pp 2685-2691 (1991).

6. Wood D, McKee P and Dames M: 'Multiple imaging and multiple-focusing Fresnel lenses with high numerical aperture', SPIE Proceedings, 1732 , paper 38 (1992).

7. Feldman M W and Welch W H: 'Iterative encoding of holographic optical elements', SPIE Proceedings, 1732 , paper 99 (1992).

8. Smyth P P, Wood D, Ritchie S and Cassidy S: 'Optical wireless: new enabling transmitter technologies', ICC'93, pp 562-566, Geneva (May 1993).

9. Wilkinson M, Cassidy S A, McKee P and Wood D: 'Novel computer designed waveguide grating structures with optimised reflection characteristics', Electron Lett, 28 , pp 1660-1661 (1992).

10. Wood D: 'Real time control using a neural network', BT Technol J, 10 , No 3, pp 69-76 (July 1992).

11. Wood D: 'Training high-speed, opto-electronic neural networks', Opt Commun, 82 , pp 236-236 (1991).

12. Barnes N M, O'Neill A W and Wood D: 'Rapid, supervised training of a two-layer, opto-electronic neural network using simulated annealing', Opt Commun, 87 , pp 203-206 (1992).

13. Goldberg D E: 'Genetic Algorithms', Addison-Wesley, New York (1989).

14. O'Neill A W: 'Genetic based training of a two-layer, optoelectronic neural network', Electron Lett, 28 , pp 47-48 (1992).

15. Cox J A, Werner T, Lee J, Nelson S, Fritz B and Bergstrom J: 'Diffraction efficiency of binary optical elements', SPIE J, 1211 (1990).

16. Farn M W and Goodman J W: 'Effect of VLSI fabrication errors on kinoform efficiency', SPIE J, 1211 (1990).

17. Webb R P and O'Neill A W: 'Optoelectronic neural networks', BT Technol J, 10 , No 3, pp 144-154 (July 1992).

18. Yariv A and Nakamura M: 'Periodic structures for integrated optics', IEEE J Quantum Electron, QE-13 , No 4, pp 233-253 (1977).

19 Sorin W V and Shaw H J: 'A single-mode fiber evanescent grating reflector', IEEE J Lightwave Technol, LT-3 , No 5, pp 1041-1043 (1985).

20. Dix C and McKee P F: 'High accuracy e-beam grating lithography for optical and optoelectronic devices', J Vac Sci & Technol B, 10 , No 6, pp 2667-2770 (1992).

21. Wilkinson M, Cassidy S A, McKee P and Wood D: 'Computer designed grating filters with tailored profiles for wavelength multiplexed systems', Proceedings ECOC'92, Berlin, 1 , paper We B9.3, pp 421-424 (October 1992).

Index